AF716122

Recent Developments in Orthogonal Polynomials

CONTEMPORARY MATHEMATICS

822

Recent Developments in Orthogonal Polynomials

AMS Special Session
Orthogonal Polynomials and their Applications
January 4–7, 2023
Boston, Massachusetts

Ahmad Barhoumi
Roozbeh Gharakhloo
Andrei Martinez-Finkelshtein
Editors

AMERICAN MATHEMATICAL SOCIETY

EDITORIAL COMMITTEE

Michael Loss, Chair

John Etnyre Angela Gibney Alex Kontorovich

2020 *Mathematics Subject Classification*. Primary 33C45, 42C05, 65F05.

Library of Congress Cataloging-in-Publication Data

Contemporary Mathematics ISSN: 0271-4132 (print); ISSN: 1098-3627 (online)
DOI: https://doi.org/10.1090/conm/822

Color graphic policy. Any graphics created in color will be rendered in grayscale for the printed version unless color printing is authorized by the Publisher. In general, color graphics will appear in color in the online version.

Copying and reprinting. Individual readers of this publication, and nonprofit libraries acting for them, are permitted to make fair use of the material, such as to copy select pages for use in teaching or research. Permission is granted to quote brief passages from this publication in reviews, provided the customary acknowledgment of the source is given.

Republication, systematic copying, or multiple reproduction of any material in this publication is permitted only under license from the American Mathematical Society. Requests for permission to reuse portions of AMS publication content are handled by the Copyright Clearance Center. For more information, please visit `www.ams.org/publications/pubpermissions`.

Send requests for translation rights and licensed reprints to `reprint-permission@ams.org`.

© 2025 by the American Mathematical Society. All rights reserved.
The American Mathematical Society retains all rights
except those granted to the United States Government.
Printed in the United States of America.

∞ The paper used in this book is acid-free and falls within the guidelines
established to ensure permanence and durability.
Visit the AMS home page at `https://www.ams.org/`

10 9 8 7 6 5 4 3 2 1 30 29 28 27 26 25

Contents

Preface

This publication serves as the proceedings volume for the Special Session on Orthogonal Polynomials and Their Applications, held on January 4–7, 2023, as part of the AMS Joint Mathematics Meetings in Boston, MA. The session featured 21 speakers at various stages of their careers, including graduate students, postdoctoral researchers, and established professors, representing a variety of institutions. The program and abstracts for the talks are available at

`https://jointmathematicsmeetings.org/jmm2023/2270_program_ss47.html`

The contributions in this volume, authored by leading experts in orthogonal polynomials and integrable systems, highlight the breadth of scientific interests within the community and its strong connections to integrable probability, numerical methods, random matrix theory, and special functions. The papers presented here contain new results, recent developments, and outline emerging trends in these areas. Many contributions are based on talks delivered at the special session, and we hope that these proceedings offer a glimpse into the rich theory and numerous applications of orthogonal polynomials.

As organizers and editors, we would like to extend our gratitude to the participants and speakers who joined us in Boston from across the continent, making the session both memorable and productive. Special thanks go to the contributors for their work and for their patience throughout the preparation of this volume. Finally, we would like to express our appreciation to Christine Thivierge from the AMS staff for her invaluable support in the production of these proceedings.

The Editors

Contemporary Mathematics
Volume **822**, 2025
https://doi.org/10.1090/conm/822/16472

The Akhiezer iteration

Cade Ballew and Thomas Trogdon

Abstract. We develop the Akhiezer iteration, a generalization of the classical Chebyshev iteration, for the inner product-free, iterative solution of indefinite linear systems using orthogonal polynomials for measures supported on multiple, disjoint intervals. The iteration applies to shifted linear solves and can then be used for efficient matrix function approximation. Using the asymptotics of orthogonal polynomials, error bounds are provided. A key component in the efficiency of the method is the ability to compute the first k orthogonal polynomial recurrence coefficients and the first k weighted Stieltjes transforms of these orthogonal polynomials in $O(k)$ complexity using a numerical Riemann–Hilbert approach. For a special class of orthogonal polynomials, the Akhiezer polynomials, the method can be sped up significantly, with the greatest speedup occurring in the two interval case where important formulae of Akhiezer are employed and the Riemann–Hilbert approach is bypassed.

1. Introduction

The Chebyshev iteration [13, 17, 20] is an inner product-free iterative method often used to solve symmetric, definite linear systems $\mathbf{Ax} = \mathbf{b}$ where an estimate on the eigenvalues of $\mathbf{A}$ is known. This iteration is based on the minimax property of the Chebyshev polynomials. More specifically, let $\hat{T}_k$ denote the (ortho)normalized kth Chebyshev polynomial shifted and scaled to an interval $[a, b]$ that contains the spectrum of $\mathbf{A}$, but not zero. Then, the polynomial $r_k(x) = \frac{\hat{T}_k(x)}{\hat{T}_k(0)}$ will be small on the spectrum of $\mathbf{A}$, and $q_k(x) = \frac{1-r_k(x)}{x}$ will satisfy $\mathbf{b} - \mathbf{A}q_k(\mathbf{A})\mathbf{b} \approx 0$ as well as a convenient recurrence relation.

A somewhat unconventional variant of this method arises from building a Chebyshev approximation to the function $f(x) = \frac{1}{x}$ on an interval containing the spectrum of $\mathbf{A}$. Specifically, f can be written in a Chebyshev-$\hat{T}$ series as[1]

$$\frac{1}{x} = \sum_{j=0}^{\infty} \left\langle \hat{T}_j, \frac{1}{\diamond} \right\rangle \hat{T}_j(x) = \sum_{j=0}^{\infty} \left(\int_a^b \frac{1}{z} \frac{\hat{T}_j(z)\mathrm{d}z}{\pi\sqrt{z-a}\sqrt{b-z}} \right) \hat{T}_j(x),$$
$$\int_a^b \frac{\hat{T}_j(z)\hat{T}_k(z)}{\pi\sqrt{z-a}\sqrt{b-z}}\mathrm{d}z = \delta_{jk},$$

This work was supported in part by NSF DMS-1945652, DMS-2306438 (TT).

[1]Here $\diamond$ refers to an independent variable, e.g., $1/\diamond$ refers to the function $x \mapsto 1/x$.

©2025 American Mathematical Society

where δ_{jk} is the Kronecker delta. This yields a series for the matrix inverse by replacing $\hat{T}_j(x)$ with the matrix polynomials $\hat{T}_j(\mathbf{A})$. Thus, the solution to the linear system can be approximated as

$$\mathbf{x} = \mathbf{A}^{-1}\mathbf{b} \approx \sum_{j=0}^{k} \left(\int_a^b \frac{\hat{T}_j(z)}{z} \frac{\mathrm{d}z}{\pi\sqrt{z-a}\sqrt{b-z}} \right) \hat{T}_j(\mathbf{A})\mathbf{b} =: \mathbf{x}_{k+1}. \tag{1}$$

This formula can be implemented as an explicit iteration, see Algorithm 1. We

Algorithm 1: Modified Chebyshev iteration

Input: $\mathbf{A}$, $\mathbf{b}$, initial guess $\mathbf{x}_0$, parameters α, c such that $\alpha - c = a$ and $\alpha + c = b$.

Set $\mathbf{b} \leftarrow \mathbf{b} - \mathbf{A}\mathbf{x}_0$.
Set $\mathbf{p}_0 = \mathbf{b}$.
Set $S_0 = \frac{1}{\sqrt{\alpha - c}\sqrt{\alpha + c}}$.
Set $\mathbf{x}_1 = S_0 \mathbf{p}_0$.
for k=1,2,... **do**
  if k=1 **then**
    Set $\mathbf{p}_1 = \frac{1}{c}(\mathbf{A}\mathbf{p}_0 - \alpha\mathbf{p}_0)$.
  else
    Set $\mathbf{p}_k = \frac{2}{c}\mathbf{A}\mathbf{p}_{k-1} - \frac{2\alpha}{c}\mathbf{p}_{k-1} - \mathbf{p}_{k-2}$.
  end
  Set $S_k = S_{k-1}\left(-\frac{\alpha}{c} - \sqrt{-1-\frac{\alpha}{c}}\sqrt{1-\frac{\alpha}{c}}\right)$.
  Set $\mathbf{x}_{k+1} = \mathbf{x}_k + 2S_k\mathbf{p}_k$.
  if converged **then**
    Set $\mathbf{x}_{k+1} \leftarrow \mathbf{x}_{k+1} + \mathbf{x}_0$.
    return $\mathbf{x}_{k+1}$
  end
end

advocate for this modification because, as we describe in this paper, it allows one to directly generalize to general classes of orthogonal polynomials. Furthermore, the asymptotics of such polynomials give error bounds for the iteration. Derivations and comparisons of the classical Chebyshev iteration and our modification are given in Appendix C.

The assumption that $\mathbf{A}$ is symmetric and definite can be relaxed provided that the eigenvalues of $\mathbf{A}$ lie sufficiently close to $[a, b]$ [**20, 21**]. In fact, given an ellipse containing the spectrum of $\mathbf{A}$, but not the origin, one can find an optimal interval $[a, b]$ for which the Chebyshev iteration converges (after possible rotation $\mathbf{A} \to \mathrm{e}^{\mathrm{i}\phi}\mathbf{A}$); however, the iteration fails if there is no such ellipse [**20**].

For instance, assume that the eigenvalues of $\mathbf{A}$ lie in $[a_1, b_1] \cup [a_2, b_2]$ where $a_1 < b_1 < 0 < a_2 < b_2$. In this case, to use the iteration (1), the shifted and scaled Chebyshev polynomials need to be replaced with polynomials with similar properties on multiple intervals to those of the Chebyshev polynomials on a single interval. In the classical view, one needs to replace r_k with polynomials that are small on $[a_1, b_1] \cup [a_2, b_2]$ and large at the origin. de Boor and Rice [**9**] explore using the minimax polynomials on $[a_1, b_1] \cup [a_2, b_2]$ for this, and they present a method to compute such polynomials. Unfortunately, such polynomials do not, in general, satisfy a three-term recurrence relation which makes generating the matrix polynomials difficult. Saad **[27]** circumvents this issue by instead using polynomials that are orthogonal with respect to a weight function w supported on $[a_1, b_1] \cup [a_2, b_2]$. Since such polynomials satisfy a three-term recurrence relation, the difficulty lies in computing the recurrence coefficients and the generalization of the integral terms in (1). Using a clever construction, for this two-interval case, Saad iteratively builds the first k of these quantities in $\mathrm{O}(k^2)$ arithmetic operations using Chebyshev polynomials defined separately on each interval. Specifically, he computes the degree k polynomial q_k that minimizes a weighted L^2 norm $\|1 - \diamond q_k(\diamond)\|_{L^2_w}$ for a chosen weight function w and uses their three-term recurrence to compute $q_k(\mathbf{A})\mathbf{b}$ iteratively. In this work we, in effect, improve upon that in **[27]** by taking $w(x) = x^2 v(x)$ where $v(x)$ is the weight function that Saad employs[2].

In **[3]**, we presented a numerical Riemann–Hilbert-based method for computing recurrence coefficients and Cauchy integrals for orthogonal polynomials with weight functions supported on multiple, disjoint intervals (see **[25, 31]** for a method for measures supported on $\mathbb{R}$). This method computes the first k recurrence coefficients and Cauchy integrals in $\mathrm{O}(k)$ arithmetic operations. These quantities can be employed analogously to (1), yielding a generalization of the Chebyshev iteration that improves the complexity of **[27]**.

In the current work, we discuss how a specific choice of weight function for the orthogonal polynomials allows the numerical method of **[3]** to be sped up significantly — reducing the constant in the $\mathrm{O}(k)$ bound. However, this can be further improved in an even more special case. In **[1]**, Akhiezer derived explicit formulae for the orthogonal polynomials with respect to a specific Chebyshev like weight function supported on two disjoint intervals. We review Akhiezer's result and demonstrate how it, along with our numerical method from **[3]**, can be used to extend (1) to an iterative method for solving $\mathbf{A}\mathbf{x} = \mathbf{b}$ where $\mathbf{A}$ has eigenvalues on, or near, multiple, disjoint intervals. We will refer to this method as *the Akhiezer iteration.* Furthermore, we discuss how the Akhiezer iteration can be extended to compute matrix functions $f(\mathbf{A})\mathbf{b}$ for suitably analytic functions f, similar to the approach in **[6]**, and implemented in optimal complexity with a provably exponential convergence rate. In Figure 1, we demonstrate the convergence of the Akhiezer iteration applied to a symmetric matrix with spectrum contained in $[-3, -1] \cup [2, 4]$.

The layout of the paper is as follows. In Section 2, we establish notation for orthogonal polynomials and their Cauchy integrals. In Section 3, we present Akhiezer's orthogonal polynomial construction and discuss how it can be used to computed recurrence coefficients and Cauchy integrals. In Section 4, we present the Akhiezer iteration and its extension to matrix functions and discuss convergence rates. In Section 5, we discuss how the Akhiezer iteration can be implemented adaptively when $\mathbf{A}$ is symmetric. In Section 6, we present some examples and applications, and in Section 7, we present our conclusions and discuss potential extensions of our work. Code used to generate the plots in this paper can be found at **[2]**.

2. Orthogonal polynomials and their Cauchy integrals

2.1. Orthogonal polynomials. For our purposes, a weight function w is a nonnegative function defined on a finite union of intervals Σ, $\Sigma \subset \mathbb{R}$, that is

[2] In **[27]**, $v(x)$ is found by using h_j constant, $\alpha_j = \beta_j = -1$ in (4).

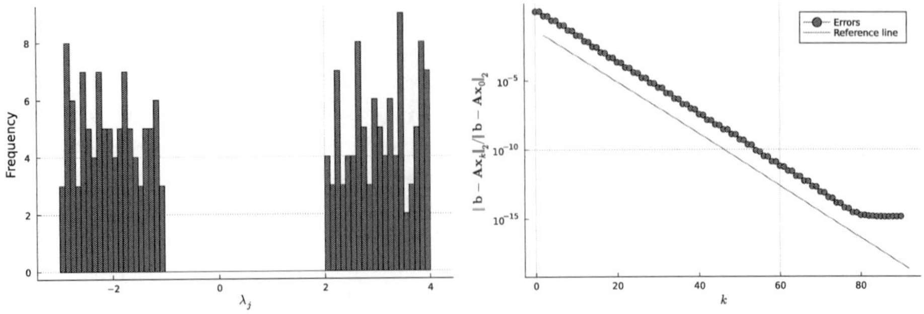

FIGURE 1. Histogram of the eigenvalues of $\mathbf{A} \in \mathbb{R}^{200\times 200}$ (left) and the convergence of the Akhiezer iteration applied to $\mathbf{Ax} = \mathbf{b}$ where $\mathbf{b} \in \mathbb{R}^{200}$ is a random Gaussian vector. The reference line that approximates the convergence rate is derived in Section 4.1.1.

continuous and positive on the interior of Σ such that $\int_\Sigma w(x)\mathrm{d}x = 1$. Consider a sequence of univariate monic polynomials $\pi_0(x), \pi_1(x), \pi_2(x), \ldots$ such that π_j has degree j for all $j \in \mathbb{N}$. These polynomials are said to be orthogonal with respect to a weight function w if

$$\langle \pi_j, \pi_k \rangle_{L^2_w(\Sigma)} = h_j \delta_{jk},$$

where $h_j > 0$, δ_{jk} is the Kronecker delta, and

$$\langle g, h \rangle_{L^2_w(\Sigma)} = \int_\Sigma g(x)\overline{h(x)} w(x)\mathrm{d}x, \quad \|g\|_{L^2_w(\Sigma)} = \sqrt{\langle g, g\rangle_{L^2_w(\Sigma)}}.$$

The orthonormal polynomials $p_0(x), p_1(x), p_2(x), \ldots$ are defined by

$$p_j(x) = \frac{1}{\sqrt{h_j}} \pi_j(x),$$

for all $j \in \mathbb{N}$. The orthonormal polynomials satisfy a symmetric three-term recurrence

$$\begin{aligned} xp_0(x) &= a_0 p_0(x) + b_0 p_1(x), \\ xp_k(x) &= b_{k-1} p_{k-1}(x) + a_k p_k(x) + b_k p_{k+1}(x), \quad k \geq 1, \end{aligned} \tag{2}$$

where $b_k > 0$ for all k. A general reference is [**28**].

2.2. Cauchy integrals. Given a curve $\Gamma \subset \mathbb{C}$ and a function $f : \Gamma \to \mathbb{C}$, the Cauchy transform $\mathcal{C}_\Gamma$ is an operator that maps f to its Cauchy integral, i.e.,

$$\mathcal{C}_\Gamma f(z) = \frac{1}{2\pi \mathrm{i}} \int_\Gamma \frac{f(s)}{s - z} \mathrm{d}s, \quad z \in \mathbb{C} \setminus \Gamma.$$

Cauchy integrals and Stieltjes transforms of orthogonal polynomials will also be of use in this work. For orthonormal polynomials $p_k(x)$ corresponding to the weight function w on $\Sigma \subset \mathbb{R}$, define

$$C_k(z) = \mathcal{C}_\Sigma[p_k w](z) = \frac{1}{2\pi \mathrm{i}} \int_\Sigma \frac{p_k(s) w(s)}{s - z} \mathrm{d}s, \quad \mathcal{S}_\Sigma[p_k w](z) = 2\pi \mathrm{i} \mathcal{C}_\Sigma[p_k w](z),$$

for $z \in \mathbb{C} \setminus \Sigma$. It is well-known that if the three-term recurrence for p_k is given by (2), these Cauchy integrals also satisfy the three-term recurrence [**24**]:

$$\begin{aligned} zC_0(z) &= a_0 C_0(z) + b_0 C_1(z) - \frac{1}{2\pi \mathrm{i}}, \\ zC_k(z) &= b_{k-1}C_{k-1}(z) + a_k C_k(z) + b_k C_{k+1}(z), \quad k \geq 1. \end{aligned} \tag{3}$$

REMARK 2.1. It is important to note that the direct numerical use of (3) is hampered by the fact that it is intrinsically unstable. Cauchy integrals decay exponentially in the complex plane, $z \notin \Sigma$, whereas the orthogonal polynomials grow exponentially. Any error, say one on the order of machine precision, will be amplified exponentially as the recurrence is run.

3. Akhiezer polynomials

In [**3**], we described a Riemann–Hilbert-based numerical method for computing orthogonal polynomials corresponding to genus g weight functions of the form

$$w(x) = \sum_{j=1}^{g+1} \mathbb{1}_{[a_j,b_j]}(x) h_j(x) \left(\sqrt{x-a_j}\right)^{\alpha_j} \left(\sqrt{b_j - x}\right)^{\beta_j}, \quad \alpha_j, \beta_j \in \{-1, 1\}, \tag{4}$$

where $[a_j, b_j]$, $j = 1, \ldots, g+1$ are disjoint and h_j is positive on $[a_j, b_j]$ and has an analytic extension to a neighborhood Ω_j of $[a_j, b_j]$. A judicious choice of h_j yields the weight function

$$w(x) \propto \mathbb{1}_\Sigma(x) \frac{\sqrt{b_{g+1} - x} \prod_{j=1}^{g+1} \sqrt{x - a_j}}{\prod_{j=1}^{g} \sqrt{x - b_j}}. \tag{5}$$

If the weight function is restricted to be of the form (5), the underlying Riemann–Hilbert problem can be simplified greatly, and the numerical method can be sped up significantly. See Appendix A for a full discussion of this. In a sense, weight functions of the form (5) are a natural extension of the Chebyshev polynomials, as they limit to a scaled and shifted Chebyshev-second-kind weight as the gaps between the intervals approach zero.

The Akhiezer polynomials are defined as the orthogonal polynomials with respect to the weight function

$$w(x) = \frac{1}{\pi} \mathbb{1}_\Sigma(x) \frac{\prod_{j=1}^{g} \sqrt{x - b_j}}{\sqrt{b_{g+1} - x} \prod_{j=1}^{g+1} \sqrt{x - a_j}}, \tag{6}$$

which is the normalized reciprocal of (5) and corresponds to a Chebyshev-first-kind-like weight [**7**]. Formulae for such polynomials can be worked out explicitly in terms of Riemann theta functions [**8**]. In the case $g = 1$, the Riemann theta functions reduce to Jacobi theta functions, and the polynomials can be computed efficiently via these theta functions. In the following, we summarize Akhiezer's construction of the polynomials from [**1**, Section 53] and derive formulae for their Cauchy integrals and recurrence coefficients.

3.1. Akhiezer's formulae and notation. Following Akhiezer's convention, we consider[3] $\Sigma = [-1, \alpha] \cup [\beta, 1]$ such that $-1 < \alpha < \beta < 1$ and define the weight

[3]This can be generalized to any configuration of two disjoint intervals by shifting and scaling the resulting polynomials as appropriate.

function

$$w(x) = \frac{1}{\pi}\mathbb{1}_\Sigma(x)\frac{\sqrt{x-\alpha}}{\sqrt{1-x}\sqrt{x+1}\sqrt{x-\beta}}. \tag{7}$$

Then, the orthonormal polynomials with respect to the weight function w can be computed according to the following procedure: First, compute the elliptic modulus[4] [**1**, Equation 6]

$$k = \sqrt{\frac{2(\beta-\alpha)}{(1-\alpha)(1+\beta)}},$$

which will serve as the elliptic modulus in all Jacobi elliptic functions that follow. Additionally, $K = K(k)$ is the complete elliptic integral:

$$K(k) = \int_0^1 \frac{\mathrm{d}t}{\sqrt{(1-t^2)(1-k^2t^2)}} = \int_0^{\pi/2} \frac{\mathrm{d}\theta}{\sqrt{1-k^2\sin^2\theta}}.$$

Then, the parameter ρ is defined so that

$$1 - 2\,\mathrm{sn}^2\,\rho = \alpha,$$

and $0 < \rho < K$ where $\mathrm{sn}(\diamond) = \mathrm{sn}(\diamond, k)$ denotes the Jacobi sn function [**1**, Equation 10]. Since Jacobi elliptic functions are defined as the inverses of elliptic integrals, ρ can be written as

$$\rho = F\left(\arcsin\sqrt{\frac{1-\alpha}{2}}, k\right),$$

where F is the incomplete elliptic integral of the first kind defined by

$$F(\phi, k) = \int_0^{\sin\phi} \frac{\mathrm{d}t}{\sqrt{(1-t^2)(1-k^2t^2)}} = \int_0^{\phi} \frac{\mathrm{d}\theta}{\sqrt{1-k^2\sin^2\theta}}.$$

Now, given $x \in \mathbb{C}$, define $u = u(x)$ such that [**1**, Equation 9]

$$x - \alpha = \frac{1-\alpha^2}{2\,\mathrm{sn}^2(u) + \alpha - 1}.$$

An explicit formula for u is

$$u = F\left(\arcsin\sqrt{\frac{(\alpha-1)(1+x)}{2(\alpha-x)}}, k\right). \tag{8}$$

Define the Jacobi theta functions

$$\theta_1(z,q) = 2\sum_{j=0}^{\infty}(-1)^j q^{\left(j+\frac{1}{2}\right)^2}\sin\left((2j+1)z\right),$$

$$\theta_4(z,q) = 1 + 2\sum_{j=1}^{\infty}(-1)^j q^{j^2}\cos\left(2jz\right).$$

In what follows, we will always take the parameter q to be given by

$$q = q(k) = \exp\left(-\pi\frac{K(\sqrt{1-k^2})}{K(k)}\right).$$

[4]In this subsection, we follow standard notation and let k be the elliptic modulus, not the iteration number of an algorithm.

Now, define the functions

$$H(z) = \theta_1\left(\frac{\pi z}{2K}, q\right), \quad \Theta(z) = \theta_4\left(\frac{\pi z}{2K}, q\right).$$

Then, the nth degree orthonormal polynomial p_n is given by [**1**]

$$p_n(x) = \frac{C_n}{2}\left(\left(\frac{H(u-\rho)}{H(u+\rho)}\right)^n \frac{\Theta(u+2n\rho)}{\Theta(u)} + \left(\frac{H(u+\rho)}{H(u-\rho)}\right)^n \frac{\Theta(u-2n\rho)}{\Theta(u)}\right), \tag{9}$$

where

$$C_n = \frac{\sqrt{2}\Theta(\rho)}{\sqrt{\Theta((2n-1)\rho)\,\Theta((2n+1)\rho)}},$$

for $n \geq 1$, $C_0 = 1$, and u is given by (8).

Akhiezer's derivation also includes a formula for the Cauchy integrals of these polynomials:

$$\mathcal{C}_\Sigma[p_n w](x) = -\frac{C_n}{2\pi \mathrm{i}}\left(\frac{H(u-\rho)}{H(u+\rho)}\right)^n \frac{\Theta(u+2n\rho)}{\Theta(u)} \frac{\sqrt{x-\alpha}}{\sqrt{x-1}\sqrt{x+1}\sqrt{x-\beta}}, \tag{10}$$

with u and C_n defined as above.

In Appendix B, we derive the following formulae for the recurrence coefficients of the Akhiezer polynomials:

(11)

$$\begin{aligned}
a_0 &= (1-\alpha^2)\left(\frac{1}{8\,\mathrm{sn}^2\rho} + \frac{\mathrm{sn}''\rho}{8\,\mathrm{sn}\,\rho\,(\mathrm{sn}'\rho)^2} + \frac{H'(2\rho)}{H(2\rho)}\frac{1}{4\,\mathrm{sn}\,\rho\,\mathrm{sn}'\rho}\right) \\
&\quad + \frac{1-\alpha^2}{2\,\mathrm{sn}\,\rho\,\mathrm{sn}'\rho}\frac{\Theta'(\rho)}{\Theta(\rho)} + \alpha, \\
a_n &= (1-\alpha^2)\left(\frac{1}{8\,\mathrm{sn}^2\rho} + \frac{\mathrm{sn}''\rho}{8\,\mathrm{sn}\,\rho\,(\mathrm{sn}'\rho)^2}\right) \\
&\quad + \frac{1-\alpha^2}{4\,\mathrm{sn}\,\rho\,\mathrm{sn}'\rho}\left(\frac{H'(2\rho)}{H(2\rho)} + \frac{\Theta'((2n+1)\rho)}{\Theta((2n+1)\rho)} - \frac{\Theta'((2n-1)\rho)}{\Theta((2n-1)\rho)}\right) + \alpha, \quad n \geq 1, \\
b_0 &= \frac{1-\alpha^2}{4\,\mathrm{sn}\,\rho\,\mathrm{sn}'\rho}\frac{H'(0)}{H(2\rho)}\sqrt{\frac{2\Theta(3\rho)}{\Theta(\rho)}}, \\
b_n &= \frac{1-\alpha^2}{4\,\mathrm{sn}\,\rho\,\mathrm{sn}'\rho}\frac{H'(0)}{H(2\rho)}\frac{\sqrt{\Theta((2n+3)\rho)\Theta((2n-1)\rho)}}{\Theta((2n+1)\rho)}, \quad n \geq 1.
\end{aligned}$$

Both the polynomials themselves and their Cauchy integrals can be efficiently computed via standard algorithms for computing Jacobi theta functions and elliptic integrals [**12**, Sections 19.36, 20.14] while the recurrence coefficients can be efficiently computed via standard algorithms for computing Jacobi elliptic [**12**, Section 22.20] and Jacobi theta functions. In particular, the parameter q is typically very small and grows very slowly as the size of the intervals shrinks, so the Jacobi theta functions can be computed by simply summing the first few terms of their respective series. However, one weakness of these formulae is that the mapped variable $u(x)$ is undefined at $x = \alpha$. While, by continuity, $u(\alpha) = \mathrm{i}K(\sqrt{1-k^2})$, the non-Lipschitz behavior of the involved functions leads to numerical instabilities when this value is used in the formulae. Thus, if one needs to compute the polynomials at this endpoint, it is better to generate them via the recurrence coefficients rather than using (9).

Otherwise, we find that these formulae are extremely efficient for evaluating the Akhiezer polynomials, their Cauchy integrals, and their recurrence coefficients in this special $g = 1$ case, outperforming by multiple orders of magnitude both our $\mathrm{O}(n)$ Riemann–Hilbert approach and the optimized $\mathrm{O}(n^2)$ algorithm given as RKPW in [**16**] and `lanczos.m` in [**15**]. Thus, these formulae should be used for all two interval problems for which one can choose the weight function to be of the form (7). For higher genus problems, choosing a weight function of the form (5) allows our Riemann–Hilbert approach to be applied efficiently at all degrees. If one wishes to consider a more general weight function of the form (4), they should consider applying the RKPW algorithm for the low degree polynomials then switching to the asymptotic version of our Riemann–Hilbert approach for higher degrees. The Cauchy integrals for the lower degree polynomials can then be obtained by applying the three-term recurrence (3) to those obtained from our Riemann–Hilbert approach. Indeed, (3) can be written as the linear system

$$\begin{pmatrix} a_0 - z & b_0 & & & \\ b_0 & a_1 - z & b_1 & & \\ & b_1 & a_2 - z & b_2 & \\ & & \ddots & \ddots & \ddots \end{pmatrix} \begin{pmatrix} C_0(z) \\ C_1(z) \\ C_2(z) \\ \vdots \end{pmatrix} = \begin{pmatrix} \frac{1}{2\pi \mathrm{i}} \\ 0 \\ 0 \\ \vdots \end{pmatrix}.$$

If for some N, $C_{N+2}(z)$ and $C_{N+1}(z)$ are known, then the truncated linear system linear system

$$\begin{pmatrix} a_0 - z & b_0 & & & \\ b_0 & a_1 - z & b_1 & & \\ & \ddots & \ddots & \ddots & \\ & & b_{N-2} & a_{N-1} - z & b_{N-1} \\ & & & b_{N-1} & a_N - z \\ & & & & b_N \end{pmatrix} \begin{pmatrix} C_0(z) \\ C_1(z) \\ \vdots \\ C_{N-1}(z) \\ C_{N-2}(z) \\ C_N(z) \end{pmatrix} = \begin{pmatrix} \frac{1}{2\pi \mathrm{i}} \\ 0 \\ \vdots \\ 0 \\ -b_N C_{N+1}(z) \\ (z - a_{N+1}) C_{N+1}(z) - b_{N+1} C_{N+2}(z) \end{pmatrix},$$

can be solved efficiently to obtain $C_0(z), \ldots, C_N(z)$. As noted previously, the forward recurrence (3) becomes more numerically unstable the further z is from Σ, so it is not, in general, a reliable way to compute Cauchy integrals.

4. The Akhiezer iteration and matrix function evaluation

4.1. The Akhiezer iteration for solving linear systems. Following [**27**], orthogonal polynomials with respect to a weight function w of the form (4) (in particular (7)) can be used to extend the classical Chebyshev iteration to solve indefinite linear systems. Assume that a square matrix $\mathbf{A}$ has eigenvalues on, or near, $\Sigma = \bigcup_{j=1}^{g+1} [a_j, b_j]$ and that $0 \notin \Sigma$. If $p_0, p_1, \ldots$ denote the orthonormal polynomials with respect to w, then a p_j-series expansion for $1/(x - z)$, for $x \in \Sigma$,

$z \notin \Sigma$, is given by

$$\frac{1}{x-z} = \sum_{j=0}^{\infty} \left\langle p_j, \frac{1}{\diamond - z} \right\rangle p_j(x) = \sum_{j=0}^{\infty} \left(\int_{\Sigma} \frac{p_j(s)}{s-z} w(s) \mathrm{d}s \right) p_j(x) \tag{12}$$

$$= \sum_{j=0}^{\infty} \mathcal{S}_{\Sigma} \left[p_j w \right] (z) p_j(x).$$

This can be extended to an iterative method for solving the linear system $\mathbf{Ax} = \mathbf{b}$ by truncating the series with $z = 0$:

$$\mathbf{x} = \mathbf{A}^{-1}\mathbf{b} = \sum_{j=0}^{\infty} \mathcal{S}_{\Sigma} \left[p_j w \right] (0) p_j(\mathbf{A}) \mathbf{b} \approx \sum_{j=0}^{k} \mathcal{S}_{\Sigma} \left[p_j w \right] (0) p_j(\mathbf{A}) \mathbf{b} =: \mathbf{x}_{k+1}. \tag{13}$$

We call the iterative implementation of this approximation *the Akhiezer iteration.* The recurrence coefficients and Cauchy integrals can be computed either by our Riemann–Hilbert approach [**3**] or by (10) and (11) in the case of the $g = 1$ Akhiezer weight. Then, the Akhiezer iteration for shifted linear solves can be implemented as in Algorithm 2.

Algorithm 2: Akhiezer iteration to approximate $(\mathbf{A} - z\mathbf{I})^{-1}\mathbf{b}$, $z \notin \Sigma$

Input: $\mathbf{A}$, $\mathbf{b}$, initial guess $\mathbf{x}_0$, and a function to compute recurrence coefficients a_k, b_k and Stieltjes transforms $S_k = \mathcal{S}_{\Sigma} \left[p_k w \right] (z)$.
Set $\mathbf{b} \leftarrow \mathbf{b} - \mathbf{A}\mathbf{x}_0$.
Set $\mathbf{p}_0 = \mathbf{b}$.
Set $\mathbf{x}_1 = S_0 \mathbf{p}_0$.
for *k=1,2,...* **do**
  if *k=1* **then**
    Set $\mathbf{p}_1 = \frac{1}{b_0}(\mathbf{A}\mathbf{p}_0 - a_0 \mathbf{p}_0)$.
  else
    Set $\mathbf{p}_k = \frac{1}{b_{k-1}}(\mathbf{A}\mathbf{p}_{k-1} - a_{k-1}\mathbf{p}_{k-1} - b_{k-2}\mathbf{p}_{k-2})$.
  end
  Set $\mathbf{x}_{k+1} = \mathbf{x}_k + S_k \mathbf{p}_k$.
  if converged **then**
    Set $\mathbf{x}_{k+1} \leftarrow \mathbf{x}_{k+1} + \mathbf{x}_0$.
    return $\mathbf{x}_{k+1}$
  end
end

Algorithm 2 can instead be utilized to build an approximation to the matrix inverse $\mathbf{A}^{-1}$ by using $\mathbf{b} = \mathbf{I}$ and removing the initial guess $\mathbf{x}_0$. Then, at each step, the algorithm requires matrix-matrix multiplication rather than matrix-vector products but is otherwise identical.

REMARK 4.1. In (12), one can differentiate with respect to z a number of times to find expansions for x^{-p} for $p \in \mathbb{N}$, provided one can compute derivatives of the Stieltjes transforms $\mathcal{S}_{\Sigma}$. The Riemann–Hilbert approach [**3**] allows for this computation in a fairly straightforward manner, but we do not implement this here.

4.1.1. *Convergence.* From [**11**], we have the following behavior of the orthogonal polynomials:

LEMMA 4.2. *Fix $\epsilon > 0$ and the set $V = \{z \in \mathbb{C}\colon |z - a_j| \geq \epsilon, |z - b_j| \geq \epsilon, j = 1, \ldots, g+1\}$. Then,*

$$p_n(z) = \hat{\delta}_n(z)\mathrm{e}^{n\mathfrak{g}(z)}, \qquad 2\pi\mathrm{i}\mathcal{C}_\Sigma\left[p_n w\right](z) = \delta_n(z)\mathrm{e}^{-n\mathfrak{g}(z)}, \quad z \in V,$$

where $\delta_n(z)$ and $\hat{\delta}_n(z)$ are uniformly bounded in both n and $z \in V$ and $\mathfrak{g}$ is the exterior Green's function with pole at infinity[5] *corresponding to Σ. Furthermore, $\mathrm{e}^{\mathfrak{g}(z)}$ is a conformal map from $\mathbb{C} \setminus \Sigma$ to the exterior of the unit disk satisfying $\operatorname{Re}\mathfrak{g}(z) = 0$ for $z \in \Sigma$.*

PROOF. By [**11**, Equations B.2, B.3] and undoing the lensing step of the Riemann–Hilbert problem deformations, the above formulae hold for $\delta_n(z)$ and $\hat{\delta}_n(z)$ depending on bounded quantities and $\prod_{j=0}^{n-1} \frac{1}{b_j \mathfrak{c}}$ where $\log \mathfrak{c}$ is the $O(1)$ term in the expansion of $\mathfrak{g}$ at infinity. By [**11**, Equations B.3, B.12],

$$\mathfrak{c}^n \prod_{j=0}^{n-1} b_j = \mathrm{O}(1),$$

as $n \to \infty$, so $\delta_n(z)$ and $\hat{\delta}_n(z)$ are bounded from above. The rest of the claims follow immediately from [**11**]. □

In the $g = 1$ case,

$$\mathrm{e}^{\mathfrak{g}(z)} = -\frac{H(u+\rho)}{H(u-\rho)},$$

in the notation of Akhiezer's formulation (9). When g is larger than one, $\mathfrak{g}(z)$ can be computed numerically as described in [**3**, Section 4.2].

To ease presentation, we introduce the following definitions.

DEFINITION 4.3. We say that a matrix is generic if it is diagonalizable and has no eigenvalues at the endpoints of Σ.

This definition is here to simplify the analysis, and our theorems require the matrix in question to be generic. With a more technical approach, one can obtain similar bounds for non-generic matrices and the algorithms in question still converge at roughly the same rate. We remark on this below.

DEFINITION 4.4. Given Σ and $\mathbf{A} \in \mathbb{C}^{n \times n}$ define

$$\nu(z; \mathbf{A}) = \max_j \operatorname{Re}\mathfrak{g}(\lambda_j) - \operatorname{Re}\mathfrak{g}(z), \quad \nu(\mathbf{A}) = \max_j \operatorname{Re}\mathfrak{g}(\lambda_j).$$

where λ_j are the eigenvalues of $\mathbf{A}$.

In the following, $\|\diamond\|_2$ will denote the Euclidean 2-norm and the induced matrix norm.

THEOREM 4.5. *Consider a density w of the form (4). If $\mathbf{A} \in \mathbb{C}^{n \times n}$ is generic and its eigenvalues λ_j all satisfy $\operatorname{Re}\mathfrak{g}(\lambda_j) < \operatorname{Re}\mathfrak{g}(z)$, then there exists $c > 0$, depending only on the eigenvalues of $\mathbf{A}$, such that*

$$E_k(z) = \left\| \sum_{j-0}^{k-1} \mathcal{S}_\Sigma\left[p_j w\right](z) p_j(\mathbf{A})\mathbf{b} - (\mathbf{A} - z\mathbf{I})^{-1}\mathbf{b} \right\|_2,$$

[5]See Appendix A for the construction of $\mathfrak{g}$. We point out that it is determined by Σ alone.

satisfies

$$E_k(z) \leq c\|\mathbf{V}\|_2 \left\|\mathbf{V}^{-1}\right\|_2 \|\mathbf{b}\|_2 \frac{\mathrm{e}^{k\nu(z;\mathbf{A})}}{1-\mathrm{e}^{\nu(z;\mathbf{A})}},$$

where $\mathbf{A}$ *is diagonalized as* $\mathbf{A} = \mathbf{V}\mathbf{\Lambda}\mathbf{V}^{-1}$, $\mathbf{\Lambda} = \operatorname{diag}(\lambda_1, \ldots, \lambda_n)$. *In particular, the Akhiezer iteration converges, at worst, at a geometric rate given by* $\mathrm{e}^{\nu(z;\mathbf{A})}$ *if* $z \notin \Sigma$ *since*

$$\|\mathbf{x}_k - (\mathbf{A} - z\mathbf{I})^{-1}\mathbf{b}\|_2 = E_k(z).$$

PROOF. We have

$$\begin{aligned} E_k(z) &= \left\| \sum_{j=0}^{k-1} \mathcal{S}_\Sigma [p_j w](z) p_j(\mathbf{A})\mathbf{b} - (\mathbf{A} - z\mathbf{I})^{-1}\mathbf{b} \right\|_2 = \left\| \sum_{j=k}^{\infty} \mathcal{S}_\Sigma [p_j w](z) p_j(\mathbf{A})\mathbf{b} \right\|_2 \\ &= \left\| \sum_{j=k}^{\infty} 2\pi \mathrm{i} \mathcal{C}_\Sigma [p_j w](z) \mathbf{V} p_j(\mathbf{\Lambda}) \mathbf{V}^{-1}\mathbf{b} \right\|_2 \\ &\leq \|\mathbf{V}\|_2 \left\|\mathbf{V}^{-1}\right\|_2 \|\mathbf{b}\|_2 \sum_{j=k}^{\infty} \max_{\ell \in \{1,\ldots,n\}} \left|2\pi \mathrm{i} \mathcal{C}_\Sigma [p_j w](z) p_j(\lambda_\ell)\right|. \end{aligned}$$

Since λ_ℓ is not at an endpoint of Σ, by Lemma 4.2,

$$\left|2\pi \mathrm{i} \mathcal{C}_\Sigma [p_j w](z) p_j(\lambda_\ell)\right| = |\delta_j(z)||\hat{\delta}_j(\lambda_\ell)| \mathrm{e}^{j \operatorname{Re}(\mathfrak{g}(\lambda_\ell) - \mathfrak{g}(z))} \leq c \mathrm{e}^{j\nu(z;\mathbf{A})},$$

for some constant c. The theorem follows from

$$\sum_{j=k}^{\infty} \mathrm{e}^{j\nu(z;\mathbf{A})} = \frac{\mathrm{e}^{k\nu(z;\mathbf{A})}}{1-\mathrm{e}^{\nu(z;\mathbf{A})}}.$$

□

REMARK 4.6. Theorem 4.5 could, in principle, be extended to a broader class of matrices. For instance, if $\mathbf{A}$ does have eigenvalues at the endpoints of Σ, the orthogonal polynomials at these endpoints are either bounded or grow at a polynomial rate. Since the convergence rate is geometric, this growth could be accounted for in the error bound without inhibiting convergence. However, one could instead perturb Σ so that these eigenvalues lie in the interior, and the iteration will suffer a small perturbation to its geometric rate of convergence.

REMARK 4.7. Jordan blocks, i.e., degenerate eigenspaces for $\mathbf{A}$, could be accounted for by investigating the derivatives of the orthogonal polynomials. Due to the Markov brothers' inequality, [**26**, Theorem 1.10], it is clear that this will not change the exponential rate of convergence. Yet, more refined estimates are possible.

4.2. The Akhiezer iteration for matrix function approximation. The Akhiezer iteration can be extended to compute matrix functions for square matrices $\mathbf{A}$ with eigenvalues on or near $\Sigma = \bigcup_{j=1}^{g+1}[a_j, b_j]$. Consider the Cauchy integral representation of a matrix function

$$f(\mathbf{A})\mathbf{b} = \frac{1}{2\pi\mathrm{i}} \int_\Gamma f(z)(z\mathbf{I} - \mathbf{A})^{-1}\mathbf{b}\,\mathrm{d}z,$$

where Γ is a counterclockwise oriented curve that encloses the spectrum of $\mathbf{A}$ and f is analytic in a region containing Γ. Applying a quadrature rule, such as a trapezoid rule [**30**], with nodes z_j and weights w_j, this can be approximated as

$$f(\mathbf{A})\mathbf{b} \approx \frac{1}{2\pi \mathrm{i}} \sum_{j=1}^{m} f(z_j) w_j (z_j \mathbf{I} - \mathbf{A})^{-1} \mathbf{b}.$$

In practice, we find that it typically suffices to take $\Gamma = \bigcup_{j=1}^{g+1} \Gamma_j$ to be a collection of circles around each $[a_j, b_j]$. In this case the trapezoid rule on each Γ_k has weights and nodes

$$w_j = \frac{2\pi \mathrm{i} r_k}{m_k} \mathrm{e}^{\mathrm{i}\theta_j^{(k)}}, \quad z_j = r_k \mathrm{e}^{\mathrm{i}\theta_j^{(k)}} + c_k, \quad j = 1, \dots, m_k,$$

where $\{\theta_j^{(k)}\}_{j=1}^{m_k}$ are evenly spaced points in $[0, 2\pi)$, r_k and c_k are the radius and center of Γ_k, respectively, and m_k is the number of quadrature nodes on Γ_k, $\sum_{j=1}^{g+1} m_j = m$. For convenience, we reorder the weights and nodes for all contours Γ_k as

$$z_j, w_j, \quad j = 1, 2, \dots, m,$$

with the first m_1 corresponding to Γ_1, the next m_2 corresponding to Γ_2, and so on.

To approximate $f(\mathbf{A})\mathbf{b}$, the resolvent is approximated by

$$(z_j \mathbf{I} - \mathbf{A})^{-1} = -\sum_{\ell=0}^{\infty} \mathcal{S}_\Sigma [p_\ell w](z_j) p_\ell(\mathbf{A}) \approx -\sum_{\ell=0}^{k-1} \mathcal{S}_\Sigma [p_\ell w](z_j) p_\ell(\mathbf{A}),$$

and the matrix function is approximated by

$$f(\mathbf{A})\mathbf{b} \approx \mathbf{f}_k := f_{k,m}(\mathbf{A})\mathbf{b} := -\sum_{\ell=0}^{k-1} \left(\sum_{j=1}^{m} f(z_j) w_j \mathcal{C}_\Sigma [p_\ell w](z_j) \right) p_\ell(\mathbf{A})\mathbf{b}. \tag{14}$$

As before, the recurrence coefficients and Cauchy integrals can be precomputed either by our Riemann–Hilbert approach [**3**] or by (10) and (11) in the case of the $g = 1$ Akhiezer weight. Then, the Akhiezer iteration for matrix function approximation can be implemented as in Algorithm 3.

REMARK 4.8. The approximation to f need not be from the discretization of a Cauchy integral. It could be from a Padé approximation, a AAA approximation [**23**], or some other form of rational approximation. As long as the approximation can be written in pole-residue form, the Akhiezer iteration can be applied. We refer the reader back to Remark 4.1 for the case when the approximation contains more than just simple poles.

4.2.1. *Convergence.*

THEOREM 4.9. *Suppose* $\mathbf{A} \in \mathbb{C}^{n\times n}$ *is generic, its eigenvalues* λ_ℓ *satisfy* $\operatorname{Re} \mathfrak{g}(\lambda_\ell) < \min_{j=1,\dots,m} \operatorname{Re} \mathfrak{g}(z_j)$, $|f(z)| \le M$ *for* $z \in \Gamma$, *and* $\sum_{j=1}^{m} |w_j| \le 2\pi L$. *Then,*

$$\begin{aligned} &\|f(\mathbf{A})\mathbf{b} - f_{k,m}(\mathbf{A})\mathbf{b}\|_2 \\ &\quad \le \frac{1}{2\pi} \|\mathbf{V}\|_2 \left\|\mathbf{V}^{-1}\right\|_2 \|\mathbf{b}\|_2 \max_{\ell \in \{1,\dots,n\}} |\rho_m(\lambda_\ell)| + LM \max_{j \in \{1,\dots,m\}} E_k(z_j), \end{aligned}$$

Algorithm 3: Akhiezer iteration for matrix function approximation

---

Input: f, $\mathbf{A}$, $\mathbf{b}$, quadrature nodes $z_1, \ldots, z_m$, quadrature weights $w_1, \ldots, w_m$, and a function to compute recurrence coefficients a_k, b_k and Cauchy integrals $C_{k,j} = \mathcal{C}_\Sigma [p_k w](z_j)$.

Set $\mathbf{f}_0 = 0$.

for k=*0,1,...* **do**

  if k=*0* **then**

    Set $\mathbf{p}_0 = \mathbf{b}$.

  else if k=*1* **then**

    Set $\mathbf{p}_1 = \frac{1}{b_0}(\mathbf{A}\mathbf{p}_0 - a_0\mathbf{p}_0)$.

  else

    Set $\mathbf{p}_k = \frac{1}{b_{k-1}}(\mathbf{A}\mathbf{p}_{k-1} - a_{k-1}\mathbf{p}_{k-1} - b_{k-2}\mathbf{p}_{k-2})$.

  end

  Set $\alpha_k = -\sum_{j=1}^m C_{k,j} f(z_j) w_j$.

  Set $\mathbf{f}_{k+1} = \mathbf{f}_k + \alpha_k \mathbf{p}_k$.

  if converged **then**

    return $\mathbf{f}_{k+1}$.

  end

end

---

where $\mathbf{A}$ *is diagonalized as* $\mathbf{A} = \mathbf{V}\mathbf{\Lambda}\mathbf{V}^{-1}$, $\mathbf{\Lambda} = \operatorname{diag}(\lambda_1, \ldots, \lambda_n)$, *and*

$$\rho_m(x) = \int_\Gamma \frac{f(z)}{z - x}\mathrm{d}z - \sum_{j=1}^m \frac{f(z_j)}{z_j - x} w_j.$$

PROOF. By the triangle inequality,

$$\|f(\mathbf{A})\mathbf{b} - f_{k,m}(\mathbf{A})\mathbf{b}\|_2 \leq \left\| f(\mathbf{A})\mathbf{b} - \frac{1}{2\pi \mathrm{i}} \sum_{j=1}^m f(z_j) w_j (z_j \mathbf{I} - \mathbf{A})^{-1}\mathbf{b} \right\|_2$$
$$+ \left\| \frac{1}{2\pi \mathrm{i}} \sum_{j=1}^m f(z_j) w_j (z_j \mathbf{I} - \mathbf{A})^{-1}\mathbf{b} - f_{k,m}(\mathbf{A})\mathbf{b} \right\|_2 .$$

Then,

$$\left\| f(\mathbf{A})\mathbf{b} - \frac{1}{2\pi \mathrm{i}} \sum_{j=1}^m f(z_j) w_j (z_j \mathbf{I} - \mathbf{A})^{-1}\mathbf{b} \right\|_2$$
$$= \left\| \frac{1}{2\pi \mathrm{i}} \mathbf{V}^{-1} \int_\Gamma f(z)(z\mathbf{I} - \mathbf{\Lambda})^{-1}\mathbf{V}\mathbf{b}\,\mathrm{d}z - \frac{1}{2\pi \mathrm{i}} \mathbf{V}^{-1} \sum_{j=1}^m f(z_j) w_j (z_j \mathbf{I} - \mathbf{\Lambda})^{-1}\mathbf{V}\mathbf{b} \right\|_2$$
$$\leq \frac{1}{2\pi} \|\mathbf{V}\|_2 \left\|\mathbf{V}^{-1}\right\|_2 \|\mathbf{b}\|_2 \max_{\ell \in \{1,\ldots,n\}} |\rho_m(\lambda_\ell)|.$$

Then, in the notation of Theorem 4.5,

$$\begin{aligned}
&\left\| \frac{1}{2\pi \mathrm{i}} \sum_{j=1}^{m} f(z_j) w_j (z_j \mathbf{I} - \mathbf{A})^{-1}\mathbf{b} - f_{k,m}(\mathbf{A})\mathbf{b} \right\|_2 \\
&\leq \frac{1}{2\pi} \sum_{j=1}^{m} \left\| f(z_j) w_j (z_j \mathbf{I} - \mathbf{A})^{-1}\mathbf{b} + f(z_j) w_j \sum_{\ell=0}^{k-1} \mathcal{S}_\Sigma \left[p_\ell w\right](z_j) p_\ell(\mathbf{A})\mathbf{b} \right\|_2 \\
&= \frac{1}{2\pi} \sum_{j=1}^{m} |f(z_j)||w_j| E_k(z_j) \leq LM \max_{j\in\{1,\ldots,m\}} E_k(z_j),
\end{aligned}$$

and the result follows. □

For a reasonable quadrature rule, $\lim_{m\to\infty} \rho_m(x) = 0$ for all x, likely exponentially, so provided that m is taken to be sufficiently large, the Akhiezer iteration converges, at worst, at an asymptotic geometric rate given by $\mathrm{e}^{\max_j \nu(z_j;\mathbf{A})}$ provided that $\mathbf{A}$ is generic and its eigenvalues λ_ℓ all satisfy $\operatorname{Re}\mathfrak{g}(\lambda_\ell) < \min_{j=1,\ldots,m} \operatorname{Re}\mathfrak{g}(z_j)$. In practice, we find that this bound is quite pessimistic, so we offer an alternative view. We first state a lemma that generalizes estimates for the Chebyshev series for functions analytic in a Bernstein ellipse.

LEMMA 4.10. *Let Γ_ϱ denote the level curve $\mathrm{e}^{\operatorname{Re}\mathfrak{g}(z)} = \varrho$ for $\varrho > 1$, and assume that f is analytic in the interior of Γ_ϱ and satisfies $|f(z)| \leq M$ for z interior to Γ_ϱ. Then, there exists a constant $C > 0$, independent of f, such that for all $\ell \in \mathbb{N}$,*

$$\left|\langle f, p_\ell \rangle_{L^2_w(\Sigma)}\right| \leq CM\varrho^{-\ell}.$$

PROOF. Define $\Omega_\eta = \{z \in \mathbb{C} \colon \varrho^\eta \leq \mathrm{e}^{\operatorname{Re}\mathfrak{g}(z)} \leq \varrho\}$ for some $0 < \eta < 1$. Choose ϵ in Lemma 4.2 to be sufficiently small such that $V \in \Omega_\eta^c$, and denote the uniform bound on $|\delta_\ell(z)|$ by D. Let ϱ' satisfy $1 < \varrho^\eta < \varrho' < \varrho$, and consider the level curve $\Gamma_{\varrho'}$. Then, by Cauchy's integral formula and Lemma 4.2,

$$\begin{aligned}
\left|\langle f, p_\ell \rangle_{L^2_w(\Sigma)}\right| &= \left| \int_\Sigma \frac{1}{2\pi \mathrm{i}} \int_{\Gamma_{\varrho'}} \frac{f(z)}{z-x} \mathrm{d}z p_\ell(x) w(x) \mathrm{d}x \right| \\
&= \left| - \int_{\Gamma_{\varrho'}} f(z) \mathcal{C}_\Sigma \left[p_\ell w\right](z) \mathrm{d}z \right| \leq |\Gamma_\varrho| M \frac{D}{2\pi} \varrho'^{-\ell},
\end{aligned}$$

using $C = \frac{1}{2\pi} |\Gamma_\varrho| D$. Since this inequality holds for all $\varrho' < \varrho$, it must hold for ϱ. □

THEOREM 4.11. *Under the assumptions of Lemma 4.10, suppose that*

$$\sum_{\ell=0}^{\infty} \varrho^{-\ell} \|p_\ell(\mathbf{A})\mathbf{b}\|_2 < \infty.$$

Then, there exist $(\varepsilon_{m,\ell})$ satisfying $|\varepsilon_{m,\ell}| \leq \|\rho_m\|_\infty := \max_{x\in\Sigma} |\rho_m(x)|$ and a constant $C > 0$ such that in any norm

$$\left\| f(\mathbf{A})\mathbf{b} - f_{k,m}(\mathbf{A})\mathbf{b} + \sum_{\ell=0}^{k-1} \varepsilon_{m,\ell} p_\ell(\mathbf{A})\mathbf{b} \right\| \leq CM \sum_{\ell=k}^{\infty} \varrho^{-\ell} \|p_\ell(\mathbf{A})\mathbf{b}\|.$$

PROOF. For a given ℓ,

$$-\sum_{j=1}^{m} f(z_j) w_j \mathcal{C}_\Sigma [p_\ell w](z_j) = \int_\Sigma p_\ell(x) \frac{1}{2\pi \mathrm{i}} \sum_{j=1}^{m} \frac{f(z_j) w_j}{z_j - x} w(x) \mathrm{d}x = \langle f_m, p_\ell \rangle_{L^2_w(\Sigma)},$$

where

$$f_m(x) = \frac{1}{2\pi \mathrm{i}} \sum_{j=1}^{m} \frac{f(z_j) w_j}{z_j - x} \approx f(x).$$

Thus,

$$f_{k,m} = \sum_{\ell=0}^{k-1} \langle f_m, p_\ell \rangle_{L^2_w(\Sigma)} p_\ell.$$

Let $\varepsilon_{m,\ell} = \langle f_m - f, p_\ell \rangle_{L^2_w(\Sigma)}$. By the Cauchy–Schwarz inequality,

$$|\varepsilon_{m,\ell}| \leq \|f_m - f\|_{L^2_w(\Sigma)} \leq \|\rho_m\|_\infty.$$

Then,

$$f(\mathbf{A})\mathbf{b} - f_{k,m}(\mathbf{A})\mathbf{b} + \sum_{\ell=0}^{k-1} \varepsilon_{m,\ell} p_\ell(\mathbf{A})\mathbf{b} = \sum_{\ell=k}^{\infty} \langle f, p_\ell \rangle_{L^2_w(\Sigma)} p_\ell(\mathbf{A})\mathbf{b}.$$

By the triangle inequality and Lemma 4.10, there exists some $C > 0$ such that

$$\left\| f(\mathbf{A})\mathbf{b} - f_{k,m}(\mathbf{A})\mathbf{b} + \sum_{\ell=0}^{k-1} \varepsilon_{m,\ell} p_\ell(\mathbf{A})\mathbf{b} \right\| \leq CM \sum_{\ell=k}^{\infty} \varrho^{-\ell} \|p_\ell(\mathbf{A})\mathbf{b}\|.$$

□

The bound from Theorem 4.11 implies that when $\mathbf{A}$ is diagonalized as $\mathbf{A} = \mathbf{V}\mathbf{\Lambda}\mathbf{V}^{-1}$,

$$\|f(\mathbf{A})\mathbf{b} - f_{k,m}(\mathbf{A})\mathbf{b}\|_2 \\ \leq \|\mathbf{V}\|_2 \|\mathbf{V}^{-1}\|_2 \|\mathbf{b}\|_2 \left[\|\rho_m\|_\infty \sum_{\ell=0}^{k-1} \|p_\ell(\mathbf{\Lambda})\|_2 + CM \sum_{\ell=k}^{\infty} \varrho^{-\ell} \|p_\ell(\mathbf{\Lambda})\|_2 \right].$$

This demonstrates that if the quadrature rule is taken such that $\|\rho_m\|_\infty$ is sufficiently small, then the convergence rate will be largely determined by the distance from Σ to the nearest singularity of f. More precisely, $\|f(\mathbf{A})\mathbf{b} - f_{k,m}(\mathbf{A})\mathbf{b}\| = \mathrm{O}(\varrho^{-k} + \|\rho_m\|_\infty)$ when the spectrum of $\mathbf{A}$ is contained in Σ, and the rate of convergence will be governed by the furthest level curve Γ_ϱ to which f can be analytically extended. If f is entire, the initial convergence is superexponential. The level curves of Lemma 4.10 can be thought of as higher genus analogs to Bernstein ellipses as they dictate the convergence rate in the same manner. A contour plot of $\mathrm{e}^{\operatorname{Re} \mathfrak{g}(z)}$ is included as part of Figure 3.

Because Theorem 4.9 gives a different bound, we have, supposing that Γ lies inside Γ_ϱ,

$$\begin{aligned}\frac{\|f(\mathbf{A})\mathbf{b}-f_{k,m}(\mathbf{A})\mathbf{b}\|_2}{\|\mathbf{V}\|_2\|\mathbf{V}^{-1}\|_2\|\mathbf{b}\|_2} &\leq \min\left\{\|\rho_m\|_\infty\sum_{\ell=0}^{k-1}\|p_\ell(\mathbf{\Lambda})\|_2+CM\sum_{\ell=k}^{\infty}\varrho^{-\ell}\|p_\ell(\mathbf{\Lambda})\|_2,\right.\\ &\qquad\left.\frac{1}{2\pi}\|\rho_m(\mathbf{\Lambda})\|_2+cLM\max_{j\in\{1,\ldots,m\}}\frac{\mathrm{e}^{k\nu(z_j;\mathbf{A})}}{1-\mathrm{e}^{\nu(z_j;\mathbf{A})}},\right\}\\ &\leq \min\left\{c'\|\rho_m\|_\infty\sum_{\ell=0}^{k-1}\mathrm{e}^{\ell\nu(\mathbf{A})}+c'CM\frac{\mathrm{e}^{k\nu(z_*;\mathbf{A})}}{1-\mathrm{e}^{\nu(z_*;\mathbf{A})}},\right.\\ &\qquad\left.\frac{1}{2\pi}\|\rho_m(\mathbf{\Lambda})\|_2+cLM\max_{j\in\{1,\ldots,m\}}\frac{\mathrm{e}^{k\nu(z_j;\mathbf{A})}}{1-\mathrm{e}^{\nu(z_j;\mathbf{A})}},\right\},\end{aligned}$$

for another constant $c'>0$ by Lemma 4.2. Here z_* is any point on Γ_ϱ.

If $\mathbf{A}$ is generic and the spectrum of $\mathbf{A}$ is contained in Σ, then $\nu(\mathbf{A})=0$ ($\|p_\ell(\mathbf{\Lambda})\|_2=O(1)$) and the first term inside the minimum is likely to be smaller than the second since the nodes z_j are inside Γ_ϱ, but, eventually, the first term will dominate the second when $\sum_{\ell=0}^{k-1}\|p_\ell(\mathbf{\Lambda})\|_2$ becomes large. The second term then implies that the errors remain small, even if this sum is large. Also, if $\mathbf{A}$ is generic but the eigenvalues of $\mathbf{A}$ are not contained in Σ, the second term here will still be small, giving useful bounds. The first term will also likely be small, but as $\nu(\mathbf{A})>0$, the sum grows exponentially with respect to k. In this case, the balance, i.e., which term gives a better bound in a given situation, will change.

4.3. Complexity. To run Algorithm 3 for k steps, using that computing a_j, b_j, S_j requires $\mathrm{O}(1)$ arithmetic operations and supposing that $T_\mathbf{A}$ is the complexity of matrix multiplication by $\mathbf{A}$, we need less than

$$kT_\mathbf{A}+7kn+km+\mathrm{O}((1+m)k)$$

arithmetic operations if $\mathbf{A}\in\mathbb{C}^{n\times n}$. Here the $\mathrm{O}((1+m)k)$ term represents the computation of a_j, b_j, S_j for $j=0,1,\ldots k$. The complexity of the first Akhiezer iteration, Algorithm 2, is found by setting $m=0$. Furthermore, if a tolerance of ϵ is desired for function approximation, we expect to set $m=\mathrm{O}(\log\epsilon^{-1})$. This is in comparison with the method in [**6**] where the complexity is

$$kT_\mathbf{A}+\mathrm{O}(k^2m+n),$$

and $m=\mathrm{O}(\epsilon^{2/7})$ is required, in general, to achieve a tolerance of ϵ.

5. Adaptivity

In the case where the matrix $\mathbf{A}$ is symmetric, i.e., it has all real eigenvalues, the asymptotic behavior of the polynomials yields an adaptive algorithm that estimates a good choice for the bands $\Sigma=\bigcup_{j=1}^{g+1}[a_j,b_j]$ then applies the iteration (13) to solve the system $\mathbf{Ax}=\mathbf{b}$. For simplicity, we will consider only the $g=1$ case $\Sigma=[a_1,b_1]\cup[a_2,b_2]$, but such an approach can be applied to the $g>1$ cases as well. Furthermore, such an approach could be used as a first step in computing a matrix function with the first step being producing an estimate of the spectrum of $\mathbf{A}$.

By Lemma 4.2, if not all eigenvalues of $\mathbf{A}$ are contained in Σ, the matrix polynomials $p_j(\mathbf{A})$, and therefore $\|p_j(\mathbf{A})\mathbf{b}\|$ as well, grow at an asymptotic rate given by $\mathrm{e}^{\max_k \operatorname{Re} \mathfrak{g}(\lambda_k)}$, where $\lambda_1, \ldots, \lambda_n$ denote the eigenvalues of $\mathbf{A}$. Thus, if the approximate growth rate is given by r, then performing root finding on $\mathrm{e}^{\operatorname{Re} g(z)} - r$ around each endpoint of Σ will yield a new set of bands that contain the eigenvalues of $\mathbf{A}$. A naive implementation of this is given as Algorithm 4. This algorithm

---

Algorithm 4: Adaptive band selection

Input: Initial guess for bands $\Sigma = [a_1, b_1] \cup [a_2, b_2]$ where $b_1 < 0 < a_2$, $\mathbf{A} \in \mathbb{C}^{n \times n}$, $\mathbf{A} = \mathbf{A}^*$, $\mathbf{b} \in \mathbb{C}^n$, parameters $\gamma_o > 1$, $0 < \gamma_i < 1$.
while Norms of matrix polynomials $\|p_j(\mathbf{A})\mathbf{b}\|$ grow as $j \to \infty$ **do**
  Compute the growth rate r of $\|p_j(\mathbf{A})\mathbf{b}\|$ as $j \to \infty$.
  Set $f(x) = \mathrm{e}^{\operatorname{Re} \mathfrak{g}(x)} - r$ for $\mathfrak{g}$ corresponding to $[a_1, b_1] \cup [a_2, b_2]$.
  Perform bisection on f on $[\gamma_o a_1, a_1]$. If bisection fails, set $a_1 \leftarrow \gamma_o a_1$; otherwise set $a_1 \leftarrow c$ where c is the computed root.
  Perform bisection on f on $[b_1, \gamma_i b_1]$. If bisection fails, set $b_1 \leftarrow \gamma_i b_1$; otherwise set $b_1 \leftarrow c$ where c is the computed root.
  Perform bisection on f on $[\gamma_i a_2, a_2]$. If bisection fails, set $a_2 \leftarrow \gamma_i a_2$; otherwise set $a_2 \leftarrow c$ where c is the computed root.
  Perform bisection on f on $[b_2, \gamma_o b_2]$. If bisection fails, set $b_2 \leftarrow \gamma_o b_2$; otherwise set $b_2 \leftarrow c$ where c is the computed root.
  Compute norms of matrix polynomials $\|p_j(\mathbf{A})\mathbf{b}\|$ for $p_0, p_1, \ldots$ corresponding to the updated bands $\Sigma = [a_1, b_1] \cup [a_2, b_2]$.
end
return $\Sigma = [a_1, b_1] \cup [a_2, b_2]$

---

ensures that the Akhiezer iteration on the outputted bands $\Sigma = [a_1, b_1] \cup [a_2, b_2]$ applied to $\mathbf{Ax} = \mathbf{b}$ will converge at the rate $\mathrm{e}^{-\operatorname{Re} \mathfrak{g}(0)}$. In practice, computing the growth rate of the matrix polynomials can be challenging due to theta function oscillations implicit in the orthogonal polynomials. One way to do this is to consider the ratio

$$\frac{|p_{n+k}(z)|}{|p_n(z)|} = \frac{|\hat{\delta}_{n+k}(z)|}{|\hat{\delta}_n(z)|} \mathrm{e}^{k \operatorname{Re} \mathfrak{g}(z)},$$

which follows from Lemma 4.2. Since $\hat{\delta}_n(z) = \mathrm{O}(1)$, for sufficiently large n and k,

$$r \approx \mathrm{e}^{\operatorname{Re} \mathfrak{g}(z)} \approx \sqrt[k]{\frac{|p_{n+k}(z)|}{|p_n(z)|}}.$$

Of course, what "sufficiently large" means in practice is not always clear, and one may need to experiment with different values of n and k for different matrices.

The factor $\mathrm{e}^{-\operatorname{Re} \mathfrak{g}(0)}$ typically increases as the sizes of the intervals increase, so the convergence rate will be suboptimal if the eigenvalues of $\mathbf{A}$ are contained in some $\Sigma' \subsetneq \Sigma$. The convergence rate can be improved by modifying Algorithm 4 so that only one endpoint is moved at a time and the new growth rate is computed after each movement. If the growth rate decreases, the algorithm proceeds to the next endpoint. Otherwise, the endpoint is reverted to its previous value and the algorithm proceeds.

An improved band estimation technique can be employed if one is willing up to compute 4 (or more generally, $2g+2$) inner products. This is done by approximating the eigenvector corresponding to the eigenvalue with the most growth. In particular, one computes $\tilde{\mathbf{y}} = p_n(\mathbf{A})\mathbf{b}$ for a sufficiently large n and normalizes $\mathbf{y} = \frac{\tilde{\mathbf{y}}}{\|\tilde{\mathbf{y}}\|}$ to avoid round-off errors. This is iterated as $\tilde{\mathbf{y}} = p_n(\mathbf{A})\mathbf{y}$, $\mathbf{y} = \frac{\tilde{\mathbf{y}}}{\|\tilde{\mathbf{y}}\|}$ until convergence. Then, the eigenvalue corresponding to the largest growth rate is approximated by the Rayleigh quotient

$$\frac{\mathbf{y}^T\mathbf{A}\mathbf{y}}{\mathbf{y}^T\mathbf{y}}.$$

The relevant endpoint is then moved to strictly contain this eigenvalue, and the algorithm iterates until all approximated eigenvalues fall within the bands. We find this approach to be the most accurate, and the eigenvalues are computed to machine precision, so the algorithm needs only 4 Rayleigh quotient computations to accurately approximate the true bands.

On the other hand, if one only cares that the iteration converges and is not concerned with the rate, a simpler approach can be applied. If one takes the initial guess for the bands to be the symmetric intervals $[-b, -a] \cup [a, b]$, convergence can be obtained without moving the interior endpoints. This follows from the fact that the iteration converges if $\operatorname{Re}\mathfrak{g}(\lambda_k) < \operatorname{Re}\mathfrak{g}(0)$ for all eigenvalues λ_k of $\mathbf{A}$ and that $\operatorname{Re}\mathfrak{g}(z)$ is maximized on $[-a, a]$ at $z = 0$. Thus, one only needs to adjust the parameter b so that $\operatorname{Re}\mathfrak{g}(\lambda_k) < \operatorname{Re}\mathfrak{g}(0)$ for all eigenvalues such that $|\lambda_k| > b$.

6. Examples and applications

Since this work is inspired by that of Saad, we first consider an example presented in [**27**, Section 8.1]. Let $\mathbf{A} \in \mathbb{R}^{200\times 200}$ be a diagonal matrix with eigenvalues spaced uniformly in $[-2, -0.5]\cup[0.5, 6]$, and let $\mathbf{b} = \mathbf{A}\mathbf{e}$ where $\mathbf{e} \in \mathbb{R}^{200}$ is the vector of all ones. We run the Akhiezer iteration using $g = 1$ Akhiezer polynomials and plot the relative residuals in Figure 2. We observe that this looks similar to the result in [**27**].

To construct a more complicated example, we consider a 200×200 real matrix with randomly generated eigenvalues in $[-2, -0.5] \cup [0.5, 6]$. We construct $\mathbf{A}$ by moving the eigenvalues off the intervals by adding a Gaussian perturbation and removing the near-diagonal structure by conjugating by an orthogonal matrix. Again, we take $\mathbf{b} = \mathbf{A}\mathbf{e}$. In Figure 3, we plot the eigenvalues of $\mathbf{A}$ on a contour map of $\operatorname{Re}\mathfrak{g}$ and the relative residuals for the first 370 iterations of our the Akhiezer iteration using $g = 1$ Akhiezer polynomials.

To demonstrate the convergence of the Akhiezer iteration for matrix function approximation, we consider the same matrix $\mathbf{A}$ as the previous example but let $\mathbf{b} \in \mathbb{R}^{200}$ be a random Gaussian vector. We let the contour be two circles: one centered around each interval with diameter 1.15 times the length of the interval. We take 200 quadrature nodes divided between the two circles proportionally to their width. First considering the entire function $f(x) = \exp(x)$, we use (14) with $g = 1$ Akhiezer polynomials to approximate $f(\mathbf{A})\mathbf{b}$ and plot the 2-norm of the difference between this approximation and $f(\mathbf{A})\mathbf{b}$ in Figure 4. We observe superexponential convergence until the iteration saturates due to rounding errors. Instead, letting $f(x) = \tanh(x)$ and then $f(x) = \exp(x)/x$, we plot the same errors in at each iteration in Figures 5 and 6 for two different matrices $\mathbf{A}$. We observe that this approach allows one to compute $f(\mathbf{A})\mathbf{b}$ for functions f that have singularities at zero, albeit at a slower rate than for functions with singularities farther away from Σ. Furthermore, the bound of Theorem 4.11 estimates the convergence rate

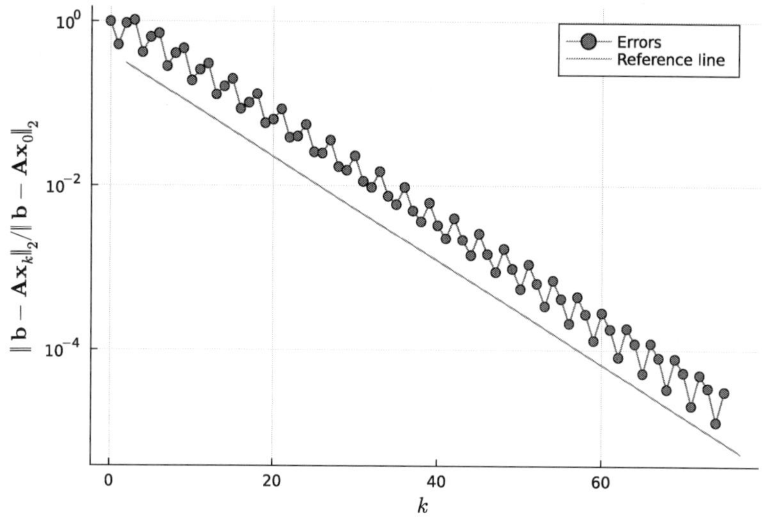

FIGURE 2. Relative residuals at each iteration of the Akhiezer iteration applied to a diagonal matrix $\mathbf{A} \in \mathbb{R}^{200\times 200}$ with eigenvalues spaced uniformly in $[-2,-0.5] \cup [0.5, 6]$ and $\mathbf{b} = \mathbf{A}\mathbf{e}$. The rate of convergence is well-approximated by the reference line $C\mathrm{e}^{-k \operatorname{Re} \mathfrak{g}(0)}$.

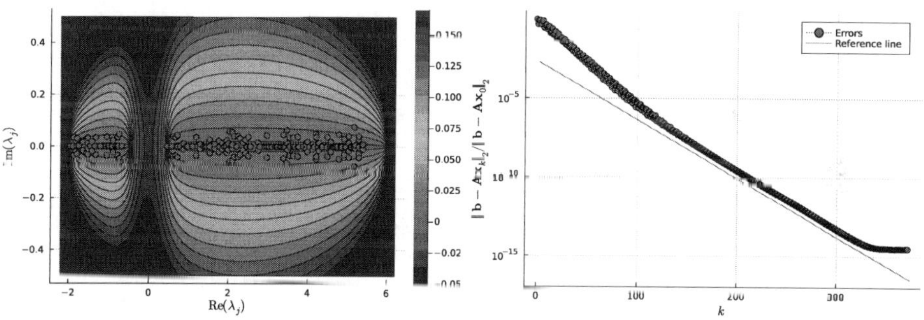

FIGURE 3. Eigenvalues of a matrix $\mathbf{A}$ superimposed on a contour plot of $\operatorname{Re} \mathfrak{g}(z)$ (left) and relative residuals at each iteration for the Akhiezer iteration applied to $\mathbf{A}\mathbf{x} = \mathbf{b} := \mathbf{A}\mathbf{e}$. The asymptotic rate of convergence is well-approximated by the reference line $C\mathrm{e}^{k\nu(0;\mathbf{A})}$.

accurately when the spectrum of $\mathbf{A}$ is contained in Σ and is pessimistic but close when this is not true.

To illustrate the advantage of having more knowledge of the spectrum, we consider $\mathbf{A}$ and $\mathbf{b}$ constructed in the same manner as the previous example but with the original eigenvalues taken to be in $\Sigma = [-2,-0.5] \cup [0.5, 0.7] \cup [5.8, 6]$. In Figure 7, we plot the convergence of the Akhiezer iteration using Akhiezer polynomials on $[-2,-0.5] \cup [0.5, 6]$ and orthogonal polynomials with the weight (5) on Σ. We find that the iteration converges to a relative residual of 10^{-10} in 177 iterations in

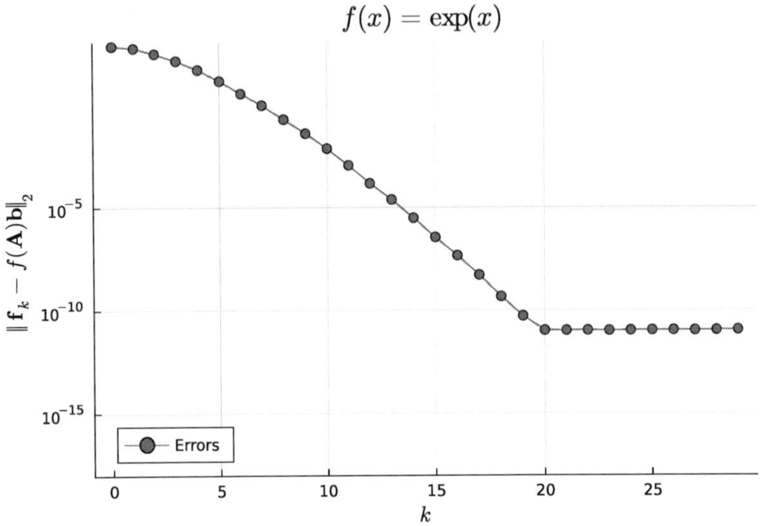

FIGURE 4. 2-norm error at each iteration for the Akhiezer iteration approximation to $f(\mathbf{A})\mathbf{b}$ where $f(x) = \exp(x)$, $\mathbf{A} \in \mathbb{R}^{200\times 200}$ is as in Figure 3, and $\mathbf{b} \in \mathbb{R}^{200}$ is a random Gaussian vector. The iteration converges superexponentially, saturating within 20 iterations.

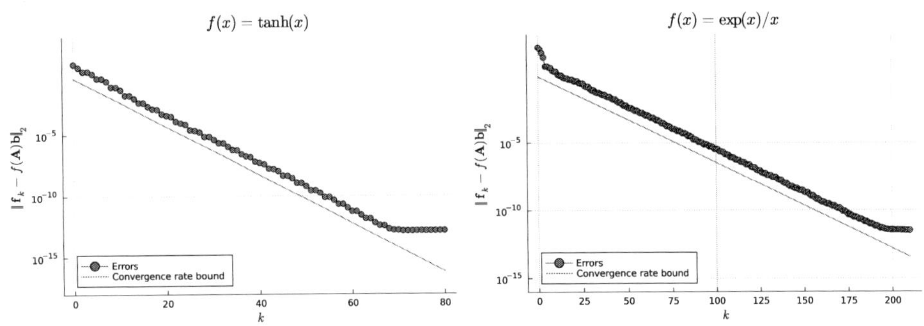

FIGURE 5. 2-norm error at each iteration for the Akhiezer iteration approximation to $f(\mathbf{A})\mathbf{b}$ where $f(x) = \tanh(x)$ (left) and $f(x) = \exp(x)/x$ (right), $\mathbf{A} \in \mathbb{R}^{200\times 200}$ is has randomly generated eigenvalues in $[-2, -0.5] \cup [0.5, 6]$, and $\mathbf{b} \in \mathbb{R}^{200}$ is a random Gaussian vector. The convergence rate bounds $C_1 \mathrm{e}^{k\nu(\mathrm{i}\pi/2;\mathbf{A})}$ (left) and $C_2 \mathrm{e}^{k\nu(0;\mathbf{A})}$ (right) approximate the true convergence rates.

the former case and 102 in the latter in spite of the unbounded behavior at the endpoints of orthogonal polynomials with the weight (5).

As a final example, we consider the boundary-value problem (BVP) on $[0, 1]$:

$$-\frac{\mathrm{d}^2}{\mathrm{d}x^2} u(x) - 30\mathrm{e}^x u(x) = x,$$
$$u(0) = u(1) = 0.$$

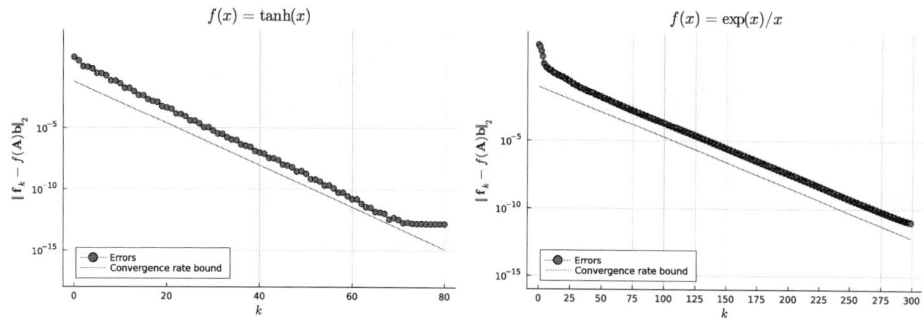

FIGURE 6. 2-norm error at each iteration for the Akhiezer iteration approximation to $f(\mathbf{A})\mathbf{b}$ where $f(x) = \tanh(x)$ (left) and $f(x) = \exp(x)/x$ (right), $\mathbf{A} \in \mathbb{R}^{200\times 200}$ is as in Figure 3, and $\mathbf{b} \in \mathbb{R}^{200}$ is a random Gaussian vector. The convergence rate bounds $C_1 \mathrm{e}^{k\nu(\mathrm{i}\pi/2;\mathbf{A})}$ (left) and $C_2 \mathrm{e}^{k\nu(0;\mathbf{A})}$ (right) approximate the true convergence rates.

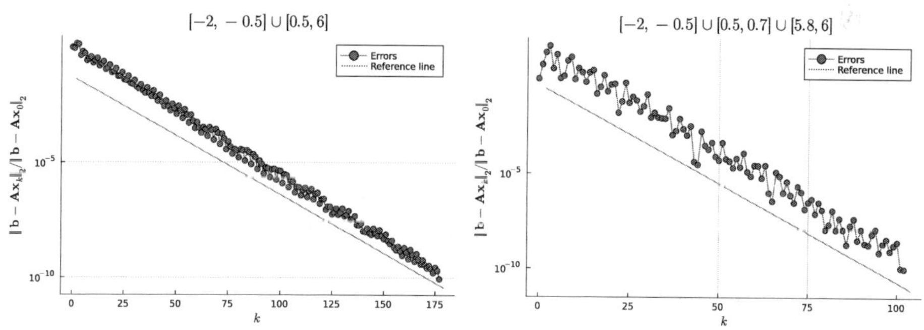

FIGURE 7. Relative residuals at each iteration for the Akhiezer iteration applied to $\mathbf{A}\mathbf{x} = \mathbf{b}$ where $\mathbf{A} \in \mathbb{R}^{200\times 200}$ has eigenvalues near $[-2,-0.5] \cup [0.5, 0.7] \cup [5.8, 6]$ and $\mathbf{b} \in \mathbb{R}^{200}$ is a random Gaussian vector with Akhiezer polynomials on $[-2,-0.5] \cup [0.5, 6]$ (left) and and orthogonal polynomials with the weight (5) on $[-2,-0.5] \cup [0.5, 0.7] \cup [5.8, 6]$ (right). The convergence rate $\mathrm{e}^{\nu(0;\mathbf{A})}$ plotted in the reference line is approximately 0.888 (left) and 0.787 (right). The iteration converges to a relative residual of 10^{-10} in 177 iterations (left) and 102 iterations (right).

Applying a centered finite difference approximation on an equally spaced grid of 100 points yields a symmetric indefinite linear system $\mathbf{A}\mathbf{x} = \mathbf{b}$ with eigenvalues on the positive real axis that are unbounded as the grid spacing is refined. To control the eigenvalues, we apply a preconditioner $\mathbf{L}^{-1}$ where $\mathbf{L}$ corresponds to a centered finite difference approximation on the same grid applied to the BVP

$$-\frac{\mathrm{d}^2}{\mathrm{d}x^2} u(x) = 0,$$
$$u(0) = u(1) = 0.$$

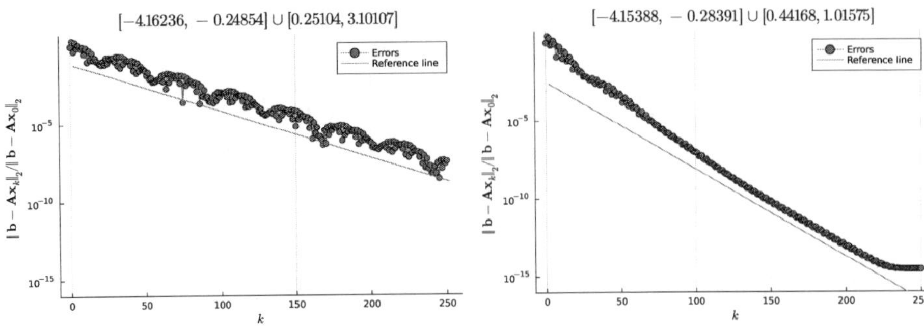

FIGURE 8. Relative residuals at each iteration for 250 iterations of the Akhiezer iteration applied to a preconditioned BVP for two different approaches for adaptive band selection (left: Algorithm 4, right: Algorithm 4, moving one endpoint at a time). The convergence rate $e^{\nu(0;\mathbf{A})}$ plotted in the reference line is approximately 0.933 (left) and 0.879 (right).

We then expect $\mathbf{L}^{-1}\mathbf{A}$ to have the majority of its eigenvalues bounded by 1 on the positive real axis with the remaining eigenvalues on the negative real axis. Thus, we apply Algorithm 4 to the system $\mathbf{L}^{-1}\mathbf{A}\mathbf{x} = \mathbf{L}^{-1}\mathbf{b}$ with an initial guess of $\Sigma = [-2, -0.5] \cup [0.5, 1]$ and parameters $\gamma_o = 5, \gamma_i = 0.7$ and obtain the bands $\Sigma = [-4.16236, -0.24854] \cup [0.25104, 3.10107]$. The true eigenvalues of $\mathbf{L}^{-1}\mathbf{A}$ fall in $[-4.14928, -0.28169] \cup [0.43062, 0.99921]$ with eigenvalues at each endpoint, so this algorithm does a good job of capturing the "worst offender" but far overshoots the right endpoint. However, applying the modification of Algorithm 4 to optimize the convergence rate described in Section 5 yields the bands $\Sigma = [-4.15388, -0.28391] \cup [0.44168, 1.01575]$; these are very close to the true bands, but note that the interval endpoint $-0.28391 < -0.28169$, and an eigenvalue lies outside the computed bands. This occurs due to difficulties in approximating the exact matrix polynomial growth rate (r in Algorithm 4) but gives just a slight perturbation to the geometric rate of convergence. Furthermore, applying the variation with inner products finds the true bands to full precision.

In Figure 8, we plot the relative residuals when applying the Akhiezer iteration to $\mathbf{L}^{-1}\mathbf{A}\mathbf{x} = \mathbf{L}^{-1}\mathbf{b}$ with the Akhiezer polynomials on $\Sigma = [-4.14823, -0.245] \cup [0.245, 3.09077]$ and the Akhiezer polynomials on

$$\Sigma = [-4.15746, -0.30738] \cup [0.42308, 1.01751]$$

for 250 iterations.

7. Discussion and open questions

First, in the current work, we have not addressed the question of adaptivity for matrices with eigenvalues off the real axis. In [**21**], this is handled for the classical Chebyshev iteration by estimating the convex hull of the spectrum via the power method and choosing the iteration parameters based on the known "optimal ellipse" containing this convex hall [**20**]. For the Akhiezer iteration, similar eigenvalue estimations could be employed, but one would instead need to look for at least two distinct regions of eigenvalues rather than the convex hull. Extending the notion of the optimal ellipse to a higher genus seems to be the primary difficulty. That is, given ellipses containing the spectrum, finding the intervals that optimize the convergence rate of the Akhiezer iteration requires a generalization of the results of [**20**] to more complex geometries based on the level curves of $\operatorname{Re}\mathfrak{g}(z)$. How to best approach this is an important and interesting question that we leave for future work.

On the other hand, the adaptivity techniques for symmetric matrices of Section 5 can be used for eigenvalue approximation, even when decoupled from the Akhiezer iteration. When the eigenvalues are contained in two disjoint intervals, the variant based on inner products computes both interior and exterior eigenvalues to full precision. In particular, given any symmetric matrix, one could compute interior eigenvalues in the interval $[b, c]$ by choosing an initial guess of $[a, b] \cup [c, d]$. Then, the algorithm would compute two consecutive eigenvalues in $[b, c]$ or leave b and/or c unchanged if there are zero/one eigenvalues in this interval. This requires at most 4 Rayleigh quotient computations, so this provides an efficient method to compute interior eigenvalues to full precision.

7.1. Other applications to orthogonal polynomials. Through the framework established here and in [**3**], an approach similar to that in [**29**] could be employed to compute the N-point Gaussian quadrature rule for a measure with density of the form (4) in $O(N)$ complexity. In particular, Akhiezer's formulae should enable this approach to yield a particularly efficient algorithm for the Akhiezer polynomials on two intervals. Unfortunately, a good application for this quadrature rule is elusive at this time.

Finally, the computations in this work could potentially be improved upon by directly using the asymptotic formulae for the orthogonal polynomials corresponding to weight functions of the form (4). These asymptotic formulae could be obtained via an approach similar to that in [**19**], and then, like the Akhiezer polynomials on two intervals, computed efficiently by simply evaluating the formulae à la [**5, 18**]. Then, our Riemann–Hilbert based numerical method would be required only for low degree polynomials before switching to evaluating the asymptotic formulae when the degree is sufficiently large. Such an approach will be explored further in future work.

Appendix A. Computing orthogonal polynomials via a Riemann–Hilbert problem

Recurrence coefficients and Cauchy integrals of weight functions of the form (5) can be computed by numerically solving a deformed version of the Fokas–Its–Kitaev (FIK) Riemann–Hilbert problem [**14**]. We first define the notation necessary for this problem.

Let $\mathfrak{g}$ denote the exterior Green's function with pole at infinity for

$$\Sigma = \bigcup_{j=1}^{g+1} [a_j, b_j].$$

More specifically, define

$$\mathfrak{g}'(z) = \frac{Q_g(z)}{R(z)}, \quad \text{where } R(z)^2 = \prod_{j=1}^{g+1} (z - a_j)(z - b_j).$$

Here, Q_g is a monic polynomial of degree g

$$Q_g(z) = z^g + \sum_{k=0}^{g-1} h_k z^k,$$

whose coefficients h_k are determined by the linear system

$$\int_{b_j}^{a_{j+1}} \sum_{k=0}^{g-1} h_k \frac{z^k}{R(z)} \mathrm{d}z = -\int_{b_j}^{a_{j+1}} \frac{z^g}{R(z)} \mathrm{d}z,$$

for $j = 1, \ldots, g$. Then, we define $\mathfrak{g}$ by normalizing at the leftmost endpoint. If some function satisfies $\psi' = \mathfrak{g}'$, then we let

$$\mathfrak{g}(z) = \psi(z) - \psi(a_1).$$

Define $\mathfrak{g}_1$ to be the coefficient in the $\mathrm{O}(z^{-1})$ term in asymptotic expansion of $\mathfrak{g}$. That is,

$$\mathfrak{g}(z) = \log \mathfrak{c} z + \frac{\mathfrak{g}_1}{z} + \ldots, \quad z \to \infty. \tag{15}$$

We also define

$$\Delta_j = \mathfrak{g}^+(z) - \mathfrak{g}^-(z) = 2\sum_{k=1}^{j} \int_{a_k}^{b_k} (\mathfrak{g}')^+(z)\mathrm{d}z \in \mathrm{i}\mathbb{R},$$

for $z \in (b_j, a_{j+1})$, $j = 1, \ldots, g$, where this value is independent of z within each of these intervals.

Define the function $\mathfrak{h}_n$ by

$$\mathfrak{h}_n(z) = R(z)\left(\sum_{j=1}^{g+1} \frac{1}{2\pi \mathrm{i}} \int_{a_j}^{b_j} \frac{A_j(n)}{R_+(s)(s-z)}\mathrm{d}s + \sum_{\ell=1}^{g} \int_{b_\ell}^{a_{\ell+1}} \frac{\log(\mathrm{e}^{n\Delta_\ell})}{R(s)(s-z)}\mathrm{d}s\right),$$

where $A_1(n), \ldots, A_{g+1}(n)$ are determined by the linear system

$$\sum_{j=1}^{g+1} A_j(n) \int_{a_j}^{b_j} \frac{s^{k-1}}{R_+(s)}\mathrm{d}s = -\sum_{\ell=1}^{g} \log(\mathrm{e}^{n\Delta_\ell}) \int_{b_\ell}^{a_{\ell+1}} \frac{s^{k-1}}{R(s)}\mathrm{d}s,$$

for $k = 1, \ldots, g+1$. Define $\mathfrak{h}_n^{(1)}$ by

$$\mathfrak{h}_n(z) = \frac{\mathfrak{h}_n^{(1)}}{z} + \mathrm{O}(z^{-2}), \quad z \to \infty.$$

Finally, for a general weight function of the form (4), define

$$w_j(x) = h_j(x)\left(\sqrt{x-a_j}\right)^{\alpha_j}\left(\sqrt{b_j - x}\right)^{\beta_j}.$$

Weight functions of the form (5) satisfy $w_j = w_0|_{(a_j,b_j)}$ for all $j = 1, \ldots, g+1$, for an analytic function $w_0 : \mathbb{C} \setminus \overline{(\mathbb{R} \setminus \Sigma)} \to \mathbb{C}$.

The deformed FIK Riemann–Hilbert problem for the unknown $\mathbf{T}_n : \mathbb{C} \setminus \Sigma \to \mathbb{C}^{2\times 2}$ is

$$\begin{aligned}
&\mathbf{T}_n \text{ is analytic in } \mathbb{C} \setminus \Sigma, \\
&\mathbf{T}_n^+(z) = \mathbf{T}_n^-(z) \begin{pmatrix} 0 & w_0(z)\mathrm{e}^{-A_j(n)} \\ -\frac{1}{w_0(z)}\mathrm{e}^{A_j(n)} & 0 \end{pmatrix}, \quad z \in (a_j, b_j), \\
&\mathbf{T}_n(z) = \mathbf{I} + \mathrm{O}(z^{-1}), \quad z \to \infty, \\
&\mathbf{T}_n(z) = \mathrm{O}\begin{pmatrix} 1 + |z-a_j|^{-\alpha_j/2} & 1 + |z-a_j|^{\alpha_j/2} \\ 1 + |z-a_j|^{-\alpha_j/2} & 1 + |z-a_j|^{\alpha_j/2} \end{pmatrix}, \quad z \to a_j, \\
&\mathbf{T}_n(z) = \mathrm{O}\begin{pmatrix} 1 + |z-b_j|^{-\beta_j/2} & 1 + |z-b_j|^{\beta_j/2} \\ 1 + |z-b_j|^{-\beta_j/2} & 1 + |z-b_j|^{\beta_j/2} \end{pmatrix}, \quad z \to b_j.
\end{aligned} \tag{16}$$

This Riemann–Hilbert problem is solved numerically via a Chebyshev collocation method, similar to that which was used in [**4**]. The solution is represented in shifted Chebyshev polynomial bases on each interval with singularity structure chosen to match the endpoint asymptotics. The jump conditions are enforced at each collocation point by computing the boundary values of the Cauchy integrals of the basis polynomials, and the expansion coefficients are approximated by the solution to a block linear system. A general reference for this type of approach is [**32**].

Given a solution to (16), the recurrence coefficients are recovered by

$$\begin{aligned}
a_n &= \left(\mathbf{T}_n^{(1)}\right)_{11} - \left(\mathbf{T}_{n+1}^{(1)}\right)_{11} - \mathfrak{h}_n^{(1)} + \mathfrak{h}_{n+1}^{(1)} - \mathfrak{g}_1, \\
b_n &= \sqrt{\left(\mathbf{T}_{n+1}^{(1)}\right)_{12}\left(\mathbf{T}_{n+1}^{(1)}\right)_{21}}.
\end{aligned} \tag{17}$$

Methods to numerically compute the first order correction terms and the constants $A_j(n)$ can be found in [**3**, Section 4]. Furthermore, if $p_0, p_1, \ldots$ denote the orthonormal polynomials with respect to the normalization of (5), the Cauchy integrals can be computed as [**3**, Equation 31]

$$\mathcal{C}_\Sigma\left[p_n w\right](z) = \left(\prod_{j=0}^{n-1} \frac{1}{b_j \mathfrak{c}}\right) (\mathbf{T}_n(z))_{12}\, \mathrm{e}^{\mathfrak{h}_n(z) - n\mathfrak{g}(z)}, \tag{18}$$

where $\mathfrak{c}$ is as defined in (15). Methods to numerically compute $\mathfrak{c}$, $\mathfrak{g}$, and $\mathfrak{h}_n$ are found in [**3**, Section 4].

To see why it suffices to solve such a simple Riemann–Hilbert problem, consider the final deformed FIK Riemann–Hilbert problem from **[3]** for the unknown $\tilde{\mathbf{S}}_n$:

$$
\begin{aligned}
&\tilde{\mathbf{S}}_n \text{ is analytic in } \mathbb{C}\setminus\left(\bigcup_{j=1}^{g+1}[a_j,b_j]\cup\bigcup_{j=1}^{g+1}C_j\right),\\
&\tilde{\mathbf{S}}_n^+(z)=\tilde{\mathbf{S}}_n^-(z)\begin{pmatrix}1 & 0\\ -\frac{1}{w_j(z)}\mathrm{e}^{2\mathfrak{h}_n(z)-2n\mathfrak{g}(z)} & 1\end{pmatrix}, \quad z\in C_j\cap\mathbb{C}^+,\\
&\tilde{\mathbf{S}}_n^+(z)=\tilde{\mathbf{S}}_n^-(z)\begin{pmatrix}1 & 0\\ \frac{1}{w_j(z)}\mathrm{e}^{2\mathfrak{h}_n(z)-2n\mathfrak{g}(z)} & 1\end{pmatrix}, \quad z\in C_j\cap\mathbb{C}^-,\\
&\tilde{\mathbf{S}}_n^+(z)=\tilde{\mathbf{S}}_n^-(z)\begin{pmatrix}0 & w_j(z)\mathrm{e}^{-A_j(n)}\\ -\frac{1}{w_j(z)}\mathrm{e}^{A_j(n)} & 0\end{pmatrix}, \quad z\in(a_j,b_j),\\
&\tilde{\mathbf{S}}_n(z)=\mathbf{I}+\mathrm{O}(z^{-1}), \quad z\to\infty,\\
&\tilde{\mathbf{S}}_n(z)=\mathrm{O}\begin{pmatrix}1+|z-a_j|^{-\alpha_j/2} & 1+|z-a_j|^{\alpha_j/2}\\ 1+|z-a_j|^{-\alpha_j/2} & 1+|z-a_j|^{\alpha_j/2}\end{pmatrix}, \quad z\to a_j,\\
&\tilde{\mathbf{S}}_n(z)=\mathrm{O}\begin{pmatrix}1+|z-b_j|^{-\beta_j/2} & 1+|z-b_j|^{\beta_j/2}\\ 1+|z-b_j|^{-\beta_j/2} & 1+|z-b_j|^{\beta_j/2}\end{pmatrix}, \quad z\to b_j,
\end{aligned}
\tag{19}
$$

where each C_j is a counterclockwise-oriented curve lying in Ω_j that encircles, but does not intersect $[a_j,b_j]$. Because weight functions of the form (5) satisfy $w_j = w_0|_{(a_j,b_j)}$ for all j and w_0 only has singularities at the endpoints of intervals, one can instead define C to be a counterclockwise-oriented circle with Σ in its interior. Then, the jump conditions on $C_1,\ldots C_{g+1}$ can be replaced with a single, identical jump condition on C.

To see why the jump conditions on C can then effectively be removed, consider a function $\mathbf{\Phi}$ that solves the Riemann–Hilbert problem:

$$
\begin{aligned}
&\mathbf{\Phi}^+(z)=\mathbf{\Phi}^-(z)\mathbf{J}(z), \quad z\in C,\\
&\mathbf{\Phi}(z)=\mathbf{I}+\mathbf{\Phi}_1 z^{-1}+\mathrm{O}(z^{-2}), \quad z\to\infty,
\end{aligned}
\tag{20}
$$

where $\mathbf{J}(z)$ is analytic exterior to C and satisfies

$$
\mathbf{J}(z)=\mathbf{I}+\begin{pmatrix}0 & 0\\ j_1 & 0\end{pmatrix}z^{-1}+\begin{pmatrix}0 & 0\\ \mathrm{O}(z^{-2}) & 0\end{pmatrix}, \quad z\to\infty.
$$

We allow $\mathbf{\Phi}$ to potentially satisfy additional jump conditions in the interior of C. We denote the interior of C by D. Define

$$
\mathbf{\Psi}(z)=\begin{cases}\mathbf{\Phi}(z) & z\in D,\\ \mathbf{\Phi}(z)\mathbf{J}(z) & z\in\mathbb{C}\setminus\overline{D}.\end{cases}
$$

Then, $\mathbf{\Psi}(z)$ satisfies the same jump conditions as $\mathbf{\Phi}$ in D, but the jump condition on C is now the identity. It has asymptotic behavior

$$
\mathbf{\Psi}(z)=\mathbf{I}+\left(\mathbf{\Phi}_1+\begin{pmatrix}0 & 0\\ j_1 & 0\end{pmatrix}\right)z^{-1}+\mathrm{O}(z^{-2}), \quad z\to\infty.
$$

We apply this to $\tilde{\mathbf{S}}_n$ where $\mathbf{J}$ is the jump on C, after the replacement of the jumps on $C_1,\ldots,C_{g+1}$ with a single jump on C. Provided the asymptotics for $\mathbf{J}$ hold as hypothesized, we can essentially ignore the jump on C, and we find the Riemann–Hilbert problem for $\mathbf{T}_n$.

Then, we note that since the $\mathrm{O}(z^{-1})$ term is all that is required to compute the recurrence coefficients (see (17)) and a_n requires only the $(1,1)$ entry of this term, a_n is directly recovered from the simplified problem for $\mathbf{T}_n$ provided that

$$\frac{1}{w_0(z)}\mathrm{e}^{2\mathfrak{h}_n(z)-2n\mathfrak{g}(z)} = \mathrm{O}(z^{-1}), \quad z \to \infty.$$

The recovery of b_n requires the additional assumption that $j_1 = 0$, i.e., $\mathbf{J}(z) = \mathbf{I} + \mathrm{O}(z^{-2})$ as $z \to \infty$. For weight functions of the form (5), $w_0(z) = \mathrm{O}(z)$ as $z \to \infty$, so

$$\frac{1}{w_0(z)}\mathrm{e}^{2\mathfrak{h}_n(z)-2n\mathfrak{g}(z)} = \mathrm{O}(z^{-1-2n}), \quad z \to \infty.$$

Thus, $j_1 = 0$ for all $n > 0$. Since b_n is recovered from $\mathbf{T}_{n+1}$, this is sufficient to compute all the recurrence coefficients for this weight. Furthermore, since multiplication by $\mathbf{J}(z)$ does not affect the $(1,2)$ entry, it is sufficient to compute the Cauchy integrals according to (18).

This analysis can be extended to weight functions w that consist of a meromorphic function multiplied by the weight (5) that are constrained to only have zeroes and poles at the endpoints of intervals. Then, the jump conditions on each C_j can still be replaced with a jump condition on C, but whether this jump condition can be thrown away depends on the asymptotic behavior of w. In particular, the Akhiezer weight (6) satisfies $w_0(z) = \mathrm{O}(z^{-1})$ as $z \to \infty$, so the full Riemann–Hilbert problem (19) is required only to compute the 0th pair of recurrence coefficients, and all others can be computed by solving the simplified problem (16).

Appendix B. Akhiezer polynomial recurrence coefficients

Assume that the nth degree orthonormal polynomial can be represented as

$$p_n(x) = k_n x^n + \ell_n x^{n-1} + \dots.$$

Then, the recurrence coefficients can be recovered by [**10**]

$$a_n = \frac{\ell_n}{k_n} - \frac{\ell_{n+1}}{k_{n+1}}, \quad b_n = \frac{k_n}{k_{n+1}}, \tag{21}$$

with the convention that $\ell_0 = 0, k_0 = 1$. To obtain such a representation, we must expand (8) around $u = -\rho$, the point corresponding to $x = \infty$ on one sheet of the Riemann surface. Noting that H is an odd function, we expand

$$\begin{aligned}
&\frac{H(u-\rho)}{H(u+\rho)} \\
&\quad = \left(-H(2\rho) + H'(2\rho)(u+\rho) + \mathrm{O}((u+\rho)^2)\right)\frac{1}{H'(0)(u+\rho)}\left(1 + \mathrm{O}((u+\rho)^2)\right)^{-1} \\
&\quad = -\frac{H(2\rho)}{H'(0)}\frac{1}{u+\rho} + \frac{H'(2\rho)}{H'(0)} + \mathrm{O}(u+\rho).
\end{aligned}$$

Similarly, Θ is an even function, and we expand

$$\begin{aligned}
&\frac{\Theta(u+2n\rho)}{\Theta(u)} \\
&\quad = \left(\Theta\left((2n-1)\rho\right) + \Theta'\left((2n-1)\rho\right)(u+\rho) + \dots\right) \frac{1}{\Theta(\rho)} \left(1 - \frac{\Theta'(\rho)}{\Theta(\rho)}(u+\rho) + \dots\right)^{-1} \\
&\quad = \frac{\Theta\left((2n-1)\rho\right)}{\Theta(\rho)} + \left(\frac{\Theta'\left((2n-1)\rho\right)}{\Theta(\rho)} + \frac{\Theta\left((2n-1)\rho\right)\Theta'(\rho)}{\Theta^2(\rho)}\right)(u+\rho) + \mathrm{O}\left((u+\rho)^2\right).
\end{aligned}$$

The other product summed in (9) is $\mathrm{O}(u+\rho)$ and will not contribute to the recurrence coefficients. To convert these expansions from $u+\rho$ to x, we utilize the relation

$$x - \alpha = \frac{1-\alpha^2}{2(\operatorname{sn}^2 u - \operatorname{sn}^2 \rho)},$$

which we rewrite as

$$(\operatorname{sn} u + \operatorname{sn} \rho)(\operatorname{sn} u - \operatorname{sn} \rho) = \frac{1-\alpha^2}{2(x-\alpha)}. \tag{22}$$

Noting that sn is an odd function, we expand

$$\begin{aligned}
&(\operatorname{sn} u + \operatorname{sn} \rho)(\operatorname{sn} u - \operatorname{sn} \rho) \\
&\quad = \left(\operatorname{sn}' \rho (u+\rho) - \frac{\operatorname{sn}'' \rho}{2}(u+\rho)^2 \dots\right)\left(-2\operatorname{sn}\rho + \operatorname{sn}'\rho(u+\rho) + \dots\right) \\
&\quad = -2\operatorname{sn}\rho \operatorname{sn}'\rho (u+\rho)\left(1 - \left(\frac{\operatorname{sn}'\rho}{2\operatorname{sn}\rho} + \frac{\operatorname{sn}''\rho}{2\operatorname{sn}'\rho}\right)(u+\rho) + \mathrm{O}\left((u+\rho)^2\right)\right).
\end{aligned}$$

Substituting this into (22), solving for $u+\rho$, and iterating,

$$\begin{aligned}
&u+\rho \\
&\quad = -\frac{1-\alpha^2}{4\operatorname{sn}\rho\operatorname{sn}'\rho}\left(1 - \left(\frac{\operatorname{sn}'\rho}{2\operatorname{sn}\rho} + \frac{\operatorname{sn}''\rho}{2\operatorname{sn}'\rho}\right)(u+\rho) + \mathrm{O}\left((u+\rho)^2\right)\right)^{-1} \frac{1}{x-\alpha} \\
&\quad = -\frac{1-\alpha^2}{4\operatorname{sn}\rho\operatorname{sn}'\rho}\left(1 - \frac{(1-\alpha^2)}{4\operatorname{sn}\rho\operatorname{sn}'\rho}\left(\frac{\operatorname{sn}'\rho}{2\operatorname{sn}\rho} + \frac{\operatorname{sn}''\rho}{2\operatorname{sn}'\rho}\right)\frac{1}{x-\alpha} + \mathrm{O}\left(\left(\frac{1}{x-\alpha}\right)^2\right)\right)\frac{1}{x-\alpha}.
\end{aligned}$$

Inverting this,

$$\frac{1}{u+\rho} = -\frac{4\operatorname{sn}\rho\operatorname{sn}'\rho}{1-\alpha^2}(x-\alpha) - \left(\frac{\operatorname{sn}'\rho}{2\operatorname{sn}\rho} + \frac{\operatorname{sn}''\rho}{2\operatorname{sn}'\rho}\right) + \mathrm{O}\left(\frac{1}{x-\alpha}\right).$$

Substituting these into the series in $u+\rho$,

$$\frac{H(u-\rho)}{H(u+\rho)} = \frac{H(2\rho)}{H'(0)}\frac{4\operatorname{sn}\rho\operatorname{sn}'\rho}{1-\alpha^2}(x-\alpha) + \frac{H(2\rho)}{H'(0)}\left(\frac{\operatorname{sn}'\rho}{2\operatorname{sn}\rho} + \frac{\operatorname{sn}''\rho}{2\operatorname{sn}'\rho}\right) + \frac{H'(2\rho)}{H'(0)} + \mathrm{O}\left(\frac{1}{x-\alpha}\right),$$

$$\frac{\Theta(u+2n\rho)}{\Theta(u)} = \frac{\Theta\left((2n-1)\rho\right)}{\Theta(\rho)} - \frac{1-\alpha^2}{4\operatorname{sn}\rho\operatorname{sn}'\rho}\left(\frac{\Theta'\left((2n-1)\rho\right)}{\Theta(\rho)} + \frac{\Theta\left((2n-1)\rho\right)\Theta'(\rho)}{\Theta^2(\rho)}\right)\frac{1}{x-\alpha} + \mathrm{O}\left(\left(\frac{1}{x-\alpha}\right)^2\right).$$

To obtain the recurrence coefficients from these expansions, denote

$$c = \frac{H(2\rho)}{H'(0)}\frac{4\operatorname{sn}\rho\operatorname{sn}'\rho}{1-\alpha^2}, \quad d = \frac{H(2\rho)}{H'(0)}\left(\frac{\operatorname{sn}'\rho}{2\operatorname{sn}\rho} + \frac{\operatorname{sn}''\rho}{2\operatorname{sn}'\rho}\right) + \frac{H'(2\rho)}{H'(0)},$$

$$\gamma_n = \frac{\Theta\left((2n-1)\rho\right)}{\Theta(\rho)},$$

$$\delta_n = -\frac{1-\alpha^2}{4\operatorname{sn}\rho\operatorname{sn}'\rho}\left(\frac{\Theta'\left((2n-1)\rho\right)}{\Theta(\rho)} + \frac{\Theta\left((2n-1)\rho\right)\Theta'(\rho)}{\Theta^2(\rho)}\right).$$

Then,

$$\begin{aligned} p_n(x) &= C_n\left(c(x-\alpha)+d+\ldots\right)^n\left(\gamma_n + \frac{\delta_n}{x-\alpha}+\ldots\right) + \mathrm{O}\left(\frac{1}{x-\alpha}\right) \\ &= C_n\gamma_n c^n x^n + C_n(n\gamma_n c^{n-1}d + \delta_n c^n - n\alpha\gamma_n c^n)x^{n-1} + \ldots, \end{aligned}$$

so

$$k_n = C_n\gamma_n c^n, \quad \ell_n = C_n(n\gamma_n c^{n-1}d + \delta_n c^n - n\alpha\gamma_n c^n).$$

Applying (21), we get that

$$\begin{aligned} a_n &= \frac{\ell_n}{k_n} - \frac{\ell_{n+1}}{k_{n+1}} = \left(\frac{nd}{c} + \frac{\delta_n}{\gamma_n} - n\alpha\right) - \left(\frac{(n+1)d}{c} + \frac{\delta_{n+1}}{\gamma_{n+1}} - (n+1)\alpha\right) \\ &= -\frac{d}{c} + \frac{\delta_n}{\gamma_n} - \frac{\delta_{n+1}}{\gamma_{n+1}} + \alpha, \\ b_n &= \frac{k_n}{k_{n+1}} = \frac{C_n}{C_{n+1}}\frac{1}{c}, \end{aligned}$$

for $n \geq 1$ and

$$\begin{aligned} a_0 &= -\frac{\ell_1}{k_1} = -\frac{d}{c} - \frac{\delta_1}{\gamma_1} + \alpha, \\ b_0 &= \frac{2}{C_1}\frac{1}{c}. \end{aligned}$$

Finally, we substitute in c, d, γ_n, δ_n and C_n and get

$$\begin{aligned}
a_0 =&(1-\alpha^2)\left(\frac{1}{8\operatorname{sn}^2\rho}+\frac{\operatorname{sn}''\rho}{8\operatorname{sn}\rho\,(\operatorname{sn}'\rho)^2}+\frac{H'(2\rho)}{H(2\rho)}\frac{1}{4\operatorname{sn}\rho\operatorname{sn}'\rho}\right)\\
&+\frac{1-\alpha^2}{2\operatorname{sn}\rho\operatorname{sn}'\rho}\frac{\Theta'(\rho)}{\Theta(\rho)}+\alpha,\\
a_n =&(1-\alpha^2)\left(\frac{1}{8\operatorname{sn}^2\rho}+\frac{\operatorname{sn}''\rho}{8\operatorname{sn}\rho\,(\operatorname{sn}'\rho)^2}\right)\\
&+\frac{1-\alpha^2}{4\operatorname{sn}\rho\operatorname{sn}'\rho}\left(\frac{H'(2\rho)}{H(2\rho)}+\frac{\Theta'((2n+1)\rho)}{\Theta((2n+1)\rho)}-\frac{\Theta'((2n-1)\rho)}{\Theta((2n-1)\rho)}\right)+\alpha, \quad n\geq 1,\\
b_0 =&\frac{1-\alpha^2}{4\operatorname{sn}\rho\operatorname{sn}'\rho}\frac{H'(0)}{H(2\rho)}\sqrt{\frac{2\Theta(3\rho)}{\Theta(\rho)}},\\
b_n =&\frac{1-\alpha^2}{4\operatorname{sn}\rho\operatorname{sn}'\rho}\frac{H'(0)}{H(2\rho)}\frac{\sqrt{\Theta((2n+3)\rho)\Theta((2n-1)\rho)}}{\Theta((2n+1)\rho)}, \quad n\geq 1.
\end{aligned}$$

Appendix C. The classical Chebyshev iteration and our modification

To compare our presentation of the Chebyshev iteration with that of [**17**], suppose that the interval $[\alpha-c,\alpha+c]$ does not contain the origin, and for simplicity suppose $\alpha-c>0$. Define the shifted and scaled Chebyshev-T polynomials by

$$\tilde{T}_k(x)=T_k\left(\frac{x-\alpha}{c}\right).$$

Based on this, define

$$r_k(x)=\frac{\tilde{T}_k(x)}{\tilde{T}_k(0)}, \quad q_k(x)=\frac{1-r_k(x)}{x}.$$

We have the recurrence

$$\begin{aligned}
&\frac{x-\alpha}{c}\tilde{T}_0(x)=\tilde{T}_1(x),\\
&2\frac{x-\alpha}{c}\tilde{T}_k(x)=\tilde{T}_{k-1}(x)+\tilde{T}_{k+1}(x), \quad k\geq 1,
\end{aligned} \tag{23}$$

so r_k must satisfy

$$\begin{aligned}
&\frac{x-\alpha}{c}r_0(x)=r_1(x)\frac{\tilde{T}_1(0)}{\tilde{T}_0(0)},\\
&2\frac{x-\alpha}{c}r_k(x)=r_{k-1}(x)\frac{\tilde{T}_{k-1}(0)}{\tilde{T}_k(0)}+r_{k+1}(x)\frac{\tilde{T}_{k+1}(0)}{\tilde{T}_k(0)}, \quad k\geq 1.
\end{aligned}$$

Using $r_k(x)=1-xq_k(x)$ and applying (23),

$$\begin{aligned}
&\frac{x-\alpha}{c}(1-xq_0(x))=\frac{\alpha}{c}xq_1(x)-\frac{\alpha}{c},\\
&2\frac{x-\alpha}{c}(1-xq_k(x))=-xq_{k-1}(x)\frac{\tilde{T}_{k-1}(0)}{\tilde{T}_k(0)}-xq_{k+1}(x)\frac{\tilde{T}_{k+1}(0)}{\tilde{T}_k(0)}-\frac{2\alpha}{c}, \quad k\geq 1.
\end{aligned}$$

This can be rearranged as

$$q_1(x) = \frac{1}{\alpha}(1 - xq_0(x) + \alpha q_0(x)),$$

$$q_{k+1}(x) = -\frac{\tilde{T}_k(0)}{\tilde{T}_{k+1}(0)}\left(\frac{2}{c}(1 - xq_k(x) + \alpha q_k(x)) + \frac{\tilde{T}_{k-1}(0)}{\tilde{T}_k(0)} q_{k-1}(x)\right), \quad k \geq 1.$$

Thus,

$$q_{k+1}(x) = -\frac{1}{\gamma_k}(r_k(x) + \alpha q_k(x) + \beta_{k-1} q_{k-1}(x)), \quad k \geq 0,$$

where

$$\gamma_0 = -\frac{1}{\alpha}, \qquad \gamma_k = \frac{c}{2}\frac{\tilde{T}_{k+1}(0)}{\tilde{T}_k(0)} = \frac{c}{2}\frac{T_{k+1}(-\alpha/c)}{T_k(-\alpha/c)}, \quad k \geq 1,$$

$$\beta_{-1} = 0, \qquad \beta_k = \frac{c}{2}\frac{\tilde{T}_k(0)}{\tilde{T}_{k+1}(0)} = \frac{c}{2}\frac{T_k(-\alpha/c)}{T_{k+1}(-\alpha/c)}, \quad k \geq 0.$$

Furthermore, the recurrence for r_k can be rearranged as

$$r_{k+1}(x) = \frac{1}{\gamma_k}(xr_k(x) - \alpha r_k(x) - \beta_{k-1} r_{k-1}(x)), \quad k \geq 0.$$

C.1. Error bounds for the residuals. The Chebyshev iteration as given in [**17**, Algorithm 1] is recovered by replacing $q_k(x)$ with $\mathbf{x}_k$ and $r_k(x)$ with $\mathbf{r}_k$ in these recurrences. That is, the iterates are defined by $\mathbf{x}_k = q_k(\mathbf{A})\mathbf{b}$, and the residuals are given by $\mathbf{r}_k = \mathbf{b} - \mathbf{A}\mathbf{x}_k = r_k(\mathbf{A})\mathbf{b}$ where $\mathbf{r}_k$ is presumably small because r_k is small on $[\alpha - c, \alpha + c]$ due to the minimax property of the Chebyshev polynomials. More specifically, for $x \in [\alpha - c, \alpha + c]$,

$$|r_k(x)| = \left|\frac{\tilde{T}_k(x)}{\tilde{T}_k(0)}\right| \leq \frac{2|\rho|^k}{1 + |\rho|^{2k}}, \quad \rho = J_+^{-1}\left(-\frac{\alpha}{c}\right),$$

where J_+^{-1} is a right inverse of the Joukowsky map, $z \mapsto (z + z^{-1})/2$, given by

$$J_+^{-1}(z) = z - \sqrt{z-1}\sqrt{z+1},$$

which maps $\mathbb{C} \setminus [-1, 1]$ to the interior of the unit disk. This results in error bounds for $\mathbf{r}_k$. An interesting aspect of this bound on r_k is that it remains bounded as $|\rho| \to 1$.

In contrast, our modification of the Chebyshev iteration implemented in Algorithm 1 relies on building an approximation to the matrix inverse out of Chebyshev polynomials. Define the normalized shifted and scaled Chebyshev-T polynomials by

$$\hat{T}_0(x) = \tilde{T}_0(x), \quad \hat{T}_k(x) = \sqrt{2}\tilde{T}_k(x), \quad k \geq 1.$$

From [**3**, Section 4.1], we have the following recurrence for $S_k = \int_{\alpha-c}^{\alpha+c} \frac{\tilde{T}_j(z)}{z} \frac{\mathrm{d}z}{\pi\sqrt{z-\alpha+c}\sqrt{\alpha+c-z}}$:

$$S_0 = \frac{1}{\sqrt{\alpha - c}\sqrt{\alpha + c}}, \quad S_k = S_{k-1} J_+^{-1}\left(\frac{-\alpha}{c}\right), \quad k \geq 1,$$

Then, (1) can be rewritten as

$$\mathbf{x}_{k+1} = S_0 \tilde{T}_0(\mathbf{A})\mathbf{b} + 2\sum_{j=1}^{k} S_j \tilde{T}_j(\mathbf{A})\mathbf{b}. \tag{24}$$

Coupling this with the recurrence (23) yields the implementation of our modification of the Chebyshev iteration in Algorithm 1.

To analyze the convergence, we utilize the series

$$\frac{1}{x} = \frac{2}{c}\frac{1}{\sqrt{\frac{\alpha}{c}-1}\sqrt{\frac{\alpha}{c}+1}}\sum_{j=0}^{\infty}{}' \rho^j \tilde{T}_j(x), \quad \rho = J_+^{-1}\left(\frac{-\alpha}{c}\right),$$

where the prime denotes that the 0th term is halved. This follows from (24) or [**22**, Equation 5.14]. Then, to estimate the error for kth iteration we use, for $x \in [\alpha - c, \alpha + c]$,

$$G_k(x) = \frac{1}{x} - \frac{2}{c}\frac{1}{\sqrt{\frac{\alpha}{c}-1}\sqrt{\frac{\alpha}{c}+1}}\sum_{j=0}^{k-1}{}' \rho^j \tilde{T}_j(x) = \frac{2}{c}\frac{1}{\sqrt{\frac{\alpha}{c}-1}\sqrt{\frac{\alpha}{c}+1}}\sum_{j=k}^{\infty} \rho^j \tilde{T}_j(x).$$

To instead investigate the residual $\mathbf{b} - \mathbf{A}\mathbf{x}_k$ of our modification, we use the three-term recurrence for $\tilde{T}_j$ to find

$$\begin{aligned} xG_k(x) &= \frac{2}{c}\frac{x}{\sqrt{\frac{\alpha}{c}-1}\sqrt{\frac{\alpha}{c}+1}}\sum_{j=k}^{\infty} \rho^j \tilde{T}_j(x) = \frac{\rho^{k-1}(\rho\tilde{T}_{k-1}(x) - \tilde{T}_k(x))}{\sqrt{\frac{\alpha}{c}-1}\sqrt{\frac{\alpha}{c}+1}} \\ &= -\frac{\rho^{k-1}(\tilde{T}_{k-1}(x) + \tilde{T}_k(x))}{\sqrt{\frac{\alpha}{c}-1}\sqrt{\frac{\alpha}{c}+1}} + \frac{\rho^{k-1}(\rho+1)\tilde{T}_{k-1}(x)}{\sqrt{\frac{\alpha}{c}-1}\sqrt{\frac{\alpha}{c}+1}}. \end{aligned}$$

From the identity $T_k(x) + T_{k-1}(x) = (x+1)V_{k-1}(x)$ where V_k is the kth Chebyshev polynomial of the third kind [**22**] and the formula for ρ, it follows that

$$\begin{aligned} xG_k(x) &= -\frac{\rho^{k-1}\left(\frac{x-\alpha}{c}+1\right)\tilde{V}_{k-1}(x)}{\sqrt{\frac{\alpha}{c}-1}\sqrt{\frac{\alpha}{c}+1}} + \frac{\rho^{k-1}(-\frac{\alpha}{c}+1)\tilde{T}_{k-1}(x)}{\sqrt{\frac{\alpha}{c}-1}\sqrt{\frac{\alpha}{c}+1}} + \rho^{k-1}\tilde{T}_{k-1}(x) \\ &= -\frac{\rho^{k-1}x\tilde{V}_{k-1}(x)}{\sqrt{\alpha-c}\sqrt{\alpha+c}} + \frac{\rho^{k-1}\sqrt{\frac{\alpha}{c}-1}(\tilde{V}_{k-1}(x) - \tilde{T}_{k-1}(x))}{\sqrt{\frac{\alpha}{c}+1}} + \rho^{k-1}\tilde{T}_{k-1}(x), \\ \tilde{V}_k(x) &= V_k\left(\frac{x-\alpha}{c}\right). \end{aligned}$$

We see that as $\alpha/c \to 1$, $\rho \to -1$, we do not enjoy the same boundedness as $r_k(x)$ above, but for $x \ll 1$, in this limit, $xG_k(x)$ remains under control. Thus, if one cares about minimizing the residual, for an ill-conditioned linear system, the classical Chebyshev iteration is likely to perform better.

C.2. Error bounds for the true error. To get a sense of the error

$$\mathbf{e}_k = \mathbf{x}_k - \mathbf{A}^{-1}\mathbf{b},$$

we compute

$$\begin{aligned} |G_k(x)| &\leq \frac{2|\rho|^k}{|\sqrt{\alpha-c}\sqrt{\alpha+c}|}\frac{1}{1-|\rho|} = \frac{2|\rho|^k}{|\sqrt{\alpha-c}\sqrt{\alpha+c}|}\frac{1}{1-\frac{\alpha}{c}-\sqrt{\frac{\alpha}{c}-1}\sqrt{\frac{\alpha}{c}+1}} \\ &= \frac{2c|\rho|^k}{|\alpha-c|\,|\sqrt{\alpha+c}|}\frac{1}{|\sqrt{\alpha+c}| - |\sqrt{c-\alpha}|}. \end{aligned}$$

The coefficient of $|\alpha - c|^{-1}|\rho|^k$ tends to one as $\alpha/c \to 1$.

Then the corresponding quantity for the classical Chebyshev iteration is

$$\left|\frac{1}{x} - q_k(x)\right| = \left|\frac{\tilde{T}_k(x)}{x\tilde{T}_k(0)}\right| \leq \frac{2|\rho|^k}{|x|(1+|\rho|^{2k})},$$

which is maximized when $x = \alpha - c$ giving

$$\left|\frac{1}{x} - q_k(x)\right| \leq \frac{2|\rho|^k}{|\alpha - c|(1+|\rho|^{2k})}.$$

Here the coefficient of $|\alpha - c|^{-1}|\rho|^k$ also tends to one as $\alpha/c \to 1$. Thus, the true error in our modification of the Chebyshev iteration performs similarly to that of the classical Chebyshev iteration in this limit when the system is ill-conditioned.

Acknowledgments

We would like to thank Alex Townsend and Walter Van Assche for pointing us to valuable references.

References

[1] N. I. Akhiezer, *Elements of the theory of elliptic functions*, Translations of Mathematical Monographs, vol. 79, American Mathematical Society, Providence, RI, 1990. Translated from the second Russian edition by H. H. McFaden, DOI 10.1090/mmono/079. MR1054205

[2] Cade Ballew and Thomas Trogdon, `https://github.com/cade-b/AkhiezerIteration`, 2023.

[3] Cade Ballew and Thomas Trogdon, *A Riemann-Hilbert approach to computing the inverse spectral map for measures supported on disjoint intervals*, Stud. Appl. Math. **152** (2024), no. 1, 31–72. MR4692723

[4] Deniz Bilman, Patrik Nabelek, and Thomas Trogdon, *Computation of large-genus solutions of the Korteweg–de Vries equation*, Phys. D **449** (2023), Paper No. 133715, 26, DOI 10.1016/j.physd.2023.133715. MR4575119

[5] I. Bogaert, *Iteration-free computation of Gauss-Legendre quadrature nodes and weights*, SIAM J. Sci. Comput. **36** (2014), no. 3, A1008–A1026, DOI 10.1137/140954969. MR3209728

[6] Jie Chen, Mihai Anitescu, and Yousef Saad, *Computing $f(A)b$ via least squares polynomial approximations*, SIAM J. Sci. Comput. **33** (2011), no. 1, 195–222, DOI 10.1137/090778250. MR2783192

[7] Yang Chen and Alexander R. Its, *A Riemann-Hilbert approach to the Akhiezer polynomials*, Philos. Trans. R. Soc. Lond. Ser. A Math. Phys. Eng. Sci. **366** (2008), no. 1867, 973–1003, DOI 10.1098/rsta.2007.2058. MR2377682

[8] Yang Chen and Nigel Lawrence, *A generalization of the Chebyshev polynomials*, J. Phys. A **35** (2002), no. 22, 4651–4699, DOI 10.1088/0305-4470/35/22/302. MR1908638

[9] Carl de Boor and John R. Rice, *Extremal polynomials with application to Richardson iteration for indefinite linear systems*, SIAM J. Sci. Statist. Comput. **3** (1982), no. 1, 47–57, DOI 10.1137/0903004. MR651866

[10] P. A. Deift, *Orthogonal polynomials and random matrices: a Riemann-Hilbert approach*, Courant Lecture Notes in Mathematics, vol. 3, New York University, Courant Institute of Mathematical Sciences, New York; American Mathematical Society, Providence, RI, 1999. MR1677884

[11] Xiucai Ding and Thomas Trogdon, *A Riemann-Hilbert approach to the perturbation theory for orthogonal polynomials: applications to numerical linear algebra and random matrix theory*, Int. Math. Res. Not. IMRN **5** (2024), 3975–4061, DOI 10.1093/imrn/rnad142. MR4714335

[12] *NIST Digital Library of Mathematical Functions*, `https://dlmf.nist.gov/`, Release 1.1.11 of 2023-09-15. F. W. J. Olver, A. B. Olde Daalhuis, D. W. Lozier, B. I. Schneider, R. F. Boisvert, C. W. Clark, B. R. Miller, B. V. Saunders, H. S. Cohl, and M. A. McClain, eds.

[13] Donald A. Flanders and George Shortley, *Numerical determination of fundamental modes*, J. Appl. Phys. **21** (1950), 1326–1332. MR40075

[14] A. S. Fokas, A. R. Its, and A. V. Kitaev, *The isomonodromy approach to matrix models in 2D quantum gravity*, Comm. Math. Phys. **147** (1992), no. 2, 395–430. MR1174420

[15] Walter Gautschi, *Orthogonal polynomials: computation and approximation*, Numerical Mathematics and Scientific Computation, Oxford University Press, New York, 2004. Oxford Science Publications. MR2061539

[16] William B. Gragg and William J. Harrod, *The numerically stable reconstruction of Jacobi matrices from spectral data*, Numer. Math. **44** (1984), no. 3, 317–335, DOI 10.1007/BF01405565. MR757489

[17] Martin H. Gutknecht and Stefan Röllin, *The Chebyshev iteration revisited*, Parallel Comput. **28** (2002), no. 2, 263–283, DOI 10.1016/S0167-8191(01)00139-9. Parallel matrix algorithms and applications (Neuchâtel, 2000). MR1889790

[18] Nicholas Hale and Alex Townsend, *Fast and accurate computation of Gauss-Legendre and Gauss-Jacobi quadrature nodes and weights*, SIAM J. Sci. Comput. **35** (2013), no. 2, A652–A674, DOI 10.1137/120889873. MR3033086

[19] Daan Huybrechs and Peter Opsomer, *Construction and implementation of asymptotic expansions for Laguerre-type orthogonal polynomials*, IMA J. Numer. Anal. **38** (2018), no. 3, 1085–1118, DOI 10.1093/imanum/drx030. MR3829155

[20] Thomas A. Manteuffel, *The Tchebychev iteration for nonsymmetric linear systems*, Numer. Math. **28** (1977), no. 3, 307–327, DOI 10.1007/BF01389971. MR474739

[21] Thomas A. Manteuffel, *Adaptive procedure for estimating parameters for the nonsymmetric Tchebychev iteration*, Numer. Math. **31** (1978/79), no. 2, 183–208, DOI 10.1007/BF01397475. MR509674

[22] J. C. Mason and D. C. Handscomb, *Chebyshev polynomials*, Chapman & Hall/CRC, Boca Raton, FL, 2003. MR1937591

[23] Yuji Nakatsukasa, Olivier Sète, and Lloyd N. Trefethen, *The AAA algorithm for rational approximation*, SIAM J. Sci. Comput. **40** (2018), no. 3, A1494–A1522, DOI 10.1137/16M1106122. MR3805855

[24] Sheehan Olver, Richard Mikaël Slevinsky, and Alex Townsend, *Fast algorithms using orthogonal polynomials*, Acta Numer. **29** (2020), 573–699, DOI 10.1017/S0962492920000045. MR4189295

[25] Sheehan Olver and Thomas Trogdon, *Numerical solution of Riemann-Hilbert problems: random matrix theory and orthogonal polynomials*, Constr. Approx. **39** (2014), no. 1, 101–149, DOI 10.1007/s00365-013-9221-3. MR3144383

[26] Theodore J. Rivlin, *An introduction to the approximation of functions*, Dover Books on Advanced Mathematics, Dover Publications, Inc., New York, 1981. Corrected reprint of the 1969 original. MR634509

[27] Youcef Saad, *Iterative solution of indefinite symmetric linear systems by methods using orthogonal polynomials over two disjoint intervals*, SIAM J. Numer. Anal. **20** (1983), no. 4, 784–811, DOI 10.1137/0720052. MR708457

[28] Gabor Szegö, *Orthogonal Polynomials*, American Mathematical Society Colloquium Publications, Vol. 23, American Mathematical Society, New York, 1939. MR077

[29] Alex Townsend, Thomas Trogdon, and Sheehan Olver, *Fast computation of Gauss quadrature nodes and weights on the whole real line*, IMA J. Numer. Anal. **36** (2016), no. 1, 337–358, DOI 10.1093/imanum/drv002. MR3463444

[30] Lloyd N. Trefethen and J. A. C. Weideman, *The exponentially convergent trapezoidal rule*, SIAM Rev. **56** (2014), no. 3, 385–458, DOI 10.1137/130932132. MR3245858

[31] Thomas Trogdon and Sheehan Olver, *A Riemann-Hilbert approach to Jacobi operators and Gaussian quadrature*, IMA J. Numer. Anal. **36** (2016), no. 1, 174–196, DOI 10.1093/imanum/dru061. MR3463437

[32] Thomas Trogdon and Sheehan Olver, *Riemann-Hilbert problems, their numerical solution, and the computation of nonlinear special functions*, Society for Industrial and Applied Mathematics (SIAM), Philadelphia, PA, 2016. MR3450072

Department of Applied Mathematics, University of Washington, Seattle, Washington 98195-3925

Email address: ballew@uw.edu

Department of Applied Mathematics, University of Washington, Seattle, Washington 98195-3925

Email address: trogdon@uw.edu

Contemporary Mathematics
Volume **822**, 2025
https://doi.org/10.1090/conm/822/16473

On Airy solutions of $\mathrm{P_{II}}$ and the complex cubic ensemble of random matrices, II

Ahmad Barhoumi, Pavel Bleher, Alfredo Deaño, and Maxim Yattselev

ABSTRACT. We describe the pole-free regions of the one-parameter family of special solutions of $\mathrm{P_{II}}$, the second Painlevé equation, constructed from the Airy functions. This is achieved by exploiting the connection between these solutions and the recurrence coefficients of orthogonal polynomials that appear in the analysis of the ensemble of random matrices corresponding to the cubic potential.

1. Introduction

This paper is a study of the so-called special function solutions of $\mathrm{P_{II}}$, the second Painlevé equation, and is a continuation of our previous work [**1**]. While generic solutions of Painlevé equations are highly transcendental, it is known that all but the first Painlevé equation admit special solutions which can be expressed in terms of elementary or classical special functions. The second Painlevé equation,

$$q''(z) = 2q^3(z) + zq(z) + \alpha, \tag{1.1}$$

possesses both rational solutions and solutions written in terms of the Airy functions for specific values of the parameter $\alpha \in \mathbb{C}$. In this work, we are interested in the latter, which are known to be meromorphic functions of $z \in \mathbb{C}$, with only simple poles. Our goal is to identify the pole-free regions for these solutions, by connecting these Airy solutions to the recurrence coefficients for orthogonal polynomials that arise in the complex cubic random matrix model, where z, properly rescaled, appears as a parameter.

It is heuristically well understood in the field of non-Hermitian orthogonal polynomials, but requires rigorous technical asymptotic analysis that we shall carry out in subsequent publications, that the pole-free regions of the recurrence coefficients as functions of the parameter correspond to the regions of the parameter plane where the attracting set of zeros of the orthogonal polynomials (as their degree

---

2020 *Mathematics Subject Classification.* Primary 15B52, 33C10, 33C47.
Key words and phrases. Airy functions, Painlevé equations, random matrix models.
The first author was partially supported by the European Research Council (ERC), Grant Agreement No. 101002013. The third author acknowledges financial support from Universidad Carlos III de Madrid (I Convocatoria para la Recualificación del Profesorado Universitario), from Grant PID2021-123969NB-I00, funded by MCIN/AEI/ 10.13039/501100011033, and from grant PID2021-122154NB-I00 from Spanish Agencia Estatal de Investigación. The research of the fourth author was supported in part by a grant from the Simons Foundation, CGM-706591.

©2025 American Mathematical Society

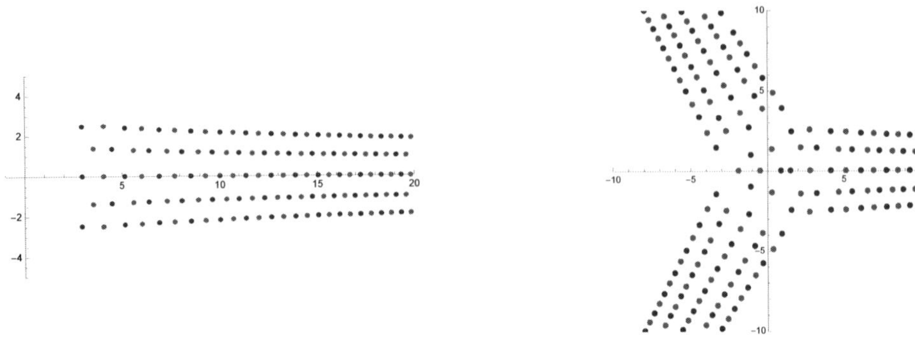

FIGURE 1. Zeros (blue) and poles (red) of $q_3(z;0)$ (left panel) and $q_3(z;\infty)$ (right panel), see (2.4)–(2.5) for the meaning of the second parameter.

tends to ∞) consists of a single Jordan arc. This is usually known in the literature as the *one–cut case*. The main goal of this work is to describe the phase diagram geometrically, that is, the partition of the parameter z space into different regions where the geometry of the zero-attracting set is qualitatively different, thus identifying the pole-free regions for the recurrence coefficients and for the respective Airy solutions of $\mathrm{P_{II}}$.

This paper is divided into three sections. In Section 2, we give a succinct but complete construction of the Airy solutions and various related functions. Section 3 is devoted to a brief description of the cubic random matrix model and the related orthogonal polynomials. There, we also state the connection between orthogonal polynomials and the Airy solutions established earlier in [**1**]. Finally, in Section 4, we prove the main result of this work describing the phase diagram for the complex cubic random matrix model. This geometric analysis is strongly connected with potential theory in the complex plane, and it is an essential step (the construction of the so-called g function) in the method of nonlinear steepest descent for asymptotics of orthogonal polynomials. This full asymptotic study will eventually allow us to prove rigorously the structure of the pole free regions for Airy solutions of $\mathrm{P_{II}}$.

2. Airy solutions of $\mathbf{P_{II}}$

It was shown by Gambier [**18**] that $\mathrm{P_{II}}$ has a one-parameter family of solutions that are expressible in terms of the Airy functions and their derivatives whenever $\alpha+1/2 = n \in \mathbb{Z}$. These are meromorphic functions in the complex plane whose poles exhibit a well-structured behavior, see Figure 1. The goal of this work is to provide a qualitative description of the pole-free regions for these solutions. Our approach proceeds through the connection with the orthogonal polynomials. Arguably one of the first successful implementations of a recasting of poles of Painlevé solutions as zeros of a related Hankel determinants was carried out in [**4**]. To understand this connection, we need a determinantal representation of these solutions, see [**29**] as well as [**14, 17**], whose derivation is explained in this section.

2.1. Airy solutions. The family of the Airy solutions of $\mathrm{P_{II}}$ is highly structured and can be generated from a single seed solution. To obtain it, compatibility with a Riccati equation is required, i.e., it is asked that $q(z)$ also satisfies

$$q'(z) = k_1(z)q^2(z) + k_2(z)q(z) + k_3(z) \tag{2.1}$$

for some functions $k_1(z), k_2(z), k_3(z)$ to be determined. Differentiating both sides of (2.1) and using (1.1) reduces the Riccati equation to an algebraic equation

$$\begin{aligned}(2 - 2k_1^2(z))q^3(z) - (k_1'(z) + 3k_1(z)k_2(z))q^2(z) - \\ (k_2'(z) + k_2^2(z) + 2k_1(z)k_3(z) - z)q(z) + \alpha - k_2(z)k_3(z) - k_3'(z) = 0,\end{aligned}$$

which must hold identically. Setting the coefficients next to the powers of $q(z)$ to zero gives that $k_1(z) = \epsilon$, $k_2(z) = 0$, $k_3(z) = \alpha z$ while $\alpha = \epsilon/2$, where $\epsilon^2 = 1$. This turns (2.1) into

$$q_\epsilon'(z) = \epsilon\big(q_\epsilon^2(z) + z/2\big). \tag{2.2}$$

Choosing $\epsilon = 1$ and making a substitution $q_1(z) = -(\log \tau_1(z))'$, linearizes (2.2) and yields

$$\tau_1''(z) + \frac{1}{2}z\tau_1(z) = 0,$$

which is a rescaled Airy equation that is solved by linear combinations

$$\tau_1(z) = C_1\mathrm{Ai}(-2^{-1/3}z) + C_2\mathrm{Bi}(-2^{-1/3}z), \tag{2.3}$$

where $C_1, C_2 \in \mathbb{C}$ are arbitrary constants. Thus, the corresponding solution of (2.2) for $\epsilon = 1$, i.e. $\alpha = 1/2$, is given by

$$q_1(z;\lambda) = -\frac{\mathrm{d}}{\mathrm{d}z}\log\left(C_1\,\mathrm{Ai}\left(-2^{-1/3}z\right) + C_2\,\mathrm{Bi}\left(-2^{-1/3}z\right)\right), \tag{2.4}$$

where $\lambda = C_2/C_1 \in \overline{\mathbb{C}}$ (note that the presence of the logarithmic derivative in the definition of $q_1(z;\lambda)$ makes it so that this is truly a one-parameter family parametrized by λ). The reader might protest that imposing the Riccati condition is unmotivated, but this does have an interpretation, see Remark 2.1 further below.

The solutions $q_n(z;\lambda)$ of (1.1) corresponding to $\alpha = n - 1/2$, $n \geq 2$, are now constructed recursively via the Bäcklund transformation

$$q_{n+1}(z;\lambda) = -q_n(z;\lambda) - \frac{2n}{2q_n^2(z;\lambda) + 2q_n'(z;\lambda) + z}, \tag{2.5}$$

see for instance [**16**, 32.7.25]. The solutions for non-positive values of n are then obtained via the symmetry $q_n(z;\lambda) = -q_{-n+1}(z;\lambda)$.

2.2. Hamiltonian structure. Each of the first six Painlevé equations can be written as a Hamiltonian system

$$\frac{\mathrm{d}q}{\mathrm{d}z} = \frac{\partial H}{\partial p} \quad \text{and} \quad \frac{\mathrm{d}p}{\mathrm{d}z} = -\frac{\partial H}{\partial q},$$

see [**23, 27, 28**]. In the case of P_{II} the Hamiltonian H_{II} is equal to

$$H_{II}(q,p,z;\alpha) = \frac{1}{2}p^2 - \left(q^2 + \frac{z}{2}\right)p - \left(\alpha + \frac{1}{2}\right)q. \tag{2.6}$$

Hence, the Hamiltonian system becomes

$$q'(z) = p(z) - q^2(z) - z/2 \quad \text{and} \quad p'(z) = 2p(z)q(z) + \alpha + 1/2. \tag{2.7}$$

Eliminating $p(z)$ from (2.7) gives (1.1), while eliminating $q(z)$ from (2.7) yields $\mathrm{P}_{\mathrm{XXXIV}}$:

$$p''(z) = \frac{(p'(z))^2 - (\alpha + 1/2)^2}{2p(z)} + 2p^2(z) - zp(z). \tag{2.8}$$

Moreover, if $q_{\alpha+1/2}(z)$ and $p_{\alpha+1/2}(z)$ solve (1.1) and (2.8), respectively, that is, as a pair solve (2.7), and

$$\sigma_{\alpha+1/2}(z) := H_{\mathrm{II}}\big(q_{\alpha+1/2}(z), p_{\alpha+1/2}(z), z; \alpha\big), \tag{2.9}$$

then this function solves S_{II}, the Jimbo-Miwa-Okamoto σ-form of P_{II}:

$$(\sigma'')^2 + 4(\sigma')^3 + 2\sigma'(z\sigma' - \sigma) = (\alpha/2 + 1/4)^2. \tag{2.10}$$

Conversely, if $\sigma_{\alpha+1/2}(z)$ solves (2.10), then the functions

$$q_{\alpha+1/2}(z) = \frac{2\sigma''_{\alpha+1/2}(z) + \alpha + 1/2}{4\sigma'_{\alpha+1/2}(z)} \quad \text{and} \quad p_{\alpha+1/2}(z) = -2\sigma'_{\alpha+1/2}(z), \tag{2.11}$$

solve (1.1) and (2.8), respectively, see [**23, 27–29**].

REMARK 2.1. Imposing the Riccati equation (2.2) when $\epsilon = -1$, i.e., $\alpha = -1/2$, is equivalent to seeking a "stationary" solution to the Hamiltonian system where $p(z) \equiv 0$. To arrive at a similar interpretation for the choice $\epsilon = 1$ requires the following modification of the Hamiltonian, cf.[1] [**29**, Remark 1.2]. Consider the change of variables

$$\hat{q}(z) := q(z), \quad \hat{p}(z) := p(z) - 2q^2(z) - z,$$

along with the Hamiltonian

$$\widehat{H}_{\mathrm{II}}(\hat{p}, \hat{q}, z; \alpha) := H_{\mathrm{II}}(q, p, z; \alpha) + q = \frac{1}{2}\hat{p}^2 + \left(\hat{q}^2 + \frac{z}{2}\right)\hat{p} - \left(\alpha - \frac{1}{2}\right)\hat{q}.$$

Then, the Hamiltonian system reads

$$\hat{q}'(z) = \hat{p}(z) + \hat{q}^2(z) + z/2 \quad \text{and} \quad \hat{p}'(z) = -2\hat{p}(z)\hat{q}(z) + \alpha - 1/2.$$

Eliminating $\hat{p}(z)$ again gives (1.1), and now we can see that imposing the Riccati equation (2.2) with $\epsilon = 1$ is equivalent to requiring $\hat{p}(z) \equiv 0$.

2.3. Tau functions. Let now $q_n(z;\lambda)$ be the solutions of (1.1) given by (2.4)–(2.5), while $p_n(z;\lambda)$ and $\sigma_n(z;\lambda)$ be the corresponding solutions of (2.8) and (2.10) obtained via (2.7) and (2.9), respectively. We may now *define* $\tau_n(z;\lambda)$, up to a multiplicative constant, via the formula

$$\sigma_n(z;\lambda) = \frac{\mathrm{d}}{\mathrm{d}z} \log \tau_n(z;\lambda), \tag{2.12}$$

where we take $\tau_0 \equiv 1$ (as $p_0 \equiv 1$ due to Remark 2.1, $\sigma_0 \equiv 0$ by (2.6) and (2.9)) and $\tau_1(z;\lambda)$ to be $\tau_1(z)$ from (2.3) for some choice C_1, C_2 such that $C_2/C_1 = \lambda$ (all such choices lead to the same function $\sigma_1(z;\lambda)$). Then

$$q_n(z;\lambda) = \frac{\mathrm{d}}{\mathrm{d}z} \log \frac{\tau_{n-1}(z;\lambda)}{\tau_n(z;\lambda)}. \tag{2.13}$$

---

[1] In the cited Remark, there is a sign error in the definition of $\widehat{H}_{\mathrm{II}}$ which we correct here.

To arrive at expression (2.13), we need the inverse of the Bäcklund transformation (2.5), see e.g. [**14**, Theorem 2],

$$q_{n-1}(z) = -q_n(z) + \frac{2(n-1)}{2q_n'(z) - 2q_n^2(z) - z}, \tag{2.14}$$

where we drop the dependence on λ for brevity. Combining (2.5), (2.14), and the first equation in (2.7), one discovers the identity

$$p_{n-1}(z) = -p_n(z) + 2q_n^2(z) + z. \tag{2.15}$$

Using (2.15), (2.14) yields the remarkable relation (cf. [**29**, Eq. $(1.13)_3$])

$$H_{\mathrm{II}}(q_{n-1}(z), p_{n-1}(z), z; n-3/2) = H_{\mathrm{II}}(q_n(z), p_n(z), z; n-3/2),$$

which, in view of (2.9) and (2.12) implies (2.13).

To derive the determinantal representation for the functions $\tau_n(z)$, observe that they satisfy the Toda equation

$$\frac{\mathrm{d}^2}{\mathrm{d}z^2}\log\tau_n(z) = K_n\frac{\tau_{n-1}(z)\tau_{n+1}(z)}{\tau_n^2(z)}, \quad K_n \in \mathbb{C}. \tag{2.16}$$

Indeed, on the one hand, (2.13) implies that

$$q_{n+1}(z) - q_n(z) = -\frac{\mathrm{d}}{\mathrm{d}z}\log\frac{\tau_{n+1}(z)\tau_{n-1}(z)}{\tau_n^2(z)}. \tag{2.17}$$

On the other hand, the Bäcklund transformation (2.5) together with the first equation in (2.7) as well as (2.11) give that

$$q_{n+1}(z) - q_n(z) = -\frac{n}{p_n(z)} + 2q_n(z) = -\frac{\mathrm{d}}{\mathrm{d}z}\log\sigma_n'(z). \tag{2.18}$$

Combining (2.17)–(2.18) with (2.12) clearly implies (2.16). Recall now that the tau functions are defined by (2.12) only up to a multiplicative constant. Given the seed functions τ_0 and τ_1 as described after (2.12), we normalize the rest of them so that (2.16) is satisfied with $K_n = 1$. This yields the representation

$$\tau_n(z;\lambda) = \det\left[\frac{\mathrm{d}^{j+k}}{\mathrm{d}z^{j+k}}\left(C_1\operatorname{Ai}\left(-2^{-1/3}z\right) + C_2\operatorname{Bi}\left(-2^{-1/3}z\right)\right)\right]_{j,k=0}^{n-1}, \tag{2.19}$$

where C_1, C_2 are the same as in (2.3), because the right-hand sides above satisfy (2.16) with $K_n = 1$ for all $n \geq 1$ according to Dodgson condensation identity[2].

The functions $\tau_n(z;\lambda)$, which are entire, have an interpretation as the isomonodromic tau function, see [**24**], that, in the context of Painlevé equations, were studied in great detail in [**23**]. There, the authors made the following observation.

Remark 2.2. The functions $\tau_n(z;\lambda)$ have only simple zeros. Indeed, if z_0 is a zero of $\tau_n(z;\lambda)$ of order p, then, in view of (2.12), $\sigma_n(z;\lambda)$ is a meromorphic function in $\mathbb{C}$ and has a simple pole at $z = z_0$ with residue p. A local analysis of (2.10) as $z \to z_0$ reveals that

$$\sigma_n(z;\lambda) = \frac{1}{z-z_0} + s_0 - \frac{z_0}{6}(z-z_0) - \frac{1}{8}(z-z_0)^2 + \cdots, \quad s_0 \in \mathbb{C},$$

which implies that $p = 1$ as claimed.

[2]Also known as Desnanot-Jacobi identity or Sylvester determinant identity.

REMARK 2.3. The functions $\tau_n(z;\lambda)$ and $\tau_{n+1}(z;\lambda)$ have no zeros in common for any $n \geq 0$. Indeed, this follows from Remark 2.2 and the fact that Toda equation (2.16) can be rewritten in the form

$$\tau_n(z;\lambda)\tau_n''(z;\lambda) - (\tau_n'(z;\lambda))^2 = \tau_{n+1}(z;\lambda)\tau_{n-1}(z;\lambda). \tag{2.20}$$

3. Cubic random matrix model

Cubic random matrix model studies statistics of $N \times N$ Hermitian matrices M drawn from the ensemble given by the probability distribution

$$\tilde{Z}_N^{-1}(u)\mathrm{e}^{-N\mathrm{Tr}(M^2/2-uM^3)}\mathrm{d}M,$$

where u is a parameter and $\tilde{Z}_N(u)$ is the normalizing constant (partition function). This model has been investigated in physical literature by Brézin, Itzykson, Parisi, and Zuber [**12**] and Bessis, Itzykson, and Zuber [**7**]. An interesting feature of the model is that its free energy possesses an asymptotic expansion in powers of N^{-2} whose coefficients are functions of u which themselves admit a Taylor series expansion near $u = 0$. It was observed that the Taylor series of the coefficient of N^{-2g} is the generating function for the number of three-valent graphs on a Riemann surface of genus g, and so this asymptotic expansion of the free energy came to be known as the *topological expansion.* The existence of a topological expansion was shown in [**15**] for matrix models with even polynomial potential. In the case of the cubic potential this expansion was obtained in [**8**] by the middle two authors of the present work, where they computed the number of three-valent graphs explicitly for $g = 0, 1$ and asymptotically for $g > 1$.

3.1. Connection to Airy solutions.

After a change of variables, see [**8**], the cubic ensemble turns into the unitary ensemble of random matrices whose partition function is formally defined by the matrix integral over the space of $N\times N$ Hermitian matrices:

$$\int_{\mathcal{H}_N} \mathrm{e}^{-N\mathrm{Tr}(-\frac{1}{3}M^3+tM)}\mathrm{d}M,$$

where $t \in \mathbb{C}$ is a complex parameter. The formal partition function of the eigenvalues of these matrices can be written as

$$\int_{-\infty}^{\infty}\cdots\int_{-\infty}^{\infty}\prod_{1\leq j<k\leq N}(s_j - s_k)^2\prod_{j=1}^{N}\mathrm{e}^{-NV(s_j;t)}\mathrm{d}s_1\ldots\mathrm{d}s_N,$$

with the polynomial potential $V(s;t)$ given by

$$V(s;t) := -\frac{1}{3}s^3 + st. \tag{3.1}$$

This expression is formal because the integrals are divergent and need regularization. To achieve it, let

$$\Gamma = \Gamma(\lambda) := \alpha_0 L_0 + \alpha_1 L_1 + \alpha_2 L_2, \tag{3.2}$$

where $L_k := \{z : z = re^{(-1+2k/3)\pi \mathrm{i}}, r \in (0,\infty)\}$ are oriented towards the origin, see Figure 2, and $\alpha_0, \alpha_1, \alpha_2 \in \mathbb{C}$ are a one-parameter family of complex numbers given in terms of the parameter λ (see (2.4)) by

$$\alpha_0 = \frac{\lambda}{\pi}, \quad \alpha_1 = -\frac{\lambda}{2\pi} + \frac{1}{2\pi\mathrm{i}}, \quad \text{and} \quad \alpha_2 = -\frac{\lambda}{2\pi} - \frac{1}{2\pi\mathrm{i}}, \tag{3.3}$$

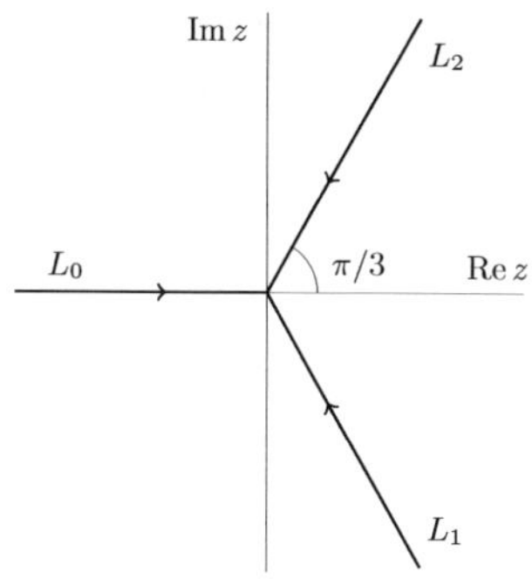

FIGURE 2. Contours L_0, L_1 and L_2 used in the construction of Γ.

when $|\lambda| < \infty$ and $\alpha_0 = 1/\pi$, $\alpha_1 = \alpha_2 = -1/2\pi$ when $\lambda = \infty$. Let

$$Z_N(t;\lambda) := \int_\Gamma \cdots \int_\Gamma \prod_{1\le j<k\le N} (s_j - s_k)^2 \prod_{j=1}^N \mathrm{e}^{-NV(s_j;t)} \mathrm{d}s_1 \ldots \mathrm{d}s_N,$$

which is well defined (the integrals are convergent) for all values $t \in \mathbb{C}$. As briefly discussed above, it was shown in [**8**] that the topological expansion of the free energy

$$F_N(t;\lambda) = \frac{1}{N^2} \log Z_N(t;\lambda)$$

is connected to the enumeration of regular graphs of degree 3 on Riemann surfaces. The partition function $Z_N(t;\lambda)$ can also be expressed as $N!D_{N-1}(t;N,\lambda)$, where

$$D_n(t;N,\lambda) := \det \left[\int_\Gamma s^{i+j} \mathrm{e}^{-NV(s;t)} \mathrm{d}s \right]_{i,j=0}^n .$$

The entries of the matrix defining $D_n(-t;N,\lambda)$ can be written as

$$\begin{aligned}\int_\Gamma s^k \mathrm{e}^{-NV(s;-t)} \mathrm{d}s &= \frac{1}{N^{k+1/3}} \frac{\mathrm{d}^k}{\mathrm{d}t^k} \left(\int_\Gamma \mathrm{e}^{-V\left(s;-N^{2/3}t\right)} \mathrm{d}s \right) \\ &= \frac{1}{N^{k+1/3}} \frac{\mathrm{d}^k}{\mathrm{d}t^k} \left(\alpha_0 \pi \operatorname{Bi}\left(-N^{2/3}t\right) + (\alpha_1 - \alpha_2)\pi \mathrm{i} \operatorname{Ai}\left(-N^{2/3}t\right) \right),\end{aligned}$$

where we used the classical integral representations, see [**16**, 9.5.4, 9.5.5],

$$\operatorname{Ai}(z) = \frac{1}{2\pi \mathrm{i}} \int_{L_1 - L_2} \mathrm{e}^{-V(s;z)} \mathrm{d}s \quad \text{and} \quad \operatorname{Bi}(z) = \frac{1}{2\pi} \left(\int_{L_0 - L_1} + \int_{L_0 - L_2} \right) \mathrm{e}^{-V(s;z)} \mathrm{d}s,$$

as well as the condition $\alpha_0 + \alpha_1 + \alpha_2 = 0$. Setting $z = (\sqrt{2}N)^{2/3} t$, we get

$$\begin{aligned}(3.4) \quad &\int_\Gamma s^k \mathrm{e}^{-NV(s;-t)} \mathrm{d}s \\ &= \frac{2^{k/3}}{N^{(k+1)/3}} \frac{\mathrm{d}^k}{\mathrm{d}z^k} \left(\alpha_0 \pi \operatorname{Bi}\left(-2^{-1/3} z\right) + (\alpha_1 - \alpha_2)\pi \mathrm{i} \operatorname{Ai}\left(-2^{-1/3} z\right) \right).\end{aligned}$$

Equation (3.4), representation (2.19) of $\tau_n(z;\lambda)$, and the Hankel structure of the determinant $D_n(-t;N,\lambda)$ now yield that

$$(3.5) \qquad D_n(-t;N,\lambda) = \frac{2^{n(n+1)/3}}{N^{(n+1)(n+2)/3}} \tau_{n+1}(z;\lambda), \quad z = (\sqrt{2}N)^{2/3} t.$$

An immediate consequence of (3.5) is that these Hankel determinants satisfy a version of the Toda equation (2.20), namely

$$D_n''(t;N,\lambda)D_n(t;N,\lambda)-(D_n'(t;N,\lambda))^2=N^2D_{n+1}(t;N,\lambda)D_{n-1}(t;N,\lambda).$$

Furthermore, if we take $n=N$, then (2.12) can be rewritten in terms of the free energy $F_N(t;\lambda)$ as

$$\sigma_N(z;\lambda)=N^2\frac{\mathrm{d}}{\mathrm{d}z}F_N\left(-(\sqrt{2}N)^{-2/3}z;\lambda\right),$$

which, of course, means that $p_N(z;\lambda)$ and $q_N(z;\lambda)$ can be expressed via $F_N(t;\lambda)$ using (2.11). The above representations can be further rewritten using the orthogonal polynomials associated to the cubic ensemble, which we do next.

3.2. Non-Hermitian orthogonal polynomials. The idea of using orthogonal polynomials in computations relating to random matrices is classical at this point, see e.g. [**26**] and references therein. Let $P_n(s;t,N,\lambda)$ be a non-identically zero polynomial of degree at most n such that

$$\int_\Gamma s^kP_n(s;t,N,\lambda)\mathrm{e}^{-NV(s;t)}\mathrm{d}s=0,\quad k\in\{0,\dots,n-1\},\tag{3.6}$$

where $\Gamma=\Gamma(\lambda)$ is as in (3.2)–(3.3) and we often shall drop the explicit dependence on λ when this dependence is not essential to us. This family of polynomials for specific values of λ was studied by Huybrechs, Kuijlaars, and Lejon in [**20, 21**] in the context of complex valued Gaussian quadrature rules, and by the authors of the present work in [**2, 9, 10**] in the context of the cubic model.

Due to the non-Hermitian character of the relations in (3.6), it might happen that polynomial satisfying (3.6) is non-unique. In this case we denote by $P_n(s;t,N,\lambda)$ the monic non-identically zero polynomial of the smallest degree; such a polynomial is always unique. It is a linear algebra exercise to verify that $\deg P_n(s;t,N,\lambda)=n$ if and only if $D_{n-1}(t;N)\neq 0$. Hence, equality (3.5) and Remark 2.3 immediately yield the following observation.

REMARK 3.1. For fixed $t\in\mathbb{C}$, $N>0$, $\lambda\in\overline{\mathbb{C}}$, and all $n\in\mathbb{N}$ it holds that

$$\deg P_n(\cdot;t,N,\lambda)=n\quad\text{or}\quad\deg P_{n+1}(\cdot;t,N,\lambda)=n+1.$$

Since $P_{n+1}(s;t,N,\lambda)=P_n(s;t,N,\lambda)$ when $\deg P_{n+1}(\cdot;t,N,\lambda)<n+1$, this means that $\deg P_{n+1}(\cdot;t,N,\lambda)\in\{n,n+1\}$ for each n.

The determinantal representation of orthogonal polynomials (see, e.g., [**37**]) yields that

$$h_n(t;N):=\int_\Gamma P_n^2(s;t,N)\mathrm{e}^{-NV(s;t)}\mathrm{d}s=\frac{D_n(t;N)}{D_{n-1}(t;N)},$$

where $D_{-1}(t;N)\equiv 1$. Since $D_n(t;N)$ is an entire function of t, each $h_n(t;N)$ is meromorphic in $\mathbb{C}$. Hence, given n, the set of the values t for which there exists $k\in\{0,\dots,n\}$ such that $h_k(t;N)=0$ is countable with no limit points in the finite plane. Outside of this set a standard argument using (3.6) shows that

$$sP_n(s;t,N)=P_{n+1}(s;t,N)+\beta_n(t;N)P_n(s;t,N)+\gamma_n^2(t;N)P_{n-1}(s;t,N),\tag{3.7}$$

and by analytic continuation (3.7) extends to those values of t for which $n+1$-st and n-th polynomials have the prescribed degrees (that is, $D_{n-1}(t;N)D_n(t;N) \neq 0$), where

$$\gamma_n^2(t;N) = \frac{h_n(t;N)}{h_{n-1}(t;N)} = \frac{D_n(t;N)D_{n-2}(t;N)}{D_{n-1}(t;N)^2}. \tag{3.8}$$

Denote by $p_{n,n-1}(t;N)$ the coefficient of $P_n(s;t,N)$ multiplying s^{n-1} and set $p_{0,-1}(t;N) := 0$. It follows from (3.7) that

$$\beta_n(t;N) = p_{n,n-1}(t;N) - p_{n+1,n}(t;N).$$

Using the determinantal form for orthogonal polynomials again and the form of the weight function in our case, we can deduce that

$$p_{n,n-1}(t;N) = \frac{1}{N}\frac{\mathrm{d}}{\mathrm{d}t}\log D_{n-1}(t;N), \tag{3.9}$$

and consequently

$$\beta_n(t;N) = \frac{1}{N}\left(\frac{\mathrm{d}}{\mathrm{d}t}\log D_{n-1}(t;N) - \frac{\mathrm{d}}{\mathrm{d}t}\log D_n(t;N)\right). \tag{3.10}$$

The following theorem has appeared in [**1**].

THEOREM 3.2. *Fix $N \geq 1$. Given $\lambda \in \overline{\mathbb{C}}$, it holds for each $n \in \mathbb{N}$ that*

$$\begin{cases} q_n(z;\lambda) &= -(N/2)^{1/3}\,\beta_{n-1}\left(-(\sqrt{2}N)^{-2/3}z;N,\lambda\right), \\ p_n(z;\lambda) &= -2(N/2)^{2/3}\,\gamma_n^2\left(-(\sqrt{2}N)^{-2/3}z;N,\lambda\right), \\ \sigma_n(z;\lambda) &= -(N/2)^{1/3}\,p_{n,n-1}\left(-(\sqrt{2}N)^{-2/3}z;N,\lambda\right), \end{cases}$$

where $q_n(z;\lambda)$ is a solution of (1.1) *given by* (2.4) *and* (2.5), *$p_n(z;\lambda)$ is the corresponding solution of* (2.8), *see* (2.7), *while $\sigma_n(z;\lambda)$ is the corresponding solution of* (2.10) *determined by* (2.9).

As we have stated before, our goal is to identify the pole-free regions and study scaling limits of the functions q_n, p_n, σ_n. Our approach goes through the connection to orthogonal polynomials stated in Theorem 3.2. In particular, combining this theorem with (3.8), (3.9), (3.10), and Remarks 2.2, 2.3 shows that the poles of the Painlevé function q_n in the z-plane correspond to zeros of D_{n-1} and D_n in the t-plane, i.e., to the values of t for which $\deg P_n = n-1$ or $\deg P_{n+1} = n$, while the poles of p_n and σ_n correspond to zeros of D_{n-1}.

The study of the asymptotic behavior of the polynomials P_n as $n \to \infty$ and for generic values of the parameter λ (i.e. for generic combination of contours, see (3.2)) is a very technical topic that we leave to subsequent publications. However, the expected outcome of such a study is well understood: if the attracting set of zeros of P_n consists of a single Jordan arc, all the zeros of P_n remain bounded and so are the quantities $\beta_{n-1}, \gamma_n, p_{n,n-1}$; otherwise polynomials P_n will have a zero exhibiting a *spurious* behavior leading to the polar singularities of $\beta_{n-1}, \gamma_n, p_{n,n-1}$.

In fact the above scheme has already been successfully carried out by us in [**1**] where we interpreted the asymptotic results of [**2, 10**] as scaling limits of $q_n(z;\lambda)$ when $\lambda \in \{0, -\mathrm{i}, \mathrm{i}\}$. These cases are essentially the same as they correspond to taking the seed function in (2.19) to be

$$\mathrm{Ai}\left(-2^{-1/3}z\right), \quad \mathrm{Ai}\left(-2^{-1/3}z\mathrm{e}^{2\pi\mathrm{i}/3}\right), \quad \text{and} \quad \mathrm{Ai}\left(-2^{-1/3}z\mathrm{e}^{-2\pi\mathrm{i}/3}\right)$$

since $\mathrm{Ai}(z)\pm\mathrm{i}\mathrm{Bi}(z) = 2\mathrm{e}^{\pm 2\pi\mathrm{i}/3}\mathrm{Ai}(z\mathrm{e}^{\mp 2\pi\mathrm{i}/3})$, see [**16**, Equation (9.2.1)]. They are also qualitatively different from the remaining ones as can be easily seen from Figure 1. Notice that these cases correspond to Γ in (3.2) for which $\alpha_0\alpha_1\alpha_2 = 0$ by (3.3). Hence, in the remaining part of the paper we shall always assume that $\alpha_0\alpha_1\alpha_2 \neq 0$.

4. Phase diagram

In this section we take the first step in understanding the pole-free regions of Airy solutions of $\mathrm{P_{II}}$ corresponding to $\lambda \notin \{0, -\mathrm{i}, \mathrm{i}\}$, or, equivalently, the behavior of the orthogonal polynomials corresponding to Γ for which $\alpha_0\alpha_1\alpha_2 \neq 0$. This step consists in describing the attracting set for the zeros of the orthogonal polynomials, or equivalently the phase diagram in the parameter space. For the rest of this section, we assume that $\alpha_0\alpha_1\alpha_2 \neq 0$.

Observe that due to the analyticity of the integrand in (3.6), the chain of integration Γ can be locally varied without changing the orthogonal polynomials. Hence, there arises a question of the identification of the contour attracting the zeros of the orthogonal polynomials. The answer to this question follows from Stahl-Gonchar-Rakhmanov theory of symmetric contours [**19, 33–35**] developed in a compact setting whose implications to unbounded contours with polynomials external fields were worked out by Kuijlaars and Silva [**25**].

4.1. Symmetric contours. Denote by $\mathcal{T}$ the collection of admissible contours defined as follows: each $T \in \mathcal{T}$ is a finite connected union of C^1-smooth Jordan arcs for which there exist $R_T > 0$ and $\epsilon_T \in (0, \pi/6)$ such that $T \setminus \{|z| \leq R_T\}$ consists of three unbounded Jordan arcs, say T_0, T_1, T_2, each connecting a point on $\{|z| = R_T\}$ to the point at infinity and satisfying $T_k \subset S_{k,\epsilon_T}$, where

$$S_{k,\epsilon} := \left\{z : \left|\arg(z) + \pi - 2k\pi/3\right| < \epsilon\right\}, \quad k \in \{0, 1, 2\}.$$

Notice that $L_0 \cup L_1 \cup L_2$, see (3.2), belongs to $\mathcal{T}$. Let $\mathcal{M}(T)$ be the space of Borel probability measures on $T \in \mathcal{T}$. The equilibrium energy of T in the external field $\operatorname{Re} V(\cdot; t)$, see (3.1), is equal to

$$\mathcal{E}_V(T) = \inf_{\nu \in \mathcal{M}(T)} \left(\iint \log \frac{1}{|s-t|} \mathrm{d}\nu(s)\mathrm{d}\nu(t) + \int \operatorname{Re} V(s; t)\mathrm{d}\nu(s) \right).$$

The infimum above is achieved by a unique minimizer, which is called the *weighted equilibrium measure* of T in the external field $\operatorname{Re} V(\cdot; t)$, see [**32**, Theorem I.1.3]. We shall denote this minimizer μ_T (of course, it is t-dependent). The support of μ_T, say J_T, is a compact subset of T. The equilibrium measure $\mu = \mu_T$ is characterized by the Euler–Lagrange variational conditions:

$$2U^\mu(z) + \operatorname{Re}\, V(z; t) \begin{cases} = \ell_T, & z \in J_T, \\ \geq \ell_T, & z \in T \setminus J_T, \end{cases} \tag{4.1}$$

for some constant ℓ_T, the Lagrange multiplier, where

$$U^\mu(z) := \int \log \frac{1}{|z-s|} \mathrm{d}\mu(s)$$

is the logarithmic potential of μ, see [**32**, Theorem I.3.3]. These notions are well defined for any $T \in \mathcal{T}$. However, it is by now well understood that one should use the contour whose equilibrium measure has support *symmetric* (with the S-*property*) in the external field $\operatorname{Re} V(\cdot; t)$. That is, it must hold that the support J_T

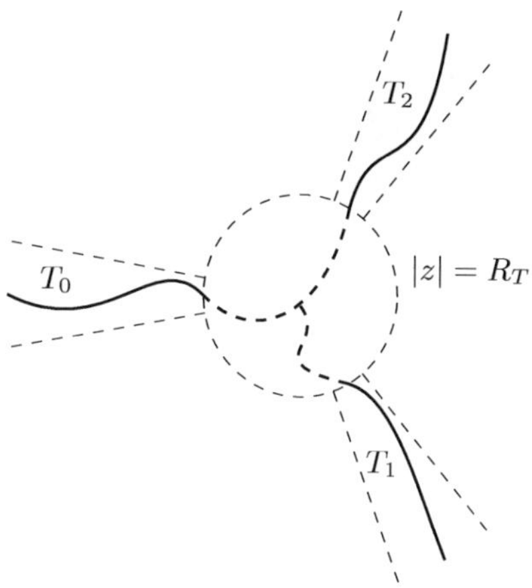

FIGURE 3. Schematic example of an admissible contour $T \in \mathcal{T}$.

consists of a finite number of open analytic arcs and their endpoints, and on each arc it must hold that

$$\frac{\partial}{\partial n_+}\left(2U^{\mu_T} + \operatorname{Re} V\right) = \frac{\partial}{\partial n_-}\left(2U^{\mu_T} + \operatorname{Re} V\right), \tag{4.2}$$

where $\partial/\partial n_+$ and $\partial/\partial n_-$ are the normal derivatives from the $(+)$- and $(-)$-side of T. It is said that a curve $T \in \mathcal{T}$ is an *S-curve* in the field $\operatorname{Re} V(\cdot;t)$, if J_T has the S-property in this field. It is also understood that geometrically J_T is comprised of *critical trajectories* of a certain quadratic differential; this was already known to Stahl (see, e.g. [**34**, Theorem 1]) and remains a useful recasting of the S-property in the context of Painlevé transcendents, see e.g. the recent work [**13**]. Recall that if $Q(z)$ is a meromorphic function, a *trajectory* (resp. *orthogonal trajectory*) of a quadratic differential $-Q(z)\mathrm{d}z^2$ is a maximal regular arc on which

$$-Q(z(s))\big(z'(s)\big)^2 > 0 \quad (\text{resp.} \quad -Q(z(s))\big(z'(s)\big)^2 < 0)$$

for any local uniformizing parameter. A trajectory is called *critical* if it is incident with a *finite critical point* (a zero or a simple pole of $-Q(z)\mathrm{d}z^2$) and it is called *short* if it is non-recurrent and incident only with finite critical points. We designate the expression *critical (orthogonal) graph of* $-Q(z)\mathrm{d}z^2$ for the totality of the critical (orthogonal) trajectories $-Q(z)\mathrm{d}z^2$. The following theorem is a specialization to $V(z;t)$ of the results of Kuijlaars and Silva [**25**, Theorems 2.3 and 2.4].

THEOREM 4.1. *Let $V(z;t)$ be as in (3.1). For each $t \in \mathbb{C}$ there exists a contour $\Gamma_t \in \mathcal{T}$ such that $\mathcal{E}_V(\Gamma_t) = \sup_{T\in\mathcal{T}} \mathcal{E}_V(T)$. Moreover,*

- *the support J_t of the equilibrium measure $\mu_t := \mu_{\Gamma_t}$ has the S-property in the external field $\operatorname{Re} V(\cdot;t)$;*
- *the function*

$$Q(z;t) = \left(\frac{V'(z;t)}{2} - \int \frac{\mathrm{d}\mu_t(s)}{z-s}\right)^2, \quad z \in \mathbb{C}\setminus J_t, \tag{4.3}$$

is a polynomial of degree 4;

- *the support J_t consists of some (possibly not all) short critical trajectories of the quadratic differential*

$$\varpi_t(z) := -Q(z;t)\mathrm{d}z^2;$$

- *the equilibrium measure μ_t is absolutely continuous with respect to Lebesgue measure and*

$$\mathrm{d}\mu_t(z) = -\frac{1}{\pi \mathrm{i}} Q_+^{1/2}(z;t)\mathrm{d}z, \quad z \in J_t, \tag{4.4}$$

where $Q^{1/2}(z;t) = \frac{1}{2}z^2 + \mathcal{O}(z)$ as $z \to \infty$ is the branch holomorphic in $\mathbb{C} \setminus J_t$.

REMARK 4.2. Recall that every $T \in \mathcal{T}$ is a finite connected union of C^1-smooth Jordan arcs. Observe also that $\mathcal{E}_V(T') \geq \mathcal{E}_V(T)$ if $T' \subseteq T$ by the very definition of minimal energy. Hence, every extremal Γ_t must consist of three semi-unbounded non-intersecting open Jordan arcs stretching out to infinity within the sectors $S_{0,\epsilon_{\Gamma_t}}, S_{1,\epsilon_{\Gamma_t}}, S_{2,\epsilon_{\Gamma_t}}$ that have a common endpoint (loops, intervals of overlap, and chains of Jordan arcs with a finite terminal point can be removed without decreasing maximality of minimal energy). Thus, $\Gamma_t = \Gamma_t^{02} \cup \Gamma_t^{12}$, where Γ_t^{ij} is a Jordan arc stretching to infinity within the sectors $S_{i,\epsilon_{\Gamma_t}}$ and $S_{j,\epsilon_{\Gamma_t}}$. Therefore,

$$\Gamma = \alpha_0 L_0 + \alpha_1 L_1 + \alpha_2 L_2 = \alpha_0(L_0 - L_2) + \alpha_1(L_1 - L_2)$$

is homologous to $\alpha_0 \Gamma_t^{02} + \alpha_1 \Gamma_t^{12}$, recall (3.2)–(3.3). Hence, we can use the latter chain in (3.6) to define orthogonal polynomials $P_n(s;t,N,\lambda)$.

REMARK 4.3. S-curve Γ_t cannot be fixed uniquely since one has freedom in choosing Γ_t away from J_t. Indeed, let

$$\mathcal{U}(z;t) := \mathrm{Re}\left(2\int_e^z Q^{1/2}(s;t)\mathrm{d}s\right) = \ell_{\Gamma_t} - \mathrm{Re}(V(z;t)) - 2U^{\mu_t}(z), \tag{4.5}$$

where $e \in J_t$ is arbitrary and the second equality follows from (4.3) (since the constant ℓ_{Γ_t} in (4.1) is the same for all connected components of J_t and the integrand is purely imaginary on J_t, the choice of e is indeed not important). It follows from the variational condition (4.1) that J_t is a part of the set $\{z : \mathcal{U}(z;t) = 0\}$ while $\Gamma_t \setminus J_t$ must belong to $\{z : \mathcal{U}(z;t) \leq 0\}$ and can be varied freely within this region. As $\mathcal{U}(z;t)$ is harmonic in $\mathbb{C} \setminus J_t$, $\Gamma_t \setminus J_t$ belongs to the closure of $\{z : \mathcal{U}(z;t) < 0\}$, which means that such variations are indeed possible (on the other hand, the arcs in J_t border $\{z : \mathcal{U}(z;t) > 0\}$ on both sides).

REMARK 4.4. The set J_t and the measure μ_t are unique for each parameter t. Indeed, let Γ_t^* and Γ_t^{**} be two S-curves corresponding to the same t with equilibrium measures μ_t^* and μ_t^{**}, respectively. Let $\epsilon > 0$ be such that both Γ_t^* and Γ_t^{**} extend to infinity within $S_{k,\epsilon}$, $k \in \{0,1,2\}$. Choose $R > 0$ large enough so that $J_t^*, J_t^{**} \subset \{|z| < R\}$ while

$$S_{k,\epsilon} \cap \{|z| \geq R\} \subset U = \{z : \mathcal{U}^*(z;t) < 0\} \cap \{z : \mathcal{U}^{**}(z;t) < 0\}.$$

According to Remark 4.3, we can modify Γ_t^* and Γ_t^{**} within U so that they intersect $\{|z| = R\}$ at the same three points. Set $\Gamma_{t,R}^* := \Gamma_t^* \cap \{|z| \leq R\}$ and define $\Gamma_{t,R}^{**}$ similarly. Since (4.1) characterizes weighted equilibrium measures, μ_t^* and μ_t^{**} remain being such measures for $\Gamma_{t,R}^*$ and $\Gamma_{t,R}^{**}$, respectively. As (4.2) also remains valid for these measures, $\Gamma_{t,R}^*$ and $\Gamma_{t,R}^{**}$ are compact S-curves. Due to Remark 4.2 and our local modification of Γ_t^* and Γ_t^{**}, $\Gamma_{t,R}^*$ and $\Gamma_{t,R}^{**}$ are homologous to each other and therefore non-Hermitian orthogonal polynomials on them must be the same. Then it follows from [**19**, Lemma 1] that the normalized zero counting measures of these polynomials must simultaneously converge weak* to μ_t^* and μ_t^{**}, which establishes the desired claim.

REMARK 4.5. It is enough to describe μ_t for $t \in O$ for any O such that $\mathbb{C} = O \cup \eta O \cup \bar{\eta} O$, where $\eta = e^{2\pi i/3}$. Indeed, it holds that $\eta T \in \mathcal{T}$ when $T \in \mathcal{T}$. Let μ be a measure on T and μ^η be the push forward measure of μ on ηT, that is, $\mu^\eta(A) = \mu(\bar{\eta} A)$ for any Borel subset A of ηT. Since

$$U^{\mu^\eta}(z) = \int \log \frac{1}{|z - \eta x|} d\mu(x) = U^\mu(\bar{\eta} z),$$

and $V(z; \bar{\eta} t) = V(\bar{\eta} z; t)$, it follows from (4.1) that the weighted equilibrium measure of ηT in the external field $\operatorname{Re} V(\cdot; \bar{\eta} t)$ is equal to μ_T^η, the push forward of the weighted equilibrium measure of T in the external field $\operatorname{Re} V(\cdot; t)$. This means that

$$\sup_{T \in \mathcal{T}} \mathcal{E}_{V(\cdot;t)}(T) = \sup_{T \in \mathcal{T}} \mathcal{E}_{V(\cdot;\bar{\eta}t)}(T)$$

and that if Γ_t is a symmetric contour from Theorem 4.1 then $\eta J_t = J_{\bar{\eta} t}$. In particular, $Q(z; \bar{\eta} t) = \eta Q(\bar{\eta} z; t)$.

4.2. Phase diagram. Both global and local structures of the critical graph of quadratic differentials have been well-studied; see e.g. [**22, 30, 36**]. Since $\deg Q(z;t) = 4$, J_t consists of one arc, two arcs, or three arcs with a common endpoint; the first case occurs when $Q(z;t)$ has two simple and one double zero, and the other two occur when $Q(z;t)$ has four simple zeros. Below, we examine when these distinct possibilities happen.

By developing the right hand side of (4.3) into a Laurent series and recalling that $Q(z;t)$ is a polynomial, we may write

$$Q(z;t) = \frac{1}{4}(z^2 - t)^2 + z + K(t), \tag{4.6}$$

for some $K(t) \in \mathbb{C}$. To find parameters t for which μ_t is supported on a single arc (often referred to as a "cut"), we start with the ansatz

$$Q(z;t) = \frac{1}{4}(z - a(t))(z - b(t))(z - c(t))^2. \tag{4.7}$$

One could identify values of $K(t)$ for which (4.6) has multiple roots by computing the discriminant, but the following approach appears more beneficial. Comparing coefficients of (4.6), (4.7) yields the system (we suppress the t-dependence for brevity)

$$\begin{cases} a + b + 2c & = 0, \\ ab + c^2 + 2(a+b)c & = -2t, \\ 2abc + (a+b)c^2 & = -4. \end{cases} \tag{4.8}$$

Letting $x := -c = (a+b)/2$ and eliminating ab from the second and third equation yields

$$x^3 - tx - 1 = 0. \tag{4.9}$$

Hence, the matter of solving the system (4.8) is now translated into that of understanding the solutions of (4.9). Precisely, we seek to identifying values of $t \in \mathbb{C}$ for

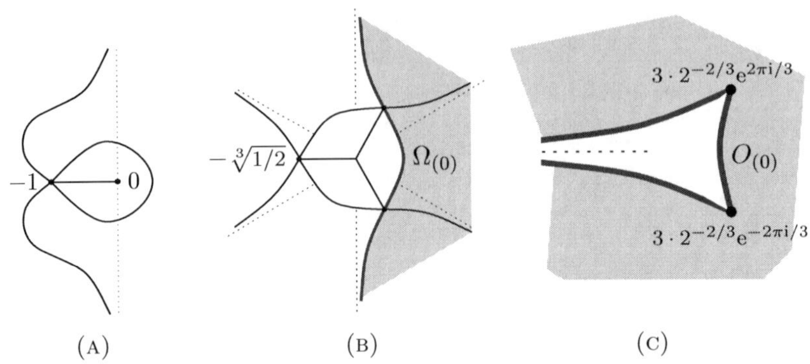

FIGURE 4. Schematic representation of (A) the critical graph $\mathcal{C}$; (B) the set Δ (solid lines) and the domain $\Omega_{(0)}$ (shaded region); (C) domain $O_{(0)}$ (shaded region).

which the critical graph of $\varpi_t(z)$, as defined by (4.7) and (4.8), admits a contour $\Gamma_t \in \mathcal{T}$ such that $\Gamma_t \subset \{z : \mathcal{U}(z;t) \leq 0\}$ and $\mathcal{U}(z;t) = 0$ holds on a subset of Γ_t of finite arclength. It was argued in [**10**, Section 5] that this condition can be imposed by requiring absence of a short trajectory connecting b to c, or

$$\operatorname{Re}\left(\int_b^c Q^{1/2}(s;t)\mathrm{d}s\right) \neq 0.$$

Thus, transitions in the critical graph correspond to choices of x for which

$$\operatorname{Re}\left(\frac{2}{3}\int_{-1}^{2x^3}\left(1+\frac{1}{s}\right)^{3/2}\mathrm{d}s\right) = 0,$$

i.e., one needed to study the critical graph of the auxiliary quadratic differential

$$-\left(1+\frac{1}{s}\right)^3\mathrm{d}s^2,$$

denoted by $\mathcal{C}$. It was shown in [**10**] that $\mathcal{C}$ consists of 5 critical trajectories emanating from -1 at the angles $2\pi k/5$, $k \in \{0,1,2,3,4\}$, one of them being $(-1,0)$, two forming a loop crossing the real line approximately at 0.635, and two approaching infinity along the imaginary axis without changing the half-plane (upper or lower), see Figure 4(A). Define

$$\Delta := \{x : 2x^3 \in \mathcal{C}\}$$

and let $\Omega_{(0)}$ to be the subset of the right-half plane bounded by three smooth subarcs of Δ as on Figure 4(B). Set $\Omega_{(\pm i)} := e^{\mp 2\pi i/3}\Omega_{(0)}$. The function $t(x)$ defined by (4.9) is holomorphic in each $\Omega_{(\tau)}$, $\tau \in \{-i, i, 0\}$, with non-vanishing derivative there. Put $O_{(\tau)} := t(\Omega_{(\tau)})$, see Figure 4(C), then the inverse map $x_\tau : O_{(\tau)} \to \Omega_{(\tau)}$ exists and is holomorphic, $\tau \in \{-i, i, 0\}$, and $O_{(\pm i)} = e^{\pm 2\pi i/3}O_{(0)}$.

Denote by O_0 the union of the unbounded components of $O_{(0)} \cap O_{(i)} \cap O_{(-i)}$ and by $O_{1,-}$ the bounded component of this intersection. Further, let $O_{1,\tau}$ be the complement of the closure of $O_{(\tau)}$ and $O_{0,\tau}$ be the connected component of O_0 that lies directly clockwise from $O_{1,\tau}$, $\tau \in \{0, i, -i\}$, see Figure 5 (numbers 0 and 1 stand for the genus of the Riemann surface of $Q^{1/2}(z;t)$). We further set $O_1 := O_{1,0} \cup O_{1,i} \cup O_{1,-i} \cup O_{1,-}$. Then the following theorem takes place.

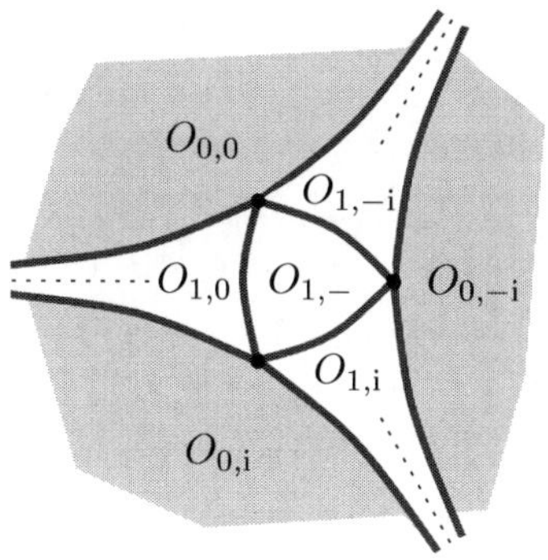

FIGURE 5. Single cut domains $O_{0,0}$, $O_{0,\mathrm{i}}$, and $O_{0,-\mathrm{i}}$; double cut domains $O_{1,0}$, $O_{1,\mathrm{i}}$, $O_{1,-\mathrm{i}}$; and three cuts with a common endpoint domain $O_{1,-}$. Boundaries of these domains correspond to single cut cases.

THEOREM 4.6. *For each* $t \in \mathbb{C}$, *let* $\Gamma_t \in \mathcal{T}$ *be as in Theorem* 4.1. *Recall that* J_t, *the support of the equilibrium measure* μ_t, *is uniquely determined. It holds that*

- *for* $t \in O_0 \cup \{t_{\mathsf{cr}}, \eta t_{\mathsf{cr}}, \bar{\eta} t_{\mathsf{cr}}\}$ *the support* J_t *consists of a single analytic Jordan arc, where* $\eta = \mathrm{e}^{2\pi\mathrm{i}/3}$ *and* $t_{\mathsf{cr}} = 3 \cdot 2^{-2/3}$;
- *for* $t \in O_{1,-}$ *the support* J_t *consists of three analytic Jordan arcs with a common endpoint;*
- *for* $t \in O_1 \setminus O_{1,-}$ *the support* J_t *consists of two analytic Jordan arcs with no endpoints in common;*
- *in the remaining cases the support* J_t *consists of two analytic Jordan arcs with a common endpoint that form a corner at it.*

The claims of Theorem 4.6 follows from Theorems 4.10–4.12 further below, in which we discuss the detailed structure of these contours Γ_t.

REMARK 4.7. According to our heuristics, Airy solutions $q_n(z;\lambda)$ must be pole-free on compact subsets of $-(\sqrt{2}n)^{2/3}O_0$ for all n large, see Theorem 3.2 where we take $n = N$ and Figure 6. We emphasize that while the locations of the poles of $q_n(-(\sqrt{2}n)^{-2/3}z;\lambda)$ depend on $\lambda \notin \{\pm\mathrm{i}, 0\}$, the region which they fill in the large n limit, O_1, does not!

4.3. One-cut cases. Below, we describe the structure of the critical trajectories of the quadratic differential $\varpi_t(z)$ as well as our preferred choice of Γ_t, where we adopt the following convention.

NOTATION. $\Gamma(z_1, z_2)$ (resp. $\Gamma[z_1, z_2]$) stands for the trajectory or orthogonal trajectory (resp. the closure of) of the differential $\varpi_t(z)$ connecting z_1 and z_2, oriented from z_1 to z_2, and $\Gamma(z, \mathrm{e}^{\mathrm{i}\theta}\infty)$ (resp. $\Gamma(\mathrm{e}^{\mathrm{i}\theta}\infty, z)$) stands for the orthogonal trajectory incident with z, approaching infinity at the angle θ, and oriented away from z (resp. oriented towards z).[3]

[3]This notation is unambiguous as the corresponding trajectories are unique for polynomial differentials as follows from Teichmüller's lemma.

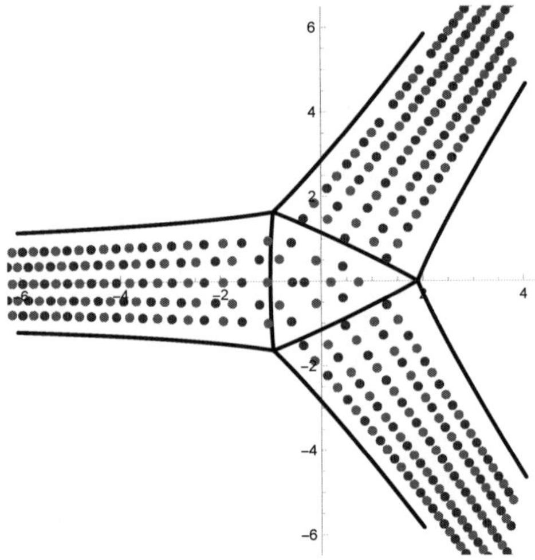

FIGURE 6. Poles (red) and zeros (blue) of $q_3(-(3\sqrt{2})^{2/3}z;\infty)$ and the curves ∂O_1 (black).

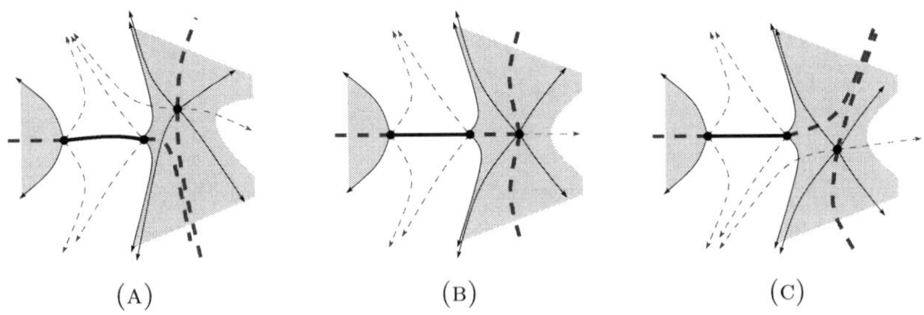

FIGURE 7. Schematic representation of the critical (solid) and critical orthogonal (dashed) graphs of $\varpi_t(z)$ when (A) $t \in O_{0,-\mathrm{i}}$, $\mathrm{Im}(t) > 0$; (B) $t \in O_{0,-\mathrm{i}}$, $\mathrm{Im}(t) = 0$; (C) $t \in O_{0,-\mathrm{i}}$, $\mathrm{Im}(t) < 0$. The bold curves represent the preferred S-curve Γ_t. Shaded region is the set $\{\mathcal{U}(z;t) < 0\}$.

REMARK 4.8. Theorem 4.10 below gives an "appropriate" choice for a chain of integration in (3.6). Each provided chain can be easily modified within the region $\{z : \mathcal{U}(z;t) < 0\}$ so that its point set is connected and corresponds to an S-curve guaranteed by Theorem 4.1. That is, our choice modifies the S-curve only within $\{z : \mathcal{U}(z;t) < 0\}$, which is possible, see Remark 4.3, and allows us to choose the chain of integration that consists solely of trajectories and orthogonal trajectories of $\varpi_t(z)$.

REMARK 4.9. Recall further that to describe S-curves for $t \notin O_1$, it is enough to describe them for $t \in \overline{O}_{0,-\mathrm{i}}$ and $t \in (\partial O_{1,-\mathrm{i}} \cap \partial O_{1,-}) \setminus \{t_{\mathsf{cr}}, \eta t_{\mathsf{cr}}\}$, see Figure 5, due to Remark 4.5.

THEOREM 4.10. *Let μ_t and $Q(z;t)$ be as in Theorem 4.1, $J_t = \mathrm{supp}(\mu_t)$. Assume that $\alpha_0\alpha_1\alpha_2 \neq 0$. When $t \in \overline{O}_{0,-\mathrm{i}}$, the polynomial $Q(z;t)$ is of the form (4.7)*

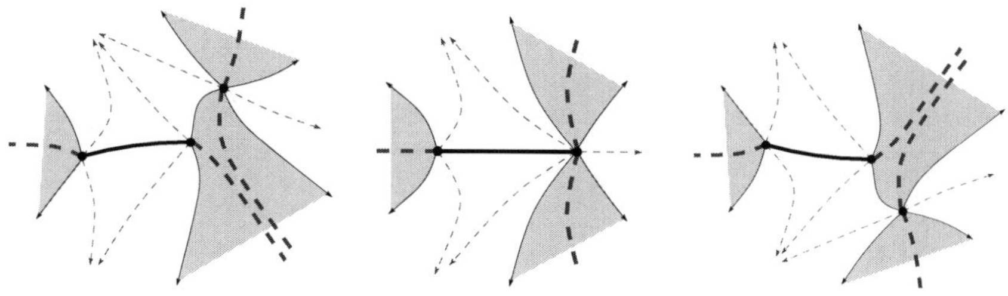

FIGURE 8. Schematic representation of the critical (solid) and critical orthogonal (dashed) graphs of $\varpi_t(z)$ when (a) $t \in \partial O_{0,-\mathrm{i}}$, $\mathrm{Im}(t) > 0$; (b) $t = 3 \cdot 2^{-2/3}$; (c) $t \in \partial O_{0,-\mathrm{i}}$, $\mathrm{Im}(t) < 0$. The bold curves represent the preferred S-curve Γ_t. Shaded region is the set $\{\mathcal{U}(z;t) < 0\}$.

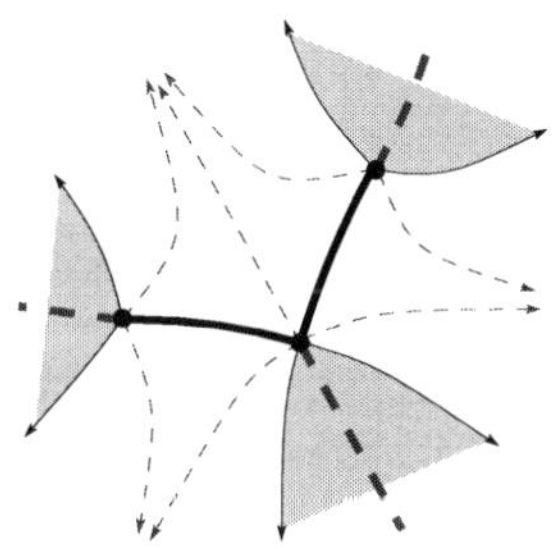

FIGURE 9. Schematic representation of the critical (solid) and critical orthogonal (dashed) graphs of $\varpi_t(z)$ when $t \in (\partial O_{1,-\mathrm{i}} \cap \partial O_{1,-}) \backslash \{t_{\mathsf{cr}}, \eta t_{\mathsf{cr}}\}$. The bold curves represent the preferred S-curve Γ_t. Shaded region is the set $\{\mathcal{U}(z;t) < 0\}$.

with $a(t)$, $b(t)$, *and* $c(t)$ *given by*

$$\begin{cases} a(t) & := x_{-\mathrm{i}}(t) - \mathrm{i}\sqrt{2}/\sqrt{x_{-\mathrm{i}}}(t), \\ b(t) & := x_{-\mathrm{i}}(t) + \mathrm{i}\sqrt{2}/\sqrt{x_{-\mathrm{i}}}(t), \\ c(t) & := -x_{-\mathrm{i}}(t), \end{cases} \tag{4.10}$$

where $\sqrt{x_{-\mathrm{i}}}(t)$ *is the branch holomorphic in* $O_{(-\mathrm{i})}$ *satisfying* $\sqrt{x_{-\mathrm{i}}}(0) = \mathrm{e}^{\pi\mathrm{i}/3}$. *In these cases* $J_t = \Gamma[a,b]$ *and the chain of integration* Γ *in* (3.6) *can be chosen as*

$$\alpha_0 \left(\Gamma\big(\mathrm{e}^{\pi\mathrm{i}}\infty, a\big) \cup J_t \cup \Gamma\big(b, \mathrm{e}^{-\pi\mathrm{i}/3}\infty\big)\right) - \alpha_2 \left(\Gamma\big(\mathrm{e}^{-\pi\mathrm{i}/3}\infty, c\big] \cup \Gamma\big(c, \mathrm{e}^{\pi\mathrm{i}/3}\infty\big)\right)$$

when $\mathrm{Im}(t) > 0$, *see Figure 7(A) and Figure 8(A); as*

$$\alpha_0 \left(\Gamma\big(\mathrm{e}^{\pi\mathrm{i}}\infty, a\big) \cup J_t \cup \Gamma(b, c]\right) + \alpha_1 \Gamma\big(\mathrm{e}^{-\pi\mathrm{i}/3}\infty, c\big) - \alpha_2 \Gamma\big(c, \mathrm{e}^{\pi\mathrm{i}/3}\infty\big)$$

when t *is real, see Figure 7(B) and Figure 8(B);*

$$\alpha_0 \left(\Gamma\big(\mathrm{e}^{\pi\mathrm{i}}\infty, a\big) \cup J_t \cup \Gamma\big(b, \mathrm{e}^{\pi\mathrm{i}/3}\infty\big)\right) + \alpha_1 \left(\Gamma\big(\mathrm{e}^{-\pi\mathrm{i}/3}\infty, c\big] \cup \Gamma\big(c, \mathrm{e}^{\pi\mathrm{i}/3}\infty\big)\right)$$

when $\mathrm{Im}(t) < 0$, *see Figure 7(C) and Figure 8(C).*

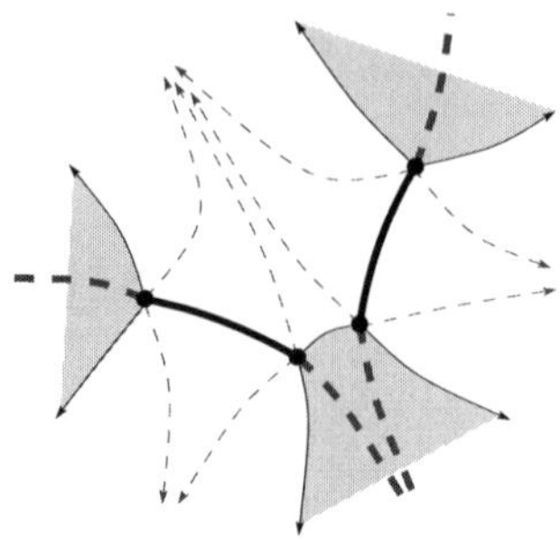

FIGURE 10. Schematic representation of the critical (solid) and critical orthogonal (dashed) graphs of $\varpi_t(z)$ when $t \in O_{1,-\mathrm{i}}$. The bold curves represent the preferred S-curve Γ_t. Shaded region is the set $\{\mathcal{U}(z;t) < 0\}$.

If $t \in (\partial O_{1,-\mathrm{i}} \cap O_{1,-}) \setminus \{t_{\mathsf{cr}}, \eta t_{\mathsf{cr}}\}$, the support $J_t = \Gamma[a,c] \cup \Gamma[c,b]$ is still a single Jordan arc and the chain of integration Γ in (3.6) *can be chosen as*

$$\alpha_0\left(\Gamma\big(\mathrm{e}^{\pi\mathrm{i}}\infty, a\big] \cup \Gamma(a,c)\right) + \alpha_1\Gamma\big(\mathrm{e}^{-\pi\mathrm{i}/3}\infty, c\big] - \alpha_2\left(\Gamma(c,b) \cup \Gamma\big[b, \mathrm{e}^{\pi\mathrm{i}/3}\infty\big)\right),$$

see Figure 9.

These claims are part of [**10**, Theorems 3.2 and 3.4], where a particular case $\lambda = -\mathrm{i}$ ($\alpha_0 = -\alpha_2 = 1/\pi\mathrm{i}$, $\alpha_1 = 0$) was analyzed. However, as one can clearly see, the obtained chains are homologous to Γ from (3.2)–(3.3). In fact, Theorem 4.10 remains valid as stated for $\lambda = -\mathrm{i}$ as well, but Remark 4.5 is not applicable to this case. So, S-curves for $\lambda = -\mathrm{i}$ and $t \in O_{(-\mathrm{i})} \setminus O_{1,-\mathrm{i}}$ are not obtained by rotation of the ones described above, see [**10**, Figure 4(a-c)].

4.4. Two-cut cases. While it is possible to analytically continue $a(t), b(t)$, and $c(t)$ in (4.10) past the boundary of $O_{0,-\mathrm{i}}$, the critical graph of $\varpi_t(z)$ as defined by (4.7) no longer admits a curve $\Gamma_t \in \mathcal{T}$ satisfying $\Gamma_t \subset \{z : \mathcal{U}(z;t) \le 0\}$, see [**10**, Theorems 3.2 and 3.4, and Figure 4]. Hence, we are forced to abandon ansatz (4.7) and have to look for $Q(z;t)$ with four simple zeros. Conditions (4.8) are now complemented by the *Boutroux conditions*:

$$\operatorname{Re}\left(\int_{z_i(t)}^{z_j(t)} Q^{1/2}(s;t)\,\mathrm{d}s\right) = 0, \quad i,j \in \{0,1,2,3\}, \tag{4.11}$$

for any continuous determination of $Q^{1/2}(z;t)$ on the path of integration, where $z_i(t)$ are the zeros of $Q(z;t)$. Relations (4.11) follow from (4.4), which tells us that the integrals are purely imaginary on the branch cuts of $Q^{1/2}(z;t)$, and (4.1), (4.5), which tell us that the integral between the cuts (if the cuts are not connected to each other) also must be purely imaginary so that the real constant ℓ_{Γ_t} is the same on all (both) cuts. Notice that both remarks made before Theorem 4.10 remain relevant to the theorem below.

THEOREM 4.11. *Let μ_t and $Q(z;t)$ be as in Theorem* 4.1, $J_t = \operatorname{supp}(\mu_t)$. *Assume that $\alpha_0\alpha_1\alpha_2 \neq 0$. When $t \in O_{1,-\mathrm{i}}$, the polynomial $Q(z;t)$ is of the form*

$$Q(z;t) = \frac{1}{4}(z - a_1(t))(z - b_1(t))(z - a_2(t))(z - b_2(t))$$

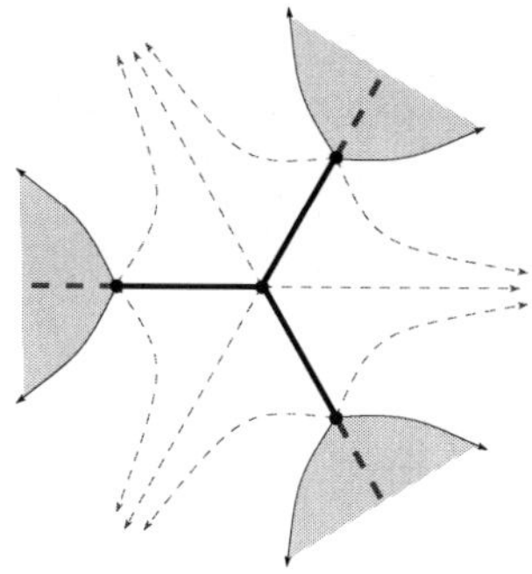

FIGURE 11. Schematic representation of the critical (solid) and critical orthogonal (dashed) graphs of $\varpi_t(z)$ when $t \in O_{1,-}$. The bold curves represent the preferred S-curve Γ_t. Shaded region is the set $\{\mathcal{U}(z;t) < 0\}$.

with $a_1(t)$*,* $b_1(t)$*,* $a_2(t)$*, and* $b_2(t)$ *all distinct. The real and imaginary parts of* $a_i(t), b_i(t)$ *are real analytic functions of* $\mathrm{Re}(t)$ *and* $\mathrm{Im}(t)$*, however, at no point of* $O_{1,-\mathrm{i}}$ *any of the functions* $a_i(t), b_i(t)$ *is complex analytic. The chain of integration* Γ *in* (3.6) *can be chosen as*

$$\alpha_0\Big(\Gamma\big(\mathrm{e}^{\pi\mathrm{i}}\infty, a_1\big) \cup \Gamma[a_1,b_1] \cup \Gamma\big(b_1, \mathrm{e}^{-\pi\mathrm{i}/3}\infty\big)\Big) - \alpha_2\Big(\Gamma\big(\mathrm{e}^{-\pi\mathrm{i}/3}\infty, a_2\big) \cup \Gamma[a_2,b_2] \cup \Gamma\big(b_2, \mathrm{e}^{\pi\mathrm{i}/3}\infty\big)\Big),$$

see Figure 10 *(this fixes our labeling of the zeros of* $Q(z;t)$*). Moreover, it holds that*

$$a_1(t), b_1(t) \to c(t^*), \quad b_1(t), a_2(t) \to c(t^*), \quad \text{and} \quad a_2(t), b_2(t) \to c(t^*)$$

as $t \to t^* \in \partial O_{1,-\mathrm{i}}$ *with* $t^* \in \partial O_{1,-}$*,* $t^* \in \partial O_{1,-\mathrm{i}}$*, and* $t^* \in \partial O_{0,0}$*, respectively.*

This is [**2**, Theorem 3.2], where the case $\lambda = -\mathrm{i}$ was analyzed. As before, one can readily see that the obtained chains are homologous to Γ from (3.2)–(3.3).

4.5. Trefoil cases. It only remains to describe what happens when $t \in O_{1,-}$. In this case the geometries of the critical and critical orthogonal graphs of $\varpi_t(z)$ are distinct from all the previous cases. Our approach essentially consists in observing that $L_0 \cup L_1 \cup L_2$ is the desired S-contour when $t = 0 \in O_{1,-}$ and arguing that the geometric structure of the critical and critical orthogonal graphs cannot change when t is varied within $O_{1,-}$.

THEOREM 4.12. *Let* μ_t *and* $Q(z;t)$ *be as in Theorem* 4.1*,* $J_t = \mathrm{supp}(\mu_t)$*. Assume that* $\alpha_0\alpha_1\alpha_2 \neq 0$*. When* $t \in O_{1,-}$*, the polynomial* $Q(z;t)$ *is of the form*

$$Q(z;t) = \frac{1}{4}(z - z_0(t))(z - z_1(t))(z - z_2(t))(z - z_3(t))$$

with $z_i(t)$*,* $i \in \{0,1,2,3\}$*, all distinct. The real and imaginary parts of* $z_i(t)$ *are real analytic functions of* $\mathrm{Re}(t)$ *and* $\mathrm{Im}(t)$*. The chain of integration* Γ *in* (3.6) *can* *be chosen as*

$$\alpha_0\left(\Gamma\left(\mathrm{e}^{\pi\mathrm{i}}\infty, z_0\right)\cup\Gamma[z_0,z_3]\right)+\alpha_1\left(\Gamma\left(\mathrm{e}^{-\pi\mathrm{i}/3}\infty, z_1\right)\cup\Gamma[z_1,z_3)\right)+\\ \alpha_2\left(\Gamma\left(\mathrm{e}^{\pi\mathrm{i}/3}\infty, z_2\right)\cup\Gamma[z_2,z_3)\right),$$

see Figure 11 *(this fixes our labeling of the zeros of $Q(z;t)$). Moreover, it holds that*

$$z_0(t)\to a(t^*),\quad z_2(t)\to b(t^*),\quad \textit{and}\quad z_1(t),z_3(t)\to c(t^*)$$

as $t\to t^*\in\partial O_{1,-\mathrm{i}}\cap\partial O_{1,-}$.

The rest of this section is dedicated to the proof of Theorem 4.12. Let us emphasize again that in the studied cases all the zeros of $Q(z;t)$ must be distinct as coincidence of any two of them leads to the ansatz (4.7), which necessarily implies that $a(t),b(t),c(t)$ are as in (4.10), where $x_{-\mathrm{i}}(t)$ can be replaced by $x_0(t)$ or $x_{\mathrm{i}}(t)$, but no combination of the critical and critical orthogonal trajectories of the corresponding differentials form a desirable curve Γ_t (all the possible configurations were already examined before, see [**10**, Theorems 3.2 and 3.4, and Figure 4]).

4.5.1. *Step 1.* Since the Jacobian of the elementary symmetric functions of the zeros of $Q(z;t)$ is its discriminant and all the zeros are distinct when $t\in O_{1,-}$, it is enough to examine dependence on t of the coefficients of $Q(z;t)$. In view of (4.6), we only need to study the free coefficient $K(t)$.

Fix $t^*\in O_{1,-}$ and let $K^*=K(t^*)$ be the corresponding constant guaranteed by Theorem 4.1 and Remark 4.4. Consider

$$Q(z;t,K):=\frac{1}{4}(z^2-t)^2+z+K$$

for t,K in some neighborhoods of t^* and K^*, respectively. We take these neighborhoods small enough so that there are disjoint neighborhoods of $z_i(t^*)$, $i\in\{0,1,2,3\}$, the zeros of $Q(z;t^*)$, each containing one zero of $Q(z;t,K)$. We label $z_i(t,K)$ the zero of $Q(z;t,K)$ that approaches $z_i(t^*)$ as $(t,K)\to(t^*,K^*)$ (these zeros are smooth functions of t and K as mentioned before). Define

$$\mathfrak{S}_{t,K}:=\left\{(z,w)\;:\;w^2=Q(z;t,K)\right\}\subset\mathbb{C}^2.$$

Since the zeros of $Q(z;t,K)$ are simple, this is an elliptic, i.e., genus 1, Riemann surface. Select a homology basis on $\mathfrak{S}_{t,K}$, say $(\boldsymbol{\alpha}_{t,K},\boldsymbol{\beta}_{t,K})$. We can do it in such a way that the natural projection $(z,w)\mapsto z$ of the cycle $\boldsymbol{\alpha}_{t,K}$ is a Jordan arc connecting $z_0(t,K)$ to $z_1(t,K)$ while the natural projection of the cycle $\boldsymbol{\beta}_{t,K}$ is a Jordan arc connecting $z_0(t,K)$ to $z_2(t,K)$. Moreover, we can arrange so that these Jordan arcs coincide for all considered (t,K) outside of small neighborhoods of $z_0(t^*)$, $z_1(t^*)$, and $z_2(t^*)$. Set

$$B_\gamma(t,K):=\frac{1}{2}\mathrm{Re}\left(\oint_{\gamma_{t,K}}w\mathrm{d}z\right),\quad \gamma\in\{\boldsymbol{\alpha},\boldsymbol{\beta}\},$$

which we consider to be functions of $\mathrm{Re}(t)$, $\mathrm{Im}(t)$, $u:=\mathrm{Re}(K)$, and $v:=\mathrm{Im}(K)$. We further make sure that the homology bases are selected so that $B_\gamma(t,K)$ are continuous functions of their parameters around (t^*,K^*). Boutroux conditions (4.11) yield that

$$B_{\boldsymbol{\alpha}}(t^*,K^*)=0\quad\text{and}\quad B_{\boldsymbol{\beta}}(t^*,K^*)=0. \tag{4.12}$$

The Cauchy-Riemann equations allow us to compute the Jacobian

$$\det\frac{\partial(B_{\boldsymbol\alpha},B_{\boldsymbol\beta})}{\partial(u,v)}=\frac{1}{4}\left(-\mathrm{Re}\left(\oint_{\boldsymbol\alpha}\frac{\mathrm{d}z}{w}\right)\mathrm{Im}\left(\oint_{\boldsymbol\beta}\frac{\mathrm{d}z}{w}\right)+\mathrm{Re}\left(\oint_{\boldsymbol\beta}\frac{\mathrm{d}z}{w}\right)\mathrm{Im}\left(\oint_{\boldsymbol\alpha}\frac{\mathrm{d}z}{w}\right)\right)$$
$$=\frac{1}{4}\mathrm{Im}\left(\oint_{\boldsymbol\alpha}\frac{\mathrm{d}z}{w}\overline{\oint_{\boldsymbol\beta}\frac{\mathrm{d}z}{w}}\right).$$

The differential $\mathrm{d}z/w$ is the unique (up to multiplication by a constant) holomorphic differential on $\mathfrak{S}_{t,K}$ and by Riemann's bilinear relation, see [**11**, Lemma 35.2], we have

$$\det\frac{\partial(B_{\boldsymbol\alpha},B_{\boldsymbol\beta})}{\partial(u,v)}<0.$$

Thus, it follows from the Implicit Function Theorem that in some neighborhood of t^* there exists a function $K^*(t)$ which is real analytic in $\mathrm{Re}(t)$ and $\mathrm{Im}(t)$ and such that the equations (4.12) hold with $(t,K^*(t))$ in this neighborhood of t^*. We still need to argue that $K(t)=K^*(t)$, To this end, we shall examine the critical graph of $\varpi_t^*(z):=Q(z;t,K^*(t))\mathrm{d}z^2$. Before we do this, let us recall relevant facts of the theory of quadratic differentials.

4.5.2. *Quadratic differentials.* Given a quadratic differential $\varpi_t^*(z)$ as above, it holds that

- two distinct (orthogonal) trajectories meet only at critical points, see [**36**, Theorem 5.5], which, in this case, are the zeros of $Q(z;t,K^*(t))$ and the point at infinity;
- no finite union of (orthogonal) trajectories can form a closed Jordan curve because $Q(z;t,K^*(t))$ is a polynomial, see [**30**, Lemma 8.3];
- trajectories of $\varpi_t^*(z)$ cannot be recurrent (dense in two-dimensional regions), see [**22**, Theorem 3.6];
- since infinity is a pole of $\varpi_t^*(z)$ order 8, the critical trajectories can approach infinity only in six distinguished directions, namely, asymptotically to the lines $\{\arg(z)=-5\pi/6+k\pi/3\}$, $k\in\{0,1,2,3,4,5\}$;
- there exists a neighborhood of infinity such that any trajectory entering it necessarily tends to infinity, see [**36**, Theorem 7.4].

Denote by $\mathcal{G}_t^*$ the critical graph of $\varpi_t^*(z)$, that is, the totality of all the critical trajectories of $\varpi_t^*(z)$. In the case of polynomial quadratic differentials, more can be said about the topological nature of $\mathbb{C}\setminus\mathcal{G}_t^*$. The complement of $\mathcal{G}_t^*$ can be written as a disjoint union of either half-plane or strip domains, see [**22**, Theorem 3.5]. Recall that a *half-plane (or end) domain* is swept by trajectories unbounded in both directions that approach infinity along consecutive critical directions. Its boundary is connected and consists of a union of two unbounded critical trajectories and a finite number (possibly zero) of short trajectories of $\varpi_t^*(z)$. The map $z\mapsto\int^z\sqrt{Q(s;t,K^*(t))}\mathrm{d}s$ maps end domains conformally onto half planes $\{z\in\mathbb{C}\mid\mathrm{Re}(z)>c\}$ for some $c\in\mathbb{R}$ that depends on the domain, and extends continuously to the boundary. Similarly, a *strip domain* is swept by trajectories unbounded in both directions, but its boundary consists of two disjoint trajectories of $\varpi_t^*(z)$, each of which is comprised of two unbounded critical trajectories and a finite number (possibly zero) of short trajectories. The map $z\mapsto\int^z\sqrt{Q(s;t,K^*(t))}\mathrm{d}s$ maps strip domains conformally onto vertical strips $\{w\in\mathbb{C}\mid c_1<\mathrm{Re}(w)<c_2\}$

for some $c_1, c_2 \in \mathbb{R}$ depending on the domain, and extends continuously to their boundaries. The number $c_2 - c_1$ is known as the *width of a strip domain* and can be calculated in terms of $\varpi_t^*(z)$ as

$$\left|\mathrm{Re}\int_p^q \sqrt{Q(s;t,K^*(t))}\mathrm{d}s\right| \tag{4.13}$$

for any two points p, q belonging to different components of the boundary of the domain.

4.5.3. *Step 2.* Equations (4.12) imply existence of three short critical trajectories of $\varpi_t^*(z)$. Indeed, let J be a Jordan arc that connects a pair of zeros of $Q(z;t,K^*(t))$ and γ be a cycle on $\mathfrak{S}_{t,K^*(t)}$ whose natural projection is J. Then it holds that

$$\mathrm{Re}\left(\int_J Q^{1/2}(s;t,K^*(t))\mathrm{d}s\right) = \pm\frac{1}{2}\mathrm{Re}\left(\oint_\gamma w\mathrm{d}s\right),$$

where $Q^{1/2}(s;t,K^*(t))$ is a continuous branch on J and the sign $+$ or $-$ depends on the choice of this branch as well as the chosen orientations for J and γ. It is well known, see [**11**, Theorem 34.3], that

$$\oint_\gamma w\mathrm{d}s = n_1\oint_{\boldsymbol{\alpha}_{t,K^*(t)}} w\mathrm{d}s + n_2\oint_{\boldsymbol{\beta}_{t,K^*(t)}} w\mathrm{d}s + 2\pi\mathrm{i}\sum m_k\,\mathrm{res}_{z_k}(w\mathrm{d}z),$$

where n_1, n_2 and m_k are integers and the sum is taken over all the residues of the differential $w\mathrm{d}z$. Since the only poles of $w\mathrm{d}z$ are located at points on top of infinity with residues ±1, the last two relations together with (4.12) imply that

$$\mathrm{Re}\left(\int_J Q^{1/2}(s;t,K^*(t))\mathrm{d}s\right) = 0 \tag{4.14}$$

for any Jordan arc J connecting any pair of zeros of $Q(s;t,K^*(t))$ (that is, this polynomial has property (4.11)). Recall now that each zero $z_i^*(t) = z_i(t,K^*(t))$ of $Q(z;t,K^*(t))$ is incident with three critical trajectories of $\varpi_t^*(z)$. If all three critical trajectories out of $z_i^*(t)$ approach infinity, then $z_i^*(t)$ must belong to a boundary of at least one strip domain. Let $z_j^*(t)$ be a different zero belonging to the other component of the boundary of this strip domain. Then it follows from (4.13) and (4.14) that the width of this strip domain is 0, which is impossible. Thus, each zero of $Q(z;t,K^*(t))$ must be coincident with at least one short trajectory. Therefore, either there is a zero connected by short trajectories to the remaining three zeros or there are at least two short trajectories connecting two pairs of zeros. In the latter case, label these zeros by $z_0^*(t), z_1^*(t)$ and $z_2^*(t), z_3^*(t)$. If the other two trajectories out of both $z_0^*(t)$ and $z_1^*(t)$ approach infinity, one of these zeros again must belong to the boundary of a strip domain with either $z_2^*(t)$ or $z_3^*(t)$ belonging to the other component of the boundary. As before, (4.13) and (4.14) yield that the width of this strip domain is 0, which, again, is impossible. Thus, in this case there also exists a third short critical trajectory.

Since short critical trajectories cannot form a loop, they form a connected set, say $\mathcal{G}_t^{\mathrm{sh}}$, that contains all four zeros of $Q(z;t,K^*(t))$ (this set is either a trefoil or a single Jordan arc). Define

$$\mathcal{U}^*(z;t) := \mathrm{Re}\left(2\int_e^z Q^{1/2}(s;t,K^*(t))\mathrm{d}s\right), \tag{4.15}$$

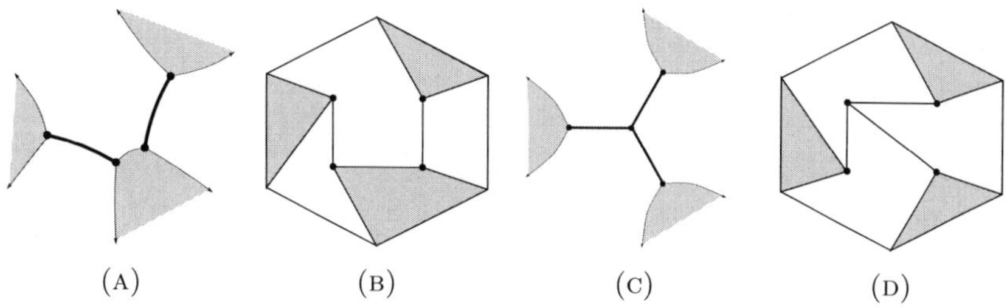

FIGURE 12. Schematic representation of $\mathcal{G}_t^*$ (A) or (C) and its associated clock diagram (B) or (D). Shaded region is the set $\{\mathcal{U}(z;t) < 0\}$.

where $e \in \mathcal{G}_t^{\rm sh}$ and $Q^{1/2}(s;t,K^*(t))$ is the branch holomorphic in $\mathbb{C} \setminus \mathcal{G}_t^{\rm sh}$ that behaves like $\frac{1}{2}z^2 + O(z)$ as $z \to \infty$. $\mathcal{U}^*(z;t)$ is a well defined harmonic function in $\mathbb{C} \setminus \mathcal{G}_t^{\rm sh}$ by (4.14) (which is independent of the choice of e). Moreover, this function is either harmonic across a given short critical trajectory in $\mathcal{G}_t^{\rm sh}$ (this happens to the "middle section" when three short critical trajectories form a single Jordan arc) or can be harmonically continued across it by $-\mathcal{U}^*(z;t)$. Since the zeros of $Q^{1/2}(s;t,K^*(t))$ converge to the zeros of $Q^{1/2}(s;t^*)$ as $t \to t^*$, we have that $\mathcal{U}^*(z;t)$ converges to $\mathcal{U}(z;t^*)$ point-wise. Using the above harmonic continuation to define a single harmonic function on a double ramified cover of this disk via concatenation of $\mathcal{U}^*(z;t)$ and $-\mathcal{U}^*(z;t)$ and applying the maximum modulus principle and [**31**, Theorem 1.3.10], we deduce that the functions $\mathcal{U}^*(z;t)$ converge to $\mathcal{U}(z;t^*)$ as $t \to t^*$ uniformly on compact subsets of $\mathbb{C}$. Furthermore, we have that

$$\mathcal{U}^*(z;t) = -\mathrm{Re}(V(z;t)) + 2\log|z/e| + \mathrm{Re}\left(2\int_e^z \left(Q^{1/2}(s;t,K^*(t)) - \frac{s^2-t}{2} - \frac{1}{s}\right)\mathrm{d}s\right), \tag{4.16}$$

where the integrand behaves as $O(z^{-2})$ at infinity and therefore the last summand above is harmonic at infinity. Thus, using (4.16) we have that the functions $\mathcal{U}^*(z;t)$ converge to $\mathcal{U}(z;t^*)$ as $t \to t^*$ uniformly on the whole sphere $\overline{\mathbb{C}}$. In particular, since the critical graph of $\varpi_t^*(z)$ is the zero level set of $\mathcal{U}^*(z;t)$, this function can be either positive and negative on different sides of a trajectory (i.e., it is harmonic across) or be positive on both sides (i.e., it is subharmonic across) since this is true for $\mathcal{U}(z;t^*)$ by (4.1) and (4.5). The latter situation happens on the part of $\mathcal{G}_t^{\rm sh}$ that serves as the branch cut for $Q^{1/2}(s;t,K^*(t))$, which we denote by J_t^*.

4.5.4. *Step 3.* Now, we would like to identify possible structures of the critical graph of $\varpi_t^*(z)$. To this end, we shall use what is known as *clock diagrams*. Introduced in [**5**, Definition 5.4], (topological) clock diagrams are a schematic way of representing the critical graph of $\varpi_t^*(z)$ that allows for a more combinatorial treatment similar in spirit to [**3**] though more concrete since we deal with a fixed polynomial potential. In our setting, a clock diagram consists of the outer hexagon whose vertices correspond to the distinguished directions at infinity; inner vertices that correspond to the finite critical points, i.e., zeros of $Q(z;t,K^*(t))$, that are connected to each other and the vertices of the hexagon by edges corresponding to the critical trajectories of $\varpi_t^*(z)$; and finally the shaded regions that correspond to

FIGURE 13. The four possible configurations of $\mathcal{G}_t^{\rm sh}$ up to rotation.

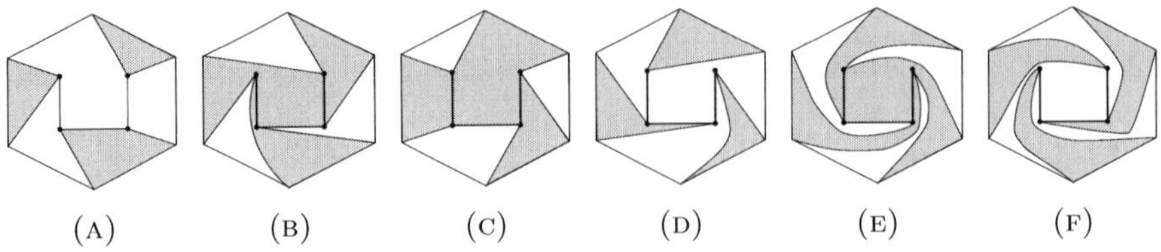

FIGURE 14. The six possible ways to connect $\mathcal{G}_t^{\rm sh}$ from Figure 13(B) to the hexagon.

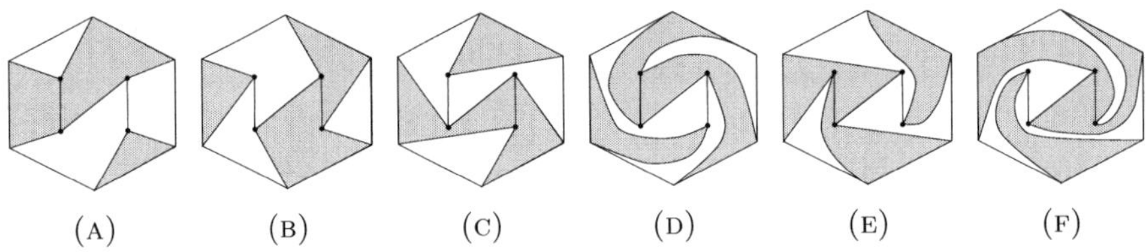

FIGURE 15. The six possible ways to connect $\mathcal{G}_t^{\rm sh}$ from Figure 13(D) to the hexagon.

the strip or end domains where $\mathcal{U}^*(z,t)<0$. It follows from (4.5) that every other edge of the hexagon must border a shaded region starting with the vertical edge on the left, which corresponds to the sector between the distinguished direction $5\pi/6$ and $7\pi/6$, see Figure 12.

Since the critical graph of $\varpi_t^*(z)$ must contain exactly three short trajectories not allowing for loops, we immediately see that there are exactly 4 possible configurations of short trajectories (up to rotation), shown on Figure 13. Each of these configurations can be attached to the boundary of the hexagon in 6 different ways since once the direction in which one trajectory goes to infinity is fixed, the others are uniquely determined by Teichmüller's Lemma, see [**36**, Theorem 14.1], see Figures 14 and 15. However, we now observe the importance of the shading. As explained after (4.16), J_t^* must border unshaded regions on both sides whereas any other trajectory must be the boundary of both a shaded and an unshaded region. Hence, of the 6 possible ways to connect the finite critical points to the hexagon, only 3 allow for a valid coloring when $\mathcal{G}_t^{\rm sh}$ is as on Figures 13(A,B) (and their rotations), see Figure 14, while there are no valid colorings when $\mathcal{G}_t^{\rm sh}$ is as on Figures 13(C,D), see Figure 15. In the sequel, we will need to talk about clock diagrams and critical graphs which are "structurally the same." To do so in our setting, we attach to a critical graph $\mathcal{G}$ with critical points z_1,z_2,z_3,z_4 the adjacency data

$$\big\{\{z_1,A_1\},\{z_2,A_2\},\{z_3,A_3\},\{z_4,A_4\}\big\},$$

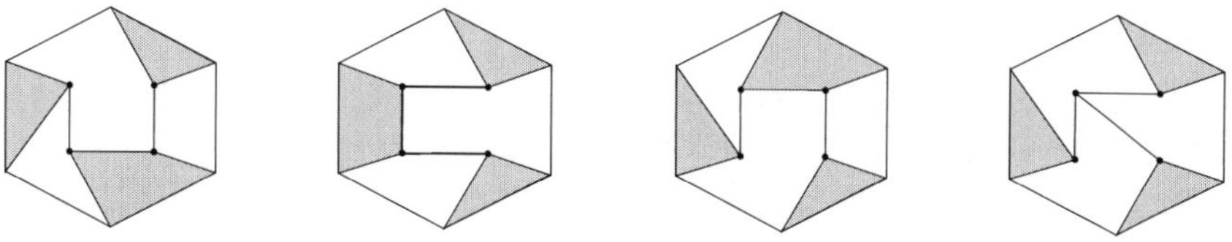
(A) (B) (C) (D)

FIGURE 16. The only four admissible clock diagrams.

where $A_i \subset \mathbb{N}$ is a finite set and $k \in A_i$ if and only if z_i is on the boundary of the end domain containing the line $\{re^{\pi k i/3} : r > R\}$ for R large enough.

DEFINITION 4.13. We say that two critical graphs are *structurally equivalent* if there exists a labeling of the critical points such that both graphs have the same adjacency data. Two clock diagrams are structurally equivalent if their corresponding critical graphs are structurally equivalent.

With this definition in mind, observe that the clock diagrams on Figures 14(A,D,F) structurally describe the same critical graphs as diagrams on Figures 16(A,B,C), respectively. Hence, if we were to draw the admissible diagrams corresponding to a rotation of $\mathcal{G}_t^{\text{sh}}$ as on Figure 13(B), they would again structurally correspond to the critical graphs depicted on Figures 16(A,B,C). As all admissible diagrams corresponding to $\mathcal{G}_t^{\text{sh}}$ as on Figure 13(A) are structurally the same, see Figure 16(D), Figure 16 depicts all possible admissible clock diagrams for the critical graph of $\varpi_t^*(z)$.

4.5.5. *Step 4.* Since $\mathcal{U}^*(z;t)$ is subharmonic in $\mathbb{C}$ and harmonic in $\mathbb{C} \setminus J_t^*$, its Laplacian is a positive measure, say $2\mu_t^*$, supported on J_t^*. It follows from (4.16), the Riesz Decomposition Theorem [**31**, Theorem 3.7.9] and Liouville's Theorem for subharmonic functions [**31**, Corollary 2.3.4] to identify the harmonic part that

$$\mathcal{U}^*(z;t) = -\text{Re}(V(z;t)) + \ell_t^* - 2U^{\mu_t^*}(z)$$

for some real constant ℓ_t^*. The measure μ_t^* has mass 1 as this is the only case when $2U^{\mu_t^*}(z)$ behaves like $-2\log|z|$ as $z \to \infty$, see (4.16). Moreover, as we have mentioned before, the very definition of $\mathcal{U}^*(z;t)$ in (4.15) implies that $-\mathcal{U}^*(z;t)$ is the harmonic continuation of $\mathcal{U}^*(z;t)$ across any subarc of J_t^*. This, however, is equivalent to saying that $\mathcal{U}^*(z;t)$ satisfies (4.2). Hence, if we add to the corresponding $\mathcal{G}_t^{\text{sh}}$, of which J_t^* is a part, three unbounded Jordan arcs that lie within $\{\mathcal{U}^*(z;t) < 0\}$ (shaded regions on Figure 16) to create an element of $\mathcal{T}$, we will obtain a symmetric contour. Recall now that the uniqueness in Remark 4.4 was not derived based on the maximization of the minimal energy, which is the defining property of J_t in Theorem 4.1, but based on being an S-curve. Hence, J_t^* is J_t and μ_t^* and μ_t. Moreover, this finishes the proof of real analytic dependence of the zeros of $Q(z;t)$ on $\text{Re}(t)$ and $\text{Im}(t)$.

Since the function $\mathcal{U}(z;t)$ converge uniformly to $\mathcal{U}(z;t^*)$ as $t \to t^*$ and the critical graph of $\varpi_t(z)$ can have only one of the four structural forms depicted on Figure 16, the critical graphs of $\varpi_t(z)$ are structurally the same as the critical graph of $\varpi_{t^*}(z)$ for all t in some small neighborhood of t^*. As any two points in $O_{1,-}$ can be joined by a path that can be covered by finitely many such neighborhoods, the structure of the critical graph of $\varpi_t(z)$ must be the same for every $t \in O_{1,-}$. It readily follows from Remark 4.5 that $K(0) = 0$ and therefore

$$\varpi_0(z) = \frac{z(z^3+4)}{4}\mathrm{d}z^2,$$

whose critical graph can be readily determined and is depicted on Figure 11. Since this structure of the critical graph uniquely determines the structure of the critical orthogonal graph, this finishes the proof of the theorem except for the last claim.

4.5.6. *Step 5.* It follows from [**6**, Theorem 5.11] that if t remains in a bounded set, so do the zeros of $Q(z;t)$. Fix $t^* \in \partial O_{1,-}$ and let $\{t_m\}$ be a sequence such that $t_m \to t^*$ as $m \to \infty$. Restricting to a subsequence if necessary, we see that there exist z_i^* such that $z_i(t_m) \to z_i^*$ as $m \to \infty$, $i \in \{0,1,2,3\}$, where $z_i(t)$ are the zeros of $Q(z;t)$. Clearly, polynomials $Q(z;t_m)$ converge uniformly on compact subsets of $\mathbb{C}$ to $Q^*(z)$, the polynomial with zeros z_i^* and leading coefficient $1/4$. With the help of uniform convergence we can conclude that (4.14) must be satisfied with $Q(z;t,K^*(t))$ replaced by $Q^*(z)$. At this point, we can repeat the steps that follow (4.14) (we will need to add clock diagrams corresponding to possible confluence of the zeros of $Q^*(z)$) to conclude that an S-curve can be constructed out of the critical and critical orthogonal trajectories of $Q^*(z)\mathrm{d}z^2$, and then appeal to Remark 4.4 to conclude that $Q^*(z) = Q(z;t^*)$.

References

[1] Ahmad Barhoumi, Pavel Bleher, Alfredo Deaño, and Maxim Yattselev. On Airy solutions of PII and complex cubic ensemble of random matrices, I. To appear in: *Orthogonal Polynomials, Special Functions and Applications - Proceedings of the 16th International Symposium, Montreal, Canada, In honor to Richard Askey.* 35, 36, 43

[2] Ahmad Barhoumi, Pavel Bleher, Alfredo Deaño, and Maxim Yattselev, *Investigation of the two-cut phase region in the complex cubic ensemble of random matrices*, J. Math. Phys. **63** (2022), no. 6, Paper No. 063303, 40, DOI 10.1063/5.0086911. MR4436195 ↑42, 43, 53

[3] Marco Bertola, *Boutroux curves with external field: equilibrium measures without a variational problem*, Anal. Math. Phys. **1** (2011), no. 2-3, 167–211, DOI 10.1007/s13324-011-0012-3. MR2853811 ↑57

[4] Marco Bertola and Thomas Bothner, *Zeros of large degree Vorob'ev-Yablonski polynomials via a Hankel determinant identity*, Int. Math. Res. Not. IMRN **19** (2015), 9330–9399, DOI 10.1093/imrn/rnu239. MR3431594 ↑36

[5] Marco Bertola and Man Yue Mo, *Commuting difference operators, spinor bundles and the asymptotics of orthogonal polynomials with respect to varying complex weights*, Adv. Math. **220** (2009), no. 1, 154–218, DOI 10.1016/j.aim.2008.09.001. MR2462838 ↑57

[6] Marco Bertola and Alexander Tovbis, *On asymptotic regimes of orthogonal polynomials with complex varying quartic exponential weight*, SIGMA Symmetry Integrability Geom. Methods Appl. **12** (2016), Paper No. 118, 50, DOI 10.3842/SIGMA.2016.118. MR3589917 ↑60

[7] Daniel Bessis, Claude Itzykson, and Jean-Bernard Zuber, *Quantum field theory techniques in graphical enumeration*, Adv. in Appl. Math. **1** (1980), no. 2, 109–157, DOI 10.1016/0196-8858(80)90008-1. MR603127 ↑40

[8] Pavel M. Bleher and Alfredo Deaño, *Topological expansion in the cubic random matrix model*, Int. Math. Res. Not. IMRN **12** (2013), 2699–2755, DOI 10.1093/imrn/rns126. MR3071662 ↑40, 41

[9] Pavel Bleher and Alfredo Deaño, *Painlevé I double scaling limit in the cubic random matrix model*, Random Matrices Theory Appl. **5** (2016), no. 2, 1650004, 58, DOI 10.1142/S2010326316500040. MR3493550 ↑42

[10] Pavel Bleher, Alfredo Deaño, and Maxim Yattselev, *Topological expansion in the complex cubic log-gas model: one-cut case*, J. Stat. Phys. **166** (2017), no. 3-4, 784–827, DOI 10.1007/s10955-016-1621-x. MR3607591 ↑42, 43, 48, 52, 54

[11] Gilbert Ames Bliss, *Algebraic functions*, Dover Publications, Inc., New York, 1966. MR203007 ↑55, 56

[12] Édouard Brézin, Claude Itzykson, Giorgio Parisi, and Jean-Bernard Zuber, *Planar diagrams*, Comm. Math. Phys. **59** (1978), no. 1, 35–51. MR471676 ↑40

[13] Robert J. Buckingham and Peter D. Miller, *Large-degree asymptotics of rational Painlevé-IV solutions by the isomonodromy method*, Constr. Approx. **56** (2022), no. 2, 233–443, DOI 10.1007/s00365-022-09586-1. MR4507087 ↑45

[14] Peter A. Clarkson, *On Airy solutions of the second Painlevé equation*, Stud. Appl. Math. **137** (2016), no. 1, 93–109, DOI 10.1111/sapm.12123. MR3529955 ↑36, 39

[15] Nicolas M. Ercolani and Kenneth D. T.-R. McLaughlin, *Asymptotics of the partition function for random matrices via Riemann-Hilbert techniques and applications to graphical enumeration*, Int. Math. Res. Not. **14** (2003), 755–820, DOI 10.1155/S1073792803211089. MR1953782 ↑40

[16] Frank W.J Olver et al. editors, NIST digital library of mathematical functions, `http://dlmf.nist.gov`. 37, 41, 44

[17] Hermann Flaschka and Alan C. Newell, *Monodromy- and spectrum-preserving deformations. I*, Comm. Math. Phys. **76** (1980), no. 1, 65–116. MR588248 ↑36

[18] Bertrand Gambier, *Sur les équations différentielles du second ordre et du premier degré dont l'intégrale générale est a points critiques fixes* (French), Acta Math. **33** (1910), no. 1, 1–55, DOI 10.1007/BF02393211. MR1555055 ↑36

[19] Andreĭ Aleksandrovich Gonchar and Evguenii A. Rakhmanov, *Equilibrium distributions and the rate of rational approximation of analytic functions* (Russian), Mat. Sb. (N.S.) **134(176)** (1987), no. 3, 306–352, 447, DOI 10.1070/SM1989v062n02ABEH003242; English transl., Math. USSR-Sb. **62** (1989), no. 2, 305–348. MR922628 ↑44, 47

[20] Daan Huybrechs, Arno Kuijlaars, and Nele Lejon, *A numerical method for oscillatory integrals with coalescing saddle points*, SIAM J. Numer. Anal. **57** (2019), no. 6, 2707–2729, DOI 10.1137/18M1221138. MR4031470 ↑42

[21] Daan Huybrechs, Arno B. J. Kuijlaars, and Nele Lejon, *Zero distribution of complex orthogonal polynomials with respect to exponential weights*, J. Approx. Theory **184** (2014), 28–54, DOI 10.1016/j.jat.2014.05.002. MR3218792 ↑42

[22] James A. Jenkins, *Univalent functions and conformal mapping*, Reihe: Moderne Funktionentheorie, Springer-Verlag, Berlin-Göttingen-Heidelberg, 1958. Ergebnisse der Mathematik und ihrer Grenzgebiete, (N.F.), Heft 18. MR96806 ↑47, 55

[23] Michio Jimbo and Tetsuji Miwa, *Monodromy preserving deformation of linear ordinary differential equations with rational coefficients. II*, Phys. D **2** (1981), no. 3, 407–448, DOI 10.1016/0167-2789(81)90021-X. MR625446 ↑37, 38, 39

[24] Michio Jimbo, Tetsuji Miwa, and Kimio Ueno, *Monodromy preserving deformation of linear ordinary differential equations with rational coefficients. I. General theory and τ-function*, Phys. D **2** (1981), no. 2, 306–352, DOI 10.1016/0167-2789(81)90013-0. MR630674 ↑39

[25] Arno B. J. Kuijlaars and Guilherme L. F. Silva, *S-curves in polynomial external fields*, J. Approx. Theory **191** (2015), 1–37, DOI 10.1016/j.jat.2014.04.002. MR3306308 ↑44, 45

[26] Madan Lal Mehta, *Random matrices*, 3rd ed., Pure and Applied Mathematics (Amsterdam), vol. 142, Elsevier/Academic Press, Amsterdam, 2004. MR2129906 ↑42

[27] Kazuo Okamoto, *Polynomial Hamiltonians associated with Painlevé equations. I*, Proc. Japan Acad. Ser. A Math. Sci. **56** (1980), no. 6, 264–268. MR581468 ↑37, 38

[28] Kazuo Okamoto, *Polynomial Hamiltonians associated with Painlevé equations. II. Differential equations satisfied by polynomial Hamiltonians*, Proc. Japan Acad. Ser. A Math. Sci. **56** (1980), no. 8, 367–371. MR596006 ↑37, 38

[29] Kazuo Okamoto, *Studies on the Painlevé equations. III. Second and fourth Painlevé equations, P_{II} and P_{IV}*, Math. Ann. **275** (1986), no. 2, 221–255, DOI 10.1007/BF01458459. MR854008 ↑36, 38, 39

[30] Christian Pommerenke, *Univalent functions*, Studia Mathematica/Mathematische Lehrbücher [Studia Mathematica/Mathematical Textbooks], Band XXV, Vandenhoeck & Ruprecht, Göttingen, 1975. With a chapter on quadratic differentials by Gerd Jensen. MR507768 ↑47, 55

[31] Thomas Ransford, *Potential theory in the complex plane*, London Mathematical Society Student Texts, vol. 28, Cambridge University Press, Cambridge, 1995, DOI 10.1017/CBO9780511623776. MR1334766 ↑57, 59

[32] Edward B. Saff and Vilmos Totik, *Logarithmic potentials with external fields*, Grundlehren der mathematischen Wissenschaften [Fundamental Principles of Mathematical Sciences], vol. 316,

Springer-Verlag, Berlin, 1997. Appendix B by Thomas Bloom, DOI 10.1007/978-3-662-03329-6. MR1485778 ↑44

[33] Herbert Stahl, *Extremal domains associated with an analytic function. I, II*, Complex Variables Theory Appl. **4** (1985), no. 4, 311–324, 325–338, DOI 10.1080/17476938508814117. MR858916 ↑44

[34] Herbert Stahl, *The structure of extremal domains associated with an analytic function*, Complex Variables Theory Appl. **4** (1985), no. 4, 339–354, DOI 10.1080/17476938508814119. MR858917 ↑44, 45

[35] Herbert Stahl, *Orthogonal polynomials with complex-valued weight function. I, II*, Constr. Approx. **2** (1986), no. 3, 225–240, 241–251, DOI 10.1007/BF01893429. MR891973 ↑44

[36] Kurt Strebel, *Quadratic differentials*, Ergebnisse der Mathematik und ihrer Grenzgebiete (3) [Results in Mathematics and Related Areas (3)], vol. 5, Springer-Verlag, Berlin, 1984, DOI 10.1007/978-3-662-02414-0. MR743423 ↑47, 55, 58

[37] Gabor Szegö, *Orthogonal polynomials*, American Mathematical Society Colloquium Publications, Vol. 23, American Mathematical Society, Providence, RI, 1959. Revised ed. MR106295 ↑42

Department of Mathematics, University of Michigan, East Hall, 530 Church Street, Ann Arbor, Michigan 48109

Current address: Department of Mathematics, Royal Institute of Technology (KTH), Stockholm, Sweden

Email address: ahmadba@kth.se

Department of Mathematical Sciences, Indiana University Indianapolis, 402 North Blackford Street, Indianapolis, Indiana 46202

Email address: pbleher@iupui.edu

Department of Mathematics, Universidad Carlos III de Madrid, Avda. de la Universidad 30, 28911 Leganés, Madrid, Spain

Email address: alfredo.deanho@uc3m.es

Department of Mathematical Sciences, Indiana University Indianapolis, 402 North Blackford Street, Indianapolis, Indiana 46202

Email address: maxyatts@iu.edu

Contemporary Mathematics
Volume **822**, 2025
https://doi.org/10.1090/conm/822/16474

Multi-orthogonal polynomials associated to a bulk queueing model

U. Fidalgo

In memoriam to the CWRU Professors of Elizabeth Meckes and Wojbor A. Woyczynski who left us during these hard years.

Abstract. We describe the asymptotic behavior of a sequence of type II multiple orthogonal polynomials whose elements belong to integral expressions for transition probability functions corresponding to a queueing model. This model has a bulk service of admitting batches with a fixed size of $\ell \in \mathbb{N}$ customers. We use the expression of the asymptotic behavior to find the corresponding system of the orthogonality measures. We also give conditions on the parameters of the model to ensure the existence of persistent states.

1. Introduction

The present work deals with the asymptotic behavior of some sequences of multi-orthogonal polynomials related to a queueing model. We use such results to find expressions for the system of measures, for which the mentioned polynomials satisfy simultaneous orthogonality relations.. In [**8**] N. Bailey considers queueing models where their services admit batches with a maximum size $\ell \in \mathbb{N} = \{1, 2, 3, \ldots\}$. That describes several situations where the customers are served in groups, for instance transportation of people in vehicles with a maximum passenger capacity. In our model the batches have a fixed size ℓ (see [**24**, Chapter 4]). We have an example of these situations in some shuttle routes where a bus only starts its trip if it completes the vehicle's capacity. This is also the behavior in certain bureaucratic processes, or what happens in some public museums where the guides only admit a fixed number of visitors and if this number is not completed they do not work. A detail description of this kind of bulk quorum queueing systems under a ℓ-policy and with a vacation schedule can be find in [**11**]. Our model could also describe a wider class of situations when the service has a low frequency and is not able to quickly satisfy customer needs to the point that there is always a queue with more than ℓ individuals waiting. In such systems what happens when the number of customers is low is not really important to the model.

Let us assume that an individual arrives at each epoch of Poisson occurrence with rate $\lambda > 0$, and the service only starts if the queue size reaches or exceeds ℓ. The service for batches is also assumed a Poisson process with parameter $\mu > 0$. In

2020 *Mathematics Subject Classification.* Primary 60H10, 60J35.
Key words and phrases. Multiple orthogonal polynomials.

©2025 American Mathematical Society

terms of Kendall's notation [**20**] used in [**24**], it is a system $M/M(\ell,\ell)/1$, which corresponds to a stationary Markov process with path function $X(t)$ whose transition probability function has entries with form

$$P_{i,j}(t) = Pr\left\{X(t+s) = j | X(s) = i\right\}, \quad (i,j) \in \mathbb{Z}_+ \times \mathbb{Z}_+ = \mathbb{Z}_+^2,$$

where $\mathbb{Z}_+ = \mathbb{N} \cup \{0\} = \{0,1,2,\ldots\}$.

According to a definition in [**4**, Chapter 5], we say that a state $i \in \mathbb{Z}_+$ is transient when the improper integral

$$\int_0^\infty P_{i,i}(t)dt < \infty \tag{1.1}$$

is convergent. The integral in (1.1) represents the expectation of the time spent by the random variable $X(t) = i$, if it starts at $X(0) = i$. When the improper integral in (1.1) diverges, we say that i is persistent. In Theorem 1.3 we give conditions on the parameters λ and μ where our model has persistent states.

Set the infinite matrix function $\mathbf{P}(t) = [P_{i,j}(t)]_{(i,j)\in\mathbb{Z}_+^2}$. Following the steps described in [**13**, Chapter XVII]), we deduce that $\mathbf{P}$ is a solution of the initial value problem

$$\begin{cases} \mathbf{P}'(t) = \mathbf{A}\mathbf{P}(t), & t \geq 0, \\ \mathbf{P}(0) = \mathbb{I}, \end{cases} \tag{1.2}$$

where $\mathbf{A} = [a_{i,j}]_{(i,j)\in\mathbb{Z}_+^2}$ has entries:

$$a_{i,j} = \begin{cases} \mu & \text{if} \quad j = i - \ell, \\ -\lambda & \text{if} \quad j = i < \ell, \\ -(\lambda+\mu) & \text{if} \quad j = i \geq \ell, \\ \lambda & \text{if} \quad j = i+1, \\ 0 & \text{if} \quad j \notin \{i+1, i, i-\ell\}, \end{cases} \qquad (i,j) \in \mathbb{Z}_+^2,$$

with schematic form:

$$\mathbf{A} = \begin{pmatrix} -\lambda & \lambda & 0 & \cdots & 0 & 0 & 0 & 0 & \\ 0 & -\lambda & \lambda & & 0 & 0 & 0 & 0 & \\ \vdots & \ddots & \ddots & \ddots & \vdots & & \vdots & \vdots & \\ 0 & 0 & \cdots & 0 & -\lambda & \lambda & 0 & 0 & \\ \mu & 0 & \cdots & 0 & 0 & -(\lambda+\mu) & \lambda & 0 & \ddots \\ 0 & \mu & & 0 & 0 & 0 & -(\lambda+\mu) & \lambda & \ddots \\ \vdots & \ddots & \ddots & & \ddots & & \ddots & \ddots & \ddots \end{pmatrix}. \tag{1.3}$$

(The first ℓ rows are marked by a brace labelled ℓ.)

From [**4**, Proposition 2.9] we deduce that there is a unique solution of the initial value problem in (1.2) which is a transition probability function. On the other hand according to the publication [**27**] by W. Sternberg this initial value problem has a unique solution with analytic entries functions on the complex plane $\mathbb{C}$, which are bounded and real valued on $\mathbb{R}_+ = [0, \infty)$. The other possible solutions have non uniform bounded entries on $\mathbb{R}_+$, hence those unbounded entries functions do not conform transition probability functions. We shall find integral expressions for the entries of a matrix $\mathbf{P}$ that is a solution of (1.2). These expressions represent integer functions, and then they conform the unique transition probability solution of the initial value problem.

The particular case of $\ell = 1$ is reduced to a Birth-and-Death process (see [**19**]) (see [**7**, Chapter III]). In [**19**], S. Karlin and J. McGregor arrived at an integral expression for the entries of $\mathbf{P}$ for a Birth and Death process. They used some aspects of the orthogonal polynomial theory. With Karlin and McGregor's approach, T. S. Chihara and M. E. H. Ismail (see [**10**]) connected sequences of orthogonal polynomials with a queueing model. Later on in [**17**] M. E. H. Ismail revisited this connection. In Karlin & Mc Gregor's case ($\ell = 1$) the matrix $\mathbf{A}$ in (1.3) is symmetrizable and it induces a symmetric operator in the Hilbert space of sequences with the usual inner product. When the coefficient are constant the operator is a compact self-adjoint operator and there is a spectral measure supported on the operator spectrum which is then contained in a bounded interval of the real line, hence Karlin and Mc Gregor's result may be expressed in terms of the orthogonal polynomials family associated to this spectral measure. In our case, when $\ell > 1$ the corresponding operator $\mathbf{A}$ is not self-adjoint, and extending these ideas is not straightforward. We are able to overcome this difficulty a family of multi-orthogonal polynomials defined by a higher order three terms of recurrence relation, which is a variation of others studied in [**5, 6, 12, 21–23**]. This kind of three terms recurrence relations does not correspond to the one that a family of standard orthogonal polynomials satisfies (see for instance [**9**]). Basically it corresponds to a recurrence relations with several terms, (more than three), that some families of multi-orthogonal polynomials satisfy (example [**1–3**]). However in this case there are only three coefficients different from zero, resulting in a three terms such that the polynomial elements of the ordered family do not appear consecutively in the expression. There is a gap between them. Observe:

Let $\{Q_n\}_{n\in\mathbb{Z}_+}$ be a sequence of polynomials generated by the recurrence relation

$$(\lambda + \mu + x)Q_n(x) = \lambda Q_{n+1}(x) + \mu Q_{n-m}(x), \quad n \geq \ell, \tag{1.4}$$

with initial conditions

$$Q_n(x) = \frac{1}{\lambda^n}(\lambda + x)^n, \quad n = 0, \ldots, \ell. \tag{1.5}$$

The conditions above (1.5) can also be written as

$$Q_0 \equiv 1, \quad Q_n(x) = \frac{1}{\lambda}(\lambda + x)Q_{n-1}, \quad n = 1, \ldots, \ell. \tag{1.6}$$

We consider a dual family of vector polynomials $\{\mathbf{q}_m = (q_{0,m}, \ldots, q_{\ell-1,m})\}_{m\in\mathbb{Z}_+}$ whose elements satisfy

$$
\begin{aligned}
&\mathbf{q}_{-1} = (0, \ldots, 0), \mathbf{q}_0 = (1, 0, \ldots, 0)\,,\\
&\mathbf{q}_1 = (0, 1, 0, \ldots, 0)\,, \ldots, \mathbf{q}_{\ell-1} = (0, \ldots, , 0, 1)\,,\\
&(\lambda + x)\mathbf{q}_m(x) = \mu\mathbf{q}_{m+\ell}(x) + \lambda\mathbf{q}_{m-1}(x), \quad m \in \{0, \ldots, \ell-1\},\\
&(\lambda + \mu + x)\mathbf{q}_m(x) = \mu\mathbf{q}_{m+\ell}(x) + \lambda\mathbf{q}_{m-1}(x), \quad m \in \{\ell, \ell+1, \ldots, \}.
\end{aligned}
\tag{1.7}
$$

In the following Theorem 1.1 we give expressions for the strong asymptotic behavior of the sequence $\{Q_n\}_{n\in\mathbb{Z}_+}$. Given two complex numbers u and v we denote by $[u, v]$ the compact segment of straight line that connects u with v. Set also an ordered pair $(\alpha, \theta) \in \mathbb{R} \times [0, 2\pi)$, and denote $\mathcal{L}(\alpha, \theta)$ the ray with slope $\tan\theta$ with endpoint at α. Let us declare the following starlike sets

$$
\begin{aligned}
\Sigma_0 &= \bigcup_{k=0}^{\ell} \left[-\lambda - \mu, -\lambda - \mu + \frac{\ell+1}{\ell}\left(\frac{\mu\lambda^\ell}{\ell}\right)^{1/(\ell+1)} \exp\frac{2\pi i k}{\ell+1}\right],\\
\Sigma_e &= \bigcup_{k=0}^{\ell} \mathcal{L}\left(-\mu - \lambda, \frac{(2k+1)\pi}{\ell+1}\right),\\
\Sigma_o &= \bigcup_{k=0}^{\ell} \mathcal{L}\left(-\mu - \lambda, \frac{2k\pi}{\ell+1}\right).
\end{aligned}
\tag{1.8}
$$

Below we sketch the shapes of these starlike set for the case $\ell = 2$ (you can find the graph corresponding to $\ell = 3$ in [**5**]).

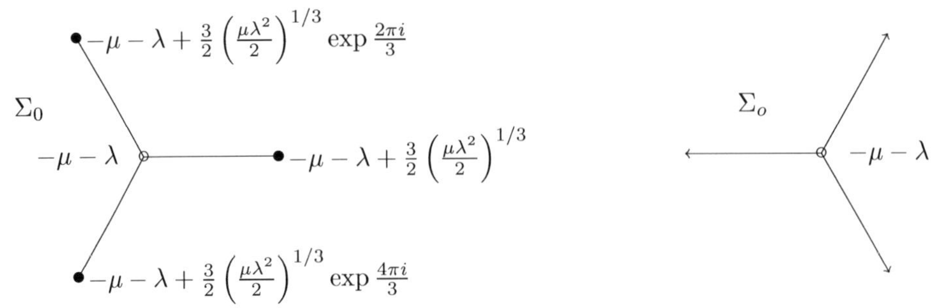

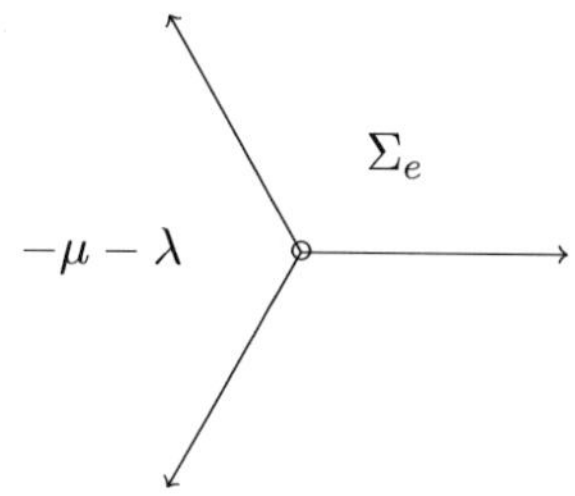

We denote

$$D_0 = \mathbb{C}\setminus\Sigma_0, \quad D_1 = \mathbb{C}\setminus(\Sigma_0\cup\Sigma_o), \quad D_j = \mathbb{C}\setminus(\Sigma_o\cup\Sigma_e), \quad 1<j<\ell,$$

and

$$D_\ell = \begin{cases} \mathbb{C}\setminus\Sigma_e & \text{if } \ell \text{ even,} \\ \\ \mathbb{C}\setminus\Sigma_o & \text{if } \ell \text{ odd.} \end{cases}$$

In the standard case of $\ell=1$, we have that $D_1 = D_0$.

Let us choose the branch solutions $\zeta_0, \zeta_1, \ldots, \zeta_\ell$ of the algebraic equation

$$\lambda\,\zeta^{\ell+1} - (z+\lambda+\mu)\,\zeta^{\ell} + \mu = 0. \tag{1.9}$$

such that

$$\zeta_j \in \mathcal{H}(D_j), \quad \zeta_j(z) = \mathcal{O}\left(\frac{1}{z}\right) \quad \text{as} \quad z\to\infty, \quad j=1,\ldots,\ell,$$

and $\zeta_0\in\mathcal{H}(D_0)$ satisfying the asymptotic behavior

$$\zeta_0(z) = \frac{z+\lambda+\mu}{\lambda} + \mathcal{O}\left(\frac{1}{z}\right) \quad \text{as} \quad z\to\infty.$$

THEOREM 1.1. *Let $\{Q_n\}_{n\in\mathbb{Z}_+}$ be the sequences of polynomials defined in (1.4). The following limit converges uniformly for any compact set*

$$K\subset D_0\setminus\bigcup_{k=1}^{\ell}\left\{-\mu-\lambda+\left[(-\lambda)^{\ell}\mu^{\ell+1}-\frac{1}{\mu}\right]\exp\left(\frac{2\,\pi\,i\,k}{\ell+1}\right)\right\}:$$

$$\lim_{n\to\infty}\frac{Q_n(z)}{\zeta_0^n(z)} = \frac{1}{1-\frac{\ell\mu}{\lambda\zeta_0^{\ell+1}(z)}}\,\frac{1-\left(-\mu\,\lambda\,\zeta_0(z)\right)^{\ell+1}}{1+\mu\,\lambda\,\zeta_0(z)}.$$

Hence for any open set O that contains

$$\bigcup_{k=1}^{\ell}\left\{-\mu-\lambda+\left[(-\lambda)^{\ell}\mu^{\ell+1}-\frac{1}{\mu}\right]\exp\left(\frac{2\,\pi\,i\,k}{\ell+1}\right)\right\}$$

there is a number $N\in\mathbb{Z}_+$ such that for all $n\geq N$ the polynomials Q_n have their roots in O. Fixed $k\in\{1,\ldots,\ell\}$, there is a domain O_k that contains the point

$$z_k = -\mu-\lambda+\left[(-\lambda)^{\ell}\mu^{\ell+1}-\frac{1}{\mu}\right]\exp\left(\frac{2\,\pi\,i\,k}{\ell+1}\right),$$

where for any open $A\subset O_k$ with $z_k\in A$, there is a large enough $\widetilde{N}\in\mathbb{N}$ such that each polynomial Q_n, $n\geq\widetilde{N}$ vanishes once in A.

A real valued function r defined on a certain set $D \subset \overline{\mathbb{C}}$ of the Riemann sphere $\overline{\mathbb{C}} = \mathbb{C} \cup \{\infty\}$ is said to be a weight if it does not change sign on D. Let δ_ζ denote the measure often called Dirac measure with a point of mass at $\zeta \in \mathbb{C}$.

THEOREM 1.2. *There exists a system of measures $(\sigma_1, \ldots, \sigma_\ell)$ such that the following bi-orthogonality relations hold*

$$\int Q_n(x) \sum_{j=1}^{\ell} q_{j,m}(x)\, d\sigma_j(x) = \delta_{n,m} = \begin{cases} 1 & \text{if } n = m \\ 0 & \text{if } n \neq m \end{cases}, \quad (n,m) \in \mathbb{Z}_+^2. \tag{1.10}$$

For each $j = 1, \ldots, \ell$, the measure σ_j, has a Radon Nikodym derivative

$$\frac{d\sigma_j}{dx} = \rho_j(x)\rho(x) + \frac{(-\mu)^j \left[1 + (-\lambda)^\ell \ell \mu^{\ell+2}\right]}{\ell+1} \sum_{k=1}^{\ell} \frac{1 - \exp \frac{2\pi i k j}{\ell+1}}{\exp \frac{2\pi i k (2j-1)}{\ell+1}} \delta_{z_k}, \tag{1.11}$$

where $\rho_1 \equiv 1$, and

$$z_k = -\mu - \lambda + \left[(-\lambda)^\ell \mu^{\ell+1} - \frac{1}{\mu}\right] \exp\left(\frac{2\pi i k}{\ell+1}\right), \quad k = 1, 2, \ldots, \ell.$$

The expression $\rho_j(x)\rho(x)$ is a function defined on the starlike set Σ_0 in (1.8) where fixed $k \in \{0, 1, \ldots, \ell\}$, for each point

$$x_k = -\mu - \lambda + t \exp \frac{2\pi i k}{m+1} \in \Sigma_0, \quad 0 \le t \le \frac{\ell+1}{\ell} \left(\frac{\mu\lambda^\ell}{\ell}\right)^{\frac{1}{\ell+1}},$$

the following symmetry holds

$$\rho_j(x_k)\, \rho(x_k) = \exp\left(-\frac{2\pi i j}{m+1}\right) \rho_j(t - \mu - \lambda)\, \rho(t - \mu - \lambda),$$

being $\rho_j(t - \mu - \lambda)\, \rho(t - \mu - \lambda)$ a weight. The functions $\rho, \rho_1, \ldots, \rho_\ell$ can be expressed explicitly in terms of the solutions of the equation (1.9).

The relation (1.10) allows us to call Q_n and $\mathbf{q}_m$ type II and type I multiple orthogonal polynomials with respect to the system of measures $(\sigma_0, \sigma_1, \ldots, \sigma_\ell)$ (Find these definitions in [**14, 25**]). When $\ell = 1$ both families coincide and they reduce to a sequence of standard orthogonal polynomials with respect to a single measure σ_0 (See [**28**]).

THEOREM 1.3. *The matrix function $\mathbf{P}(t) = [P_{n,r}]_{(n,m) \in \mathbb{Z}_+^2}$ whose entries are*

$$P_{n,m}(t) = \int e^{xt} Q_n(x) \sum_{j=1}^{\ell} q_{j,m}(x)\, d\sigma_j(x), \qquad (n,m) \in \mathbb{Z}_+^2, \tag{1.12}$$

is the unique solution of the initial value problem (1.2), that conform a transition probability function. In addition, the states $n \in \{0, 1, \ldots, \ell - 1\}$ are persistent if there exists at least a value of $k \in \{1, \ldots \ell\}$ where the parameters μ and λ satisfy

$$\left[(-\lambda)^\ell \mu^{\ell+1} - \frac{1}{\mu}\right] \cos\left(\frac{2\,\pi\, k}{\ell+1}\right) > \mu + \lambda. \tag{1.13}$$

When ℓ is odd, condition (1.13) is reduced to $1 + (\lambda\,\mu)^\ell \mu^2 > \lambda\mu + \mu^2$.

This paper is organized as follows: we prove Theorem 1.1, Theorem 1.2, and Theorem 1.3 in Section 4. Previously, in Section 2 and Section 3, we analyzed some auxiliary algebraic and asymptotic aspects that are necessary in the proofs of the main results. In Section 2 we introduce and analyze algebraic properties two auxiliary family of monic polynomials $\{L_n\}_{n\in\mathbb{Z}_+}$ and $\{T_n\}_{n\in\mathbb{Z}_+}$. The elements of the first sequence are obtained after a change of variable, and the second one is constructed with the same recurrence relation that $\{L_n\}_{n\in\mathbb{Z}_+}$ satisfies, but with different initial conditions. the sequence corresponds to an easy case of the families study in some publications such as [**1**–**3**]. Section 3 we analyze specifically the family $\{T_n\}_{n\in\mathbb{Z}_+}$. We study its zeros distribution of its polynomials to find later an expression of their strong asymptotic behavior. Such expression is then used to find another one corresponding to the sequence $\{L_n\}_{n\in\mathbb{Z}_+}$. With this last strong asymptotic analysis, using the inverse change of variable, we are able to proof Theorem 1.1 and also describe the measures that appear in the integral expressions of Theorem 1.2 and Theorem 1.3.

2. Type II polynomials. Some algebraic aspects

In this Section we review concepts and results which are necessary to show Theorem 1.1, whose proof is carried out in Section 4. Some of these ideas are brought from [**5**,**6**]. In order to make easier our work, from the family of polynomials $\{Q_n\}_{n\in\mathbb{Z}_+}$ defined by the recurrence relation (1.4) and initial conditions (1.5), we introduce another sequence of polynomials $\{L_n\}_{n\in\mathbb{Z}_+}$ with the change of variables $z = x + \lambda + \mu$:

$$L_n(z) = \lambda^n Q_n(z - \lambda - \mu), \quad n \in \mathbb{Z}_+. \tag{2.1}$$

Taking into account the relations (1.4) and (1.5) we have that

$$z\, L_n(z) = L_{n+1}(z) + \mu\lambda^\ell L_{n-\ell}, \quad n \geq \ell, \tag{2.2}$$

with initial conditions

$$L_n(z) = (z - \mu)^n, \quad n = 0, \dots, \ell. \tag{2.3}$$

We also introduce a third family of polynomials $\{T_n\}_{n\in\mathbb{Z}_+}$ satisfying the recurrence relation (2.2)

$$z\, T_n(z) = T_{n+1}(z) + \mu\lambda^\ell T_{n-\ell}, \quad n \geq \ell, \tag{2.4}$$

with different initial conditions

$$T_n(z) = z^n, \quad n = 0, \dots, \ell - 1. \tag{2.5}$$

The polynomials in $\{T_n\}_{n\in\mathbb{Z}_+}$ can be found as a particular case of the sequences analyzed in the publications [**5**, **6**, **21**–**23**]. In Section 3 we use some particular properties of our case to arrive to specific results that the elements of such sequence $\{T_n\}_{n\in\mathbb{Z}_+}$ satisfy.

Consider the linear operator $\mathbf{L}$ acting on the space of sequences ℓ_2, which is determined by the infinite non-symmetric matrix $\mathbf{L} = [k_{i,j}]_{(i,j)\in\mathbb{Z}_+^2}$ that has entries:

$$k_{i,j} = \begin{cases} \mu\lambda^\ell & \text{if } \; j = i-\ell, \\ \mu & \text{if } \; j = i < \ell, \\ 1 & \text{if } \; j = i+1, \\ 0 & \text{if } \; j \notin \{i+1, i, i-\ell\}, \end{cases} \qquad (i,j) \in \mathbb{Z}_+^2,$$

schematically:

$$\mathbf{L} = \left(\begin{array}{cccccccccc} \mu & 1 & 0 & \cdots & 0 & 0 & 0 & 0 & \\ 0 & \mu & 1 & & 0 & 0 & 0 & 0 & \\ \vdots & \ddots & \ddots & \ddots & \vdots & & \vdots & \vdots & \\ 0 & 0 & \cdots & 0 & \mu & 1 & 0 & 0 & \\ \mu\lambda^\ell & 0 & \cdots & 0 & 0 & 0 & 1 & 0 & \ddots \\ 0 & \mu\lambda^\ell & & 0 & 0 & 0 & 0 & 1 & \ddots \\ \vdots & \ddots & \ddots & & \ddots & & \ddots & \ddots & \ddots \end{array} \right) \begin{array}{l} \left.\begin{array}{l} \\ \\ \\ \\ \end{array}\right\} \ell \\ \\ \\ \\ \end{array} .$$

Set $\{\mathbf{e}_j\}_{j\in\mathbb{Z}_+}$ the standard basis of ℓ_2, then the operator $\mathbf{L}$ is defined as follows

$$\mathbf{L}^\top \mathbf{e}_j = \begin{cases} \mathbf{e}_{j+1} + \mu\mathbf{e}_j, & j \in \{0,1,2,\ldots,\ell-1\}, \\ \mathbf{e}_{j+1} + \mu\lambda^\ell\mathbf{e}_{j-\ell}, & j \geq \ell, \end{cases} \tag{2.6}$$

where $\top$ denotes the operator of matrix transposition.

Given a polynomial $p(x) = b_\kappa\, x^\kappa + \cdots + b_1\, x + b_0$ with an arbitrary degree κ we denote the operator $p(\mathbf{L})$ satisfying that

$$p(\mathbf{L}^\top)\mathbf{u} = \left(b_\kappa \mathbf{L}^{\top\kappa} + \cdots + b_1\,\mathbf{L}^\top + b_0\right)\mathbf{u} = b_\kappa\mathbf{L}^{\top\kappa}\mathbf{u} + \cdots + b_1\,\mathbf{L}^\top\mathbf{u} + b_0\mathbf{u}, \quad \mathbf{u} \in \ell_2.$$

LEMMA 2.1. *Consider the sequence of polynomials* $\{L_n\}_{n\in\mathbb{Z}_+}$ *satisfying the recurrence relation* (2.2) *and initial conditions* (2.3). *Then*

$$\mathbf{e}_n = L_n(\mathbf{L}^\top)\,\mathbf{e}_0, \quad n \in \mathbb{Z}_+. \tag{2.7}$$

PROOF. For each $n \in \{0, 1, \dots, \ell - 1\}$ we have that $L_n(\mathbf{L}^\top) = (\mathbf{L}^\top - \mu)^n$. This means that

$$L_0(\mathbf{L}^\top)\,\mathbf{e}_0 = \left(\mathbf{L}^\top - \mu\right)^0 \mathbf{e}_0 = \mathbf{e}_0,$$

$$L_1(\mathbf{L}^\top)\,\mathbf{e}_0 = \left(\mathbf{L}^\top - \mu\right)\mathbf{e}_0 = \mathbf{e}_1 + \mu\mathbf{e}_0 - \mu\mathbf{e}_0 = \mathbf{e}_1,$$

$$L_2(\mathbf{L}^\top)\,\mathbf{e}_0 = (\mathbf{L}^\top - \mu)^2\mathbf{e}_0 = \left(\mathbf{L}^\top - \mu\right)\mathbf{e}_1 = \mathbf{e}_2 + \mu\mathbf{e}_1 - \mu\mathbf{e}_1 = \mathbf{e}_2,$$

$$\vdots$$

$$L_{\ell-1}(\mathbf{L}^\top)\,\mathbf{e}_0 = (\mathbf{L}^\top - \mu)^{\ell-1}\mathbf{e}_0$$

$$= (\mathbf{L}^\top - \mu)^{\ell-2}\mathbf{e}_1 = \cdots = \left(\mathbf{L}^\top - \mu\right)\mathbf{e}_{\ell-2} = \mathbf{e}_{\ell-1}.$$

We now consider $n \geq \ell - 1$. For each $k = 1, \dots, n$, assume that $L_k(\mathbf{L}^\top)\,\mathbf{e}_0 = \mathbf{e}_k$. By the recurrence relation (2.2), we have that

$$L_{n+1}(\mathbf{L}^\top)\mathbf{e}_0 = \left[\mathbf{L}^\top L_n(\mathbf{K}^\top) - \mu\lambda^\ell L_{n-\ell}(\mathbf{L}^\top)\right]\mathbf{e}_0.$$

hence

$$L_{n+1}(\mathbf{L}^\top)\,\mathbf{e}_0 - \mathbf{e}_{n+1} + \mu\lambda^\ell\mathbf{e}_{j-\ell} - \mu\lambda^\ell\mathbf{e}_{n-\ell}] = \mathbf{e}_{n+1}.$$

This completes the proof by induction. □

LEMMA 2.2. *Set $j \in \{1, \dots, \ell\}$. For each $\nu \in \{0, 1, \dots, j - 1\}$*

$$(\mathbf{L} - \mu\mathbb{I})^\nu\mathbf{e}_{j-1} = \sum_{k=0}^{\nu}\binom{\nu}{k}\left(\mu\lambda^\ell\right)^{\nu-k}\mathbf{e}_{(\nu-k)\ell+j-1-k}. \tag{2.8}$$

PROOF. We proceed by induction. In case $\nu = 0$ according to (2.6) we have that

$$(\mathbf{L} - \mu\mathbb{I})^0\mathbf{e}_{j-1} = \mathbf{e}_{j-1}.$$

We assume that the statement (2.8) holds for a $\nu \in \{0, 1, \dots j - 2\}$, and then we prove it for $\nu + 1$.

$$(\mathbf{L} - \mu\mathbb{I})^{\nu+1}\mathbf{e}_{j-1} = (\mathbf{L} - \mu\mathbb{I})(\mathbf{L} - \mu\mathbb{I})^\nu\mathbf{e}_{j-1}$$

$$= (\mathbf{L} - \mu\mathbb{I})\sum_{k=0}^{\nu}\binom{\nu}{k}\left(\lambda\mu^\ell\right)^{\nu-k}\mathbf{e}_{(\nu-k)\ell+j-1-k}$$

$$= \sum_{k=0}^{\nu}\binom{\nu}{k}\left(\mu\lambda^\ell\right)^{\nu-k}(\mathbf{L} - \mu\mathbb{I})\,\mathbf{e}_{(\nu-k)\ell+j-1-k}$$

$$= \sum_{k=0}^{\nu}\binom{\nu}{k}\left(\mu\lambda^\ell\right)^{\nu-k}\left(\mathbf{e}_{(\nu-k)\ell+j-1-k-1} + \mu\lambda^\ell\mathbf{e}_{(\nu-k+1)\ell+j-1-k}\right) = \mathbf{e}_{j-1-\nu-1}$$

$$+ \sum_{k=1}^{\nu}\left(\binom{\nu}{k-1} + \binom{\nu}{k}\right)\left(\lambda\mu^\ell\right)^{\nu-k+1}\mathbf{e}_{(\nu-k+1)\ell+j-1-k} + \left(\lambda\mu^\ell\right)^{\nu+1}\mathbf{e}_{j-1-\nu-1}.$$

Using Pascal's identity we obtain that

$$(\mathbf{L}-\mu\mathbb{I})^{\nu+1}\mathbf{e}_{j-1}=\sum_{k=0}^{\nu+1}\binom{\nu+1}{k}\left(\mu\lambda^{\ell}\right)^{\nu+1-k}\mathbf{e}_{(\nu+1-k)\ell+j-1-k}.$$

This completes the proof by induction. □

LEMMA 2.3. *The zeros of the polynomials L_n are all contained in a subset of*

$$F=\left\{z\in\mathbb{C}:|z|\leq 1+\mu\lambda^{\ell}\right\}\cup\left\{z\in\mathbb{C}:|z-\mu|\leq 1+\mu\lambda^{\ell}\right\},$$

which implies that the linear operator $\mathbf{L}$ *has a bounded spectrum*

$$\sigma_{\mathbf{L}}=\bigcap_{j=1}^{\infty}\overline{\bigcup_{n=j}^{\infty}\{x:L_n(x)=0\}}\subset F.$$

PROOF. Set $(L_0,\ldots,L_{n-1})^{\top}$, the vector $\mathbf{e}_n=(0,0\ldots,0,1)^{\top}$ with n components, and the $n\times n$ matrix block $\mathbf{L}_n=[k_{i,j}]_{(i,j)\in\{0,1,\ldots,n-1\}^2}$ of the infinite matrix form corresponding to $\mathbf{L}$, where for each $i\in\{1,\ldots,n-1\}$

$$k_{i,i+1}=1,\quad k_{i+\ell,i}=\mu\lambda^{\ell},\quad k_{i,i}=\gamma,\quad i\in\{0,\ldots,m\}\quad\text{otherwise}\quad k_{i,j}=0.$$

From the recurrence relations in (2.2) and (2.3) we observe that $\deg L_n=n$ then L_n has n zeros. Let x_k, $k\in\{1,2,\ldots,n\}$ denote the zeros of L_n. The relations (2.2) and (2.3) can be rewritten as follows

$$z\mathbf{L}_n(z)=\mathbf{L}_n(z)+L_n(z)\mathbf{e}_n\implies x_k\mathbf{L}_n(x_k)=\mathbf{L}_n(x_k),$$

which implies that x_k, $k=1,\ldots,n$, are the n eigenvalues of $\mathbf{L}_n$. From Gershgorin Disc Theorem (see [**29**]), either $|x_k-\mu\leq 1+\mu\lambda^{\ell}$ or $|x_k|\leq 1+\mu\lambda^{\ell}$, $k=1,\ldots,n$. Then $\mathbf{L}$ has a bounded spectrum. This completes the proof. □

Let us consider the ℓ resolvent functions of the linear operator $\mathbf{L}$, (see for instance [**18, 25**]). They are complex valued functions with domains on the largest sets whose expressions are well defiined:

$$f_j(z)=(z\mathbb{I}-\mathbf{L})^{-1}\mathbf{e}_{j-1}\cdot\mathbf{e}_0,\quad j=1,\ldots,\ell. \tag{2.9}$$

Hence their domains are the set $\overline{\mathbb{C}}\setminus\sigma_{\mathbf{L}}$, where $\overline{\mathbb{C}}=\mathbb{C}\cup\{\infty\}$ is the Riemann sphere.

LEMMA 2.4. *Fix* $j\in\{1,\ldots,\ell\}$. *The resolvent function* f_j *is holomorphic on* $\overline{\mathbb{C}}\setminus\sigma_{\mathbf{L}}$ *(denote* $f_j(\cdot)\in\mathcal{H}(\overline{\mathbb{C}}\setminus\sigma_{\mathbf{L}})$*) and also behaves*

$$f_j(z)=\mathcal{O}\left(\frac{1}{z^j}\right)\quad as\quad z\to\infty. \tag{2.10}$$

PROOF. By construction $f_j\in\mathcal{H}(\overline{\mathbb{C}}\setminus\sigma_{\mathbf{L}})$. According to Lemma 2.3 we have that $\sigma_{\mathbf{L}}\subset\left\{z\in\mathbb{C}:|z|\leq 1+\lambda\mu^{\ell}\right\}\cup\left\{z\in\mathbb{C}:|z-\mu|\leq 1+\lambda\mu^{\ell}\right\}$ is bounded. Then applying Neumann's series inversion (see for instance [**26**]) we obtain that:

$$(z\mathbb{I}-\mathbf{L})^{-1}\mathbf{u}=((z-\mu)\mathbb{I}-(\mathbf{L}-\mu\mathbb{I}))^{-1}\mathbf{u}=\sum_{\nu=0}^{\infty}\frac{(\mathbf{L}-\mu\mathbb{I})^{\nu}\mathbf{u}}{(z-\mu)^{\nu+1}},\quad \mathbf{u}\in\ell_2,$$

to obtain the following expression that holds uniformly in any compact set contained in the complement of smallest disk of $\overline{\mathbb{C}}$ that contains $\sigma_{\mathbf{L}}$ (let $D(\mathbf{L})$ denote this set)

$$f_j(z)=\sum_{\nu=0}^{\infty}\frac{(\mathbf{L}-\mu\mathbb{I})^{\nu}\mathbf{e}_{j-1}\cdot\mathbf{e}_0}{(z-\mu)^{\nu+1}}.$$

From (2.8) in Lemma 2.2 for each $\nu \in \{0, \ldots, j-2\}$, we have that

$$(\mathbf{L} - \mu \mathbb{I})^{\nu} \mathbf{e}_{j-1} \cdot \mathbf{e}_0 = \left(\sum_{k=0}^{\nu} \binom{\nu}{k} \left(\lambda \mu^{\ell}\right)^{\nu-k} \mathbf{e}_{(\nu-k)\ell+j-1-k} \right) \cdot \mathbf{e}_0 = 0,$$

hence

$$f_j(z) = \sum_{\nu=j-1}^{\infty} \frac{(\mathbf{L} - \mu \mathbb{I})^{\nu} \mathbf{e}_{j-1} \cdot \mathbf{e}_0}{(z-\mu)^{\nu+1}} = \frac{1}{(z-\mu)^{j-1}} \sum_{\nu=0}^{\infty} \frac{(\mathbf{L} - \mu \mathbb{I})^{\nu+j-1} \mathbf{e}_{j-1} \cdot \mathbf{e}_0}{(z-\mu)^{\nu+1}},$$

which proves the statement. □

Let $\{c_{\nu,j}\}_{\nu \in \mathbb{Z}_+}$, $j = 1, \ldots, \ell$ denote the sequence of moments whose elements are defined by

$$c_{\nu,j} = \mathbf{L}^{\nu} \mathbf{e}_{j-1} \cdot \mathbf{e}_0, \quad \nu \in \mathbb{Z}_+, \quad j = 1, \ldots, \ell. \tag{2.11}$$

LEMMA 2.5. *Given* $j \in \{1, \ldots, \ell\}$

$$c_{\nu,j} = (\mathbf{L})^{\nu} \mathbf{e}_{j-1} \cdot \mathbf{e}_0 = 0, \quad \nu = 0, \ldots, j-1, \quad j = 1, \ldots, \ell.$$

PROOF. Fix $j \in \{1, \ldots, \ell\}$. Since $\sigma_{\mathbf{L}}$ is bounded we apply Neumann's series inversion to $(z\mathbb{I} - \mathbf{L})^{-1} \mathbf{e}_j$ such that at any compact set contained in $D(\mathbf{L})$, the following equality holds uniformly

$$f_j(z) = \sum_{\nu=0}^{\infty} \frac{\mathbf{L}^{\nu} \mathbf{e}_{j-1} \cdot \mathbf{e}_0}{z^{\nu+1}} = \sum_{\nu=0}^{\infty} \frac{c_{\nu,j}}{z^{\nu+1}}.$$

Taking now into account (2.10) we obtain that the first $j-1$th coefficients $c_{\nu,j}$ must vanish. This completes the proof. □

Let us introduce the second kind polynomials

$$L_{n,j}(z) = (z\mathbb{I} - \mathbf{L})^{-1} \mathbf{e}_{j-1} \cdot \left(L_n(z\mathbb{I}) - L_n(\mathbf{L}^{\top})\right) \mathbf{e}_0, \quad j = 1, \ldots, \ell. \tag{2.12}$$

For convenience we agree to call $L_{n,0} \equiv L_n$, $n \in \mathbb{Z}_+$.

LEMMA 2.6. *Let $n \in \mathbb{Z}_+$ and $j \in \{1, \ldots, \ell\}$ be given, with $n = \kappa\ell + s$, $(\kappa, s) \in \mathbb{Z}_+ \times \{0, 1, \ldots, \ell-1\}$. The resolvent functions in (2.9) corresponding to the linear operator $\mathbf{L}$ and the polynomial (2.12) satisfy the identity*

$$L_n(z) f_j(z, \mathbf{L}) - L_{n,j}(z) = (z\mathbb{I} - \mathbf{L})^{-1} \mathbf{e}_{j-1} \cdot L_n(\mathbf{L}^{\top}) \mathbf{e}_0, \tag{2.13}$$

where

$$(z\mathbb{I} - \mathbf{L})^{-1} \mathbf{e}_{j-1} \cdot L_n(\mathbf{L}^{\top}) \mathbf{e}_0 \in \mathcal{H}\left(\overline{\mathbb{C}} \setminus \sigma_{\mathbf{L}}\right), \tag{2.14}$$

and

$$(z\mathbb{I} - \mathbf{L})^{-1} \mathbf{e}_{j-1} \cdot L_n(\mathbf{L}^{\top}) \mathbf{e}_0 = \mathcal{O}\left(\frac{1}{z^{\tau(n,j)}}\right) \quad \text{as} \quad z \to \infty \tag{2.15}$$

with

$$\tau(n,j) = \begin{cases} \kappa + \ell + j - s + 1 & \text{if} \quad j \in \{1, \ldots, s\}, \\ \kappa + j - s + 1 & \text{if} \quad j \in \{s+1, \ldots, \ell\}. \end{cases}$$

PROOF. Fix $j \in \{1, 2, \ldots, \ell\}$. The expression in (2.12) yields the identity (2.13). From Lemma 2.4 we obtain $f_j \in \mathcal{H}\left(\overline{\mathbb{C}} \setminus \sigma_{\mathbf{L}}\right)$. Hence taking into account (2.13) we show (2.14).

Since $\sigma_{\mathbf{L}}$ is bounded we apply again Neumann's series for operator inversion to $(z\mathbb{I} - \mathbf{L})^{-1}$, and we obtain the following equality uniformly on any compact set of $D(\mathbf{L})$. We have that

$$(z\mathbb{I} - \mathbf{L})^{-1} \mathbf{e}_{j-1} \cdot L_n(\mathbf{L}^\top) \mathbf{e}_0 = \sum_{\nu=0}^{\infty} \frac{\mathbf{L}^\nu \mathbf{e}_{j-1} \cdot L_n(\mathbf{L}^\top) \mathbf{e}_0}{z^{\nu+1}}, \quad j = 1, \ldots, \ell.$$

Using the identity in (2.7) , we obtain that

$$(z\mathbb{I} - \mathbf{L})^{-1} \mathbf{e}_{j-1} \cdot L_n(\mathbf{L}^\top) \mathbf{e}_0 = \sum_{\nu=0}^{\infty} \frac{\mathbf{L}^\nu \mathbf{e}_{j-1} \cdot \mathbf{e}_n}{z^{\nu+1}}.$$

Recall (2.8) in Lemma 2.2, then

$$\begin{aligned}(z\mathbb{I} - \mathbf{L})^{-1} \mathbf{e}_{j-1} \cdot L_n(\mathbf{L}^\top) \mathbf{e}_0 &= \sum_{\nu=0}^{\infty} \frac{1}{z^{\nu+1}} \sum_{k=0}^{\nu} \binom{\nu}{k} \left(\mu\lambda^\ell\right)^{\nu-k} \mathbf{e}_{(\nu-k)\ell+j-1-k} \cdot \mathbf{e}_n \\ &= \sum_{\nu=0}^{\infty} \frac{1}{z^{\nu+1}} \sum_{k=0}^{\nu} \binom{\nu}{k} \left(\mu\lambda^\ell\right)^{\nu-k} \mathbf{e}_{(\nu-k)\ell+j-1-k} \cdot \mathbf{e}_{\kappa\ell+s} \\ &= \sum_{\nu=0}^{\infty} \frac{1}{z^{\nu+1}} \sum_{k=0}^{\nu} \binom{\nu}{k} \left(\mu\lambda^\ell\right)^{\nu-k} \delta_{(\nu-k)\ell+j-1-k,\kappa\ell+s}.\end{aligned}$$

Let us start by analyzing the cases $j \in \{1, \ldots, s\}$. In the summation above, the first term that does not vanish, ocurres when $\nu - k - 1 = s$ and $\ell + j - 1 - k = s$, hence $\nu = \kappa + \ell + j - s$. This implies that

$$(z\mathbb{I} - \mathbf{L})^{-1} \mathbf{e}_{j-1} \cdot L_n(\mathbf{L}^\top) \mathbf{e}_0 = \mathcal{O}\left(\frac{1}{z^{\kappa+\ell+j-s+1}}\right) \quad \text{as} \quad z \to \infty.$$

When $j \in \{s+1, \ldots, \ell\}$, the first term that is different from zero, corresponds to $\nu - k - 1 = \kappa$ and $j - 1 - k$, so $\nu = \kappa + j - s$ and

$$(z\mathbb{I} - \mathbf{L})^{-1} \mathbf{e}_{j-1} \cdot L_n(\mathbf{L}^\top) \mathbf{e}_0 = \sum_{\nu=0}^{\infty} \frac{\mathbf{L}^\nu \mathbf{e}_{j-1} \cdot \mathbf{e}_n}{z^{\nu+1}} = \mathcal{O}\left(\frac{1}{z^{\kappa+j-s+1}}\right), \text{ as } z \to \infty.$$

The last two expressions complete the proof. □

LEMMA 2.7. *Fix $j \in \{1, \ldots, \ell\}$. The sequence of polynomials $\{L_{n,j}\}_{n \in \mathbb{Z}_+}$ in (2.12) satisfy the following recurrence relations*

$$zL_{n,j}(z) = L_{n+1,j}(z) + \mu\lambda^\ell L_{n-\ell,j}(z), \tag{2.16}$$

with the initial conditions

$$L_{n,j} \equiv 0, \quad n = 0, 1, \ldots, j-1 \quad \text{and} \quad L_{n,j}(z) = (z-\mu)^{n-j}, \quad n = j, \ldots, \ell. \tag{2.17}$$

PROOF. For each $j \in \{1, \ldots, \ell\}$, consider (2.13)

$$L_{n,j}(z) = L_n(z) f_j(z) - (z\mathbb{I} - \mathbf{L})^{-1} \mathbf{e}_{j-1} \cdot L_n(\mathbf{L}^\top) \mathbf{e}_0.$$

Let us recall Lemma 2.4. We have that $f_j \in \mathcal{H}\left(\overline{\mathbb{C}} \setminus \sigma_{\mathbf{L}}\right)$. Combining this equality above with (2.14) and (2.15) in Lemma 2.6, we conclude that for any $n \in \{0, \dots, j-1\}$ the polynomial

$$\begin{aligned} L_{n,j}(z) &= L_n(z) f_j(z, \mathbf{L}) - (z\mathbb{I} - \mathbf{L})^{-1} \mathbf{e}_{j-1} \cdot L_n(\mathbf{L}^\top)\, \mathbf{e}_0 \\ &= \mathcal{O}\left(\frac{1}{z}\right) + \mathcal{O}\left(\frac{1}{z^{\tau(n,j)}}\right) = \mathcal{O}\left(\frac{1}{z}\right) \quad \text{as} \quad z \to \infty, \end{aligned}$$

which yields $L_{n,j} \equiv 0$, $n = 0, \dots, j-1$. When $n = j$ we have that the polynomial $L_{j,j}$ is bounded in $\overline{\mathbb{C}}$, hence its a constant. From (2.8) in Lemma 2.2 we obtain that

$$\lim_{z \to \infty} L_{j,j}(z) = \lim_{z \to \infty} L_j(z) f_j(z) = 1.$$

We now consider $n \in \{j, \dots, \ell - 1\}$ and (2.3) to obtain

$$\begin{aligned} zL_{n,j}(z) &= z\,(z\mathbb{I} - \mathbf{L})^{-1} \mathbf{e}_{j-1} \cdot \left(L_n(z\mathbb{I}) - L_n(\mathbf{L}^\top)\right) \mathbf{e}_0 \\ &= (z\mathbb{I} - \mathbf{L})^{-1} \mathbf{e}_{j-1} \cdot \left(zL_n(z\mathbb{I}) - zL_n(\mathbf{L}^\top)\right) \mathbf{e}_0 \\ &= (\mathbb{I} - \mathbf{L})^{-1} \mathbf{e}_{j-1} \cdot \left(zL_n(z\mathbb{I}) - (z\mathbb{I} - \mathbf{L}^\top) L_n(\mathbf{L}^\top) - \mathbf{L}^\top L_n(\mathbf{L}^\top)\right) \mathbf{e}_0 \\ &= (\mathbb{I} - \mathbf{L})^{-1} \mathbf{e}_{j-1} \cdot \left(L_{n+1}(z\mathbb{I}) - L_{n+1}(\mathbf{L}^\top)\right) \mathbf{e}_0 \\ &+\mu\,(\mathbb{I} - \mathbf{L})^{-1} \mathbf{e}_{j-1} \cdot \left(L_n(z\mathbb{I}) - L_n(\mathbf{L}^\top)\right) \mathbf{e}_0 = L_{n+1,j}(z) + \mu L_{n,j}(z). \end{aligned}$$

Since $n \geq j$ we obtain (2.17).

Let us prove (2.16). Set $n \geq \ell$ and recall that (2.2)

$$\begin{aligned} z\,L_{n,j}(z) &- z\,(z\mathbb{I} - \mathbf{L})^{-1} \mathbf{e}_{j-1} \cdot \left(L_n(z\mathbb{I}) \quad L_n(\mathbf{L}^\top)\right) \mathbf{e}_0 \\ &= (z\mathbb{I} - \mathbf{L})^{-1} \mathbf{e}_{j-1} \cdot \left(z\mathbb{I}L_n(z\mathbb{I}) - z\mathbb{I}L_n(\mathbf{L}^\top)\right) \mathbf{e}_0 \\ &= (z\mathbb{I} - \mathbf{L})^{-1} \mathbf{e}_{j-1} \cdot \left(z\mathbb{I}L_n(z\mathbb{I}) - (z\mathbb{I} - \mathbf{L}^\top) L_n(\mathbf{L}^\top) - \mathbf{L}^\top L_n(\mathbf{L}^\top)\right) \mathbf{e}_0 \\ &= L_{n+1,j}(z) + \mu\lambda^\ell L_{n-\ell,j}(z). \end{aligned}$$

This proves (2.16). □

REMARK 2.8. The vector rational function

$$\left(\frac{L_{n,1}}{L_n}, \dots, \frac{L_{n,\ell}}{L_n}\right), \quad n \in \mathbb{Z}_+ \tag{2.18}$$

is often called Hermite-Padé approximant associated to the system of functions $(f_1(z), \dots, f_\ell(z))$ and the multi-index $\mathbf{n} = \left(\overbrace{\kappa+1, \dots, \kappa+1}^{s}, \kappa, \dots, \kappa\right) \in \mathbb{Z}_+^\ell$ with ℓ_1-norm $|\mathbf{n}| = n = \ell\,\kappa + s$.

3. On the sequence $\{T_n\}_{n\in\mathbb{Z}_+}$

This section concerns about the sequence of polynomials $\{T_n\}_{n\in\mathbb{Z}_+}$ that satisfy the recurrence relation (2.4) with initial conditions (2.5). This sequence is an easier version of the families considered in [**5, 6, 21**–**23**], then we are able to find more details in our results, which are used in the proof of Theorem 1.1 in Section 4.

LEMMA 3.1. *For each $n \in \mathbb{Z}_+$ write $n = d(\ell+1) + r$ where $d \in \mathbb{Z}_+$ and $r \in \{0, 1, \dots, \ell\}$. Then there exists a polynomial h_n with degree d such that $T_n(z) =$*

$z^r h_n(z^{\ell+1})$. The polynomial elements of the sequence $\{h_n\}_{n\in\mathbb{Z}_+}$ satisfy the following initial conditions and recurrence relations, respectively

$$h_0 = h_1 = \cdots = h_\ell = 1 \tag{3.1}$$

and when $d \geq 1$

$$h_{n+1}(z) = \begin{cases} h_n(z) - \mu\lambda^\ell h_{n-\ell}(z) & \text{if } r \in \{0, 1, \ldots, \ell - 1\}, \\ z\, h_n(z) - \mu\lambda^\ell h_{n-\ell}(z) & \text{if } r = \ell, \end{cases} \tag{3.2}$$

and such that

1. $(-1)^d h_n(0) > 0$, $n \in \mathbb{Z}_+$.
2. *The d zeros of h_n are positive and simple.*
3. *Given an arbitrary set $n = d(\ell + 1) + r$, denote $x_{n,1} < x_{n,2} < \cdots < x_{n,d}$ the zeros of h_n. Fix now n such as $r = \ell$. Then for each $j \in \{1, \ldots, d\}$*

$$0 < x_{n+1,j} < \cdots < x_{n+\ell+1,j} < x_{n,j} < x_{n+1,j+1} < \cdots < x_{n+\ell+1,j+1} < x_{n,j+1}.$$

Proof. The proof is divided in two parts. First we show the recurrence relation in (3.2) and the initial conditions (3.1). In the second part we prove theother statements, using the mentioned recurrence relation.

1st part: The initial conditions (2.5) make evident the statement for $n \in \{0, 1, \ldots, \ell\}$. Observe that

$$T_n(z) = z^n \implies h_n \equiv 1 \quad \text{and} \quad r = n,$$

which proves (3.1). Let us use the induction method. Given $n = d(\ell + 1) + r \geq \ell$ we assume that the statement holds for every $\widetilde{n} \in \{n - \ell, n - \ell + 1, \ldots, n\}$. Observe that

$$n - \ell = d(\ell + 1) + r - \ell = d(\ell + 1) - (\ell + 1) + r + 1 = (d - 1)(\ell + 1) + r + 1.$$

Let us analyze two different cases

(1) **Case** $r = 0, 1, \ldots, \ell - 1$. We have that $n + 1 = d(\ell + 1) + r + 1$, with $d \in \mathbb{Z}_+$, $r + 1 \in \{1, \ldots, \ell\}$.

$$T_{n+1}(z) = zT_n(z) - \mu\lambda^\ell T_{n-m} = zz^r h_n(z^{\ell+1}) - \mu\lambda^\ell z^{r+1} h_{n-\ell}(z^{\ell+1}),$$

then

$$\begin{aligned} T_{n+1}(z) &= z^{r+1} h_n(z^{\ell+1}) - \mu\lambda^\ell z^{r+1} h_{n-\ell}(z^{\ell+1}) \\ &= z^{r+1} \left(h_n(z^{\ell+1}) - \mu\lambda^\ell h_{n-\ell}(z^{\ell+1})\right) = z^{r+1} h_{n+1}(z^{\ell+1}). \end{aligned}$$

We see that $\deg h_{n+1} = \deg h_n$. We also observe that the equality

$$z^{r+1} \left(h_n(z^{\ell+1}) - \mu\lambda^\ell h_{n-\ell}(z^{\ell+1})\right) = z^{r+1} h_{n+1}(z^{\ell+1})$$

implies that

$$h_{n+1}(z) = h_n(z) - \mu\lambda^\ell h_{n-\ell}(z).$$

This completes this part of the proof for cases $r = 0, 1, \ldots, \ell - 1$.

(2) **Case** $r = \ell$. Since $n = d(\ell + 1) + \ell$ then

$$n + 1 = d(\ell + 1) + \ell + 1 = (d + 1)(\ell + 1)$$

and

$$n - m = d(\ell + 1) + \ell - \ell = d(\ell + 1).$$

Hence

$$T_{n+1}(z) = zT_n(z) - \mu\lambda^\ell T_{n-\ell} = zz^\ell h_n(z^{\ell+1}) - \mu\lambda^\ell z^{\ell+1}h_{n-\ell}(z^{\ell+1})$$
$$= z^{\ell+1}h_n(z^{\ell+1}) - \mu\lambda^\ell h_{n-\ell}(z^{\ell+1})) = h_{n+1}(z^{\ell+1}).$$

This implies that $\deg h_{n+1} = \deg h_n + 1 = d+1$, and the expression

$$z^{\ell+1}\left(h_n(z^{\ell+1} - \mu\lambda^\ell h_{n-\ell}(z^{\ell+1})\right) = h_{n+1}(z^{\ell+1})$$

yields

$$h_{n+1}(z) = z\,h_n(z) - \mu\lambda^\ell h_{n-\ell}(z).$$

This completes the proof of the first part.

2nd part; We use an induction method. When $n \in \{0,\ldots,\ell\}$, $d=0$ and from (3.1) the first two statements (1) and (2) are trivially satisfied. Consider $d=1$. From recurrence relations in (3.2) we have that

$$h_{\ell+1}(z) = z\,h_\ell - \mu\lambda^\ell\,h_0 = z - \mu\lambda^\ell,$$

$$h_{\ell+2}(z) = h_{\ell+1} - \mu\lambda^\ell\,h_1 = z - 2\mu\lambda^\ell,$$

$$h_{\ell+3}(z) = h_{\ell+2} - \mu\lambda^\ell\,h_2 = z - 3\mu\lambda^\ell,$$

$$\vdots$$

$$h_{2\ell+1}(z) = h_{2\ell} - \mu\lambda^\ell\,h_\ell = z - (\ell+1)\mu\lambda^\ell,$$

hence

$$x_{\ell+k,1} = k\,\mu\lambda^\ell > 0, \quad k = 1,2,\ldots,\ell+1.$$

Observe that if $d-1=0$ $h_\ell = 1 > 0$. This implies that the two statements are trivially true

$$0 < x_{\ell+1,1} < x_{\ell+2,1} < \cdots < x_{2\ell+1,1}.$$

We assume that the three properties (1), (2), and (3) when $d \geq 1$ are true. Consider $n = d(\ell+1)+\ell$. Combining the recurrence relations in (3.2) for each $k = 1,2,\ldots,\ell+1$ we have that

$$h_{n+k}(z) = zh_n(z) - \lambda\mu^\ell\left(h_{n-\ell}(z) + h_{n-\ell+1}(z) + \cdots + h_{n-\ell+k-1}(z)\right). \tag{3.3}$$

From assumption we have that the polynomial

$$\mu\lambda^\ell(-1)^d\left(h_{n-\ell}(z) + h_{n-\ell+1}(z) + \cdots + h_{n-\ell+k-1}(z)\right) > 0,$$

hence the equation (3.3) yields

$$h_{n+k}(0) = -\mu\lambda^\ell\left(h_{n-\ell} + h_{n-\ell+1} + \cdots + h_{n-\ell+k-1}\right)(0) < 0.$$

The property (1) is already proved. Let us show (2) and (3). Note that

$$h_{n+k}(x_{n,j}) = -\mu\lambda^\ell\left(h_{n-\ell} + h_{n-\ell+1} + \cdots + h_{n-\ell+k-1}\right)(x_{n,j}), \quad j = 1,2,\ldots,d.$$

Then the polynomial h_{n+k} also changes sign at a point between two consecutive zeros of h_n. Observe that all the polynomials h_n are monic. Let us take into account that

$$\deg h_{n+k} = \deg \mu\lambda^\ell\left(h_{n-\ell} + h_{n-\ell+1} + \cdots + h_{n-\ell+k-1}\right) + 2.$$

This implies that

$$\lim_{z\to-\infty} h_{n+k}(z) = \lim_{z\to-\infty} \lambda\mu^{\ell}\left(h_{n-\ell}+h_{n-\ell+1}+\cdots+h_{n-\ell+k-1}\right)(z) = \pm\infty.$$

Assuming that the polynomial $\mu\lambda^{\ell}\left(h_{n-\ell}+h_{n-\ell+1}+\cdots+h_{n-\ell+k-1}\right)(z)$ does not vanish before $x_{n,1}$, and taking into account that the polynomials $h_{n+k}(x_{n,1})$ and $\mu\lambda^{\ell}\left(h_{n-\ell}+h_{n-\ell+1}+\cdots+h_{n-\ell+k-1}\right)(x_{n,1})$

$$h_{n+k}(x_{n,1}) = -\mu\lambda^{\ell}\left(h_{n-\ell}+h_{n-\ell+1}+\cdots+h_{n-\ell+k-1}\right)(x_{n,1}),$$

have different sing, hence $h_{n,k}$ changes sing once before $x_{n,1}$. We have proved that given $k\in\{2,\ldots,\ell+1\}$, $x_{n,k-1}<x_{n+1,k}<x_{n,k}$ and $x_{n+1,1}<x_{n,1}$.

In order to complete the proof we only need to show that given $k\in\{1,\ldots,\ell\}$, $x_{n+k,j}<x_{n+k+1,j}$. Consider again the recurrence relations in (3.2). We have that

$$h_{n+k+1}(z) = h_{n+k}(z) - \mu\lambda^{\ell} h_{n-\ell+k}(z).$$

Let us recall the assumption that the polynomial $h_{n-\ell}$ changes sign between two consecutive zeros of h_{n+k}, then h_{n+k+1} also interlace zeros with h_{n+k}. Since $h_{n-\ell+k}$ does not vanishes after $x_{n+k,d}$ and taking into account that both h_{n+k} and $h_{n-\ell+k}$ are monic polynomials and since they have different sign at $x_{n+k,d}$, we have that h_{n+k+1} vanishes ones after $x_{n+k,d}$. Considering (1), we conclude that the smallest zeros must be positive. This completes the proof. □

Combining Lemma 2.3 and Lemma 3.1 we have that given any $n=d(\ell+1)+r$ its corresponding polynomial $T_n(z)=z^r h_n(z^{\ell+1})$ has a zero of order r at the origin and $d(\ell+1)$ simple roots shared in a starlike set Σ with

$$\Sigma \subset \bigcup_{k=0}^{\ell}\left[0, \left(1+\mu\lambda^{\ell}\right)^{1/\ell+1}\exp\left(\frac{2\pi i k}{\ell+1}\right)\right].$$

There are d different real zeros

$$x_{n,j}\in\left[0,1+\mu\lambda^{\ell}\right],\quad j=1,2,\ldots,d,$$

and for each $k\in\{1,\ldots,\ell\}$, T_n vanishes at

$$z_{n,j,k} = x_{n,j}^{1/(\ell+1)}\exp\left(\frac{2\pi i k}{\ell+1}\right),\quad j=1,\ldots,d,\quad k=0,1,\ldots,\ell.$$

Studying the strong asymptomatic behavior of the polynomials T_n we prove that the set

$$S_0 = \bigcup_{k=0}^{\ell}\left[0, a\exp\frac{2\pi i k}{\ell+1}\right],\quad a=\frac{\ell+1}{\ell}\sqrt[\ell+1]{\ell\,\mu\,\lambda^{\ell}}. \tag{3.4}$$

attracts the points $z_{n,j,k}$, $j=1,2,\ldots,d$, $k=0,1,\ldots,\ell$, $n\in\mathbb{Z}_+$.

In order to analyze the strong asymptotic behavior of the sequence $\{T_n\}_{n\in\mathbb{Z}_+}$ we need to give some properties of the certain solutions $\omega_0,\ldots,\omega_\ell$ corresponding to the algebraic equation

$$\omega^{\ell+1} - z\omega^{\ell} + \mu\lambda^{\ell} = 0. \tag{3.5}$$

This equation is well studied in [**6**]. Since the roots of T_n belong to a starlike set Σ, we choose the branch for $\omega_0,\ldots,\omega_m$ whose cuts are as small as possible starlike sets. Let us declare the following starlike sets that we add to the one defined in (3.4)

$$S_0 = \bigcup_{k=0}^{\ell} \left[0,\, a \exp \frac{2\pi i k}{\ell+1}\right],$$

$$(3.6) \qquad S_e = \bigcup_{k=0}^{\ell} \mathcal{L}\left(0,\, \frac{(2k+1)\pi}{\ell+1}\right),$$

$$S_o = \bigcup_{k=0}^{\ell} \mathcal{L}\left(0,\, \frac{2k\pi}{\ell+1}\right).$$

We denote

$$\Omega_0 = \mathbb{C} \setminus S_0, \quad \Omega_1 = \mathbb{C} \setminus (S_0 \cup S_0), \quad \Omega_j = \mathbb{C} \setminus (S_o \cup S_e), \quad 1 < j < \ell$$

and

$$\Omega_\ell = \begin{cases} \mathbb{C} \setminus S_e & \text{if } \ell \text{ even,} \\ \mathbb{C} \setminus S_o & \text{if } \ell \text{ odd.} \end{cases}$$

For each $j \in \{0, 1, \dots, \ell\}$, ω_j we choose $\omega_j \in \mathcal{H}(\Omega_j)$ and

$$(3.7) \qquad \omega_0(z) = z + \mathcal{O}(1) \quad \text{and} \quad \omega_j(z) = \mathcal{O}\left(\frac{1}{z^{1/m}}\right), \quad \text{as} \quad z \to \infty.$$

From [**6**, Proposition 1] we have that

$$(3.8) \qquad |\omega_0| \geq |\omega_1| \geq \cdots \geq |\omega_\ell|, \quad \text{on } \mathbb{C}.$$

In Ω_0 the inequalities are strict.

REMARK 3.2. Observe that after the transformation

$$\omega(z) = \lambda \zeta(z - \mu - \lambda)$$

we obtain (3.5) from (1.9).

We are now ready to state an expression for the strong asymptotic behavior of the family of polynomials T_n, $n \in \mathbb{Z}_+$.

LEMMA 3.3. *Let $\{T_n\}_{n\in\mathbb{Z}_+}$ be the sequence of polynomials defined by the recurrence relation* (2.4) *with initial conditions* (2.5)*. The following equality holds uniformly for any compact set $K \subset \Omega_0$*

$$(3.9) \qquad \lim_{n\to\infty} \frac{T_n(z)}{\omega_0(z)} = \frac{1}{1 - \frac{\ell \lambda^\ell \mu}{\omega_0^{\ell+1}(z)}}.$$

Consequently for any open set D that contains S_0 there is a number $N \in \mathbb{Z}_+$ such that for all $n \geq N$ the polynomial T_n has its roots in D.

PROOF. From Lemma 3.1 we conclude that the zeros of T_n lie in a starlike set that contains S_0. Actually, if we consider the equality in (3.9), the boundary of D could be as close to S_0 as we want, hence S_0 attracts the roots of the polynomials T_n, $n \in \mathbb{Z}_+$. This means that $S_0 = \sigma_{\mathbf{T}}$.

For each $n \in \mathbb{Z}_+$ we write T_n as follows

$$T_n(z) = \left(a_0\omega_0^{n+\ell} + a_1\omega_1^{n+\ell} + \cdots + a_\ell\omega_\ell^{n+\ell}\right)(z), \quad z \in \bigcap_{j=0}^{\ell} \Omega_j,$$

whose coefficients satisfy the initial conditions

$$a_0\omega_0^n + a_1\omega_1^n + \cdots + a_\ell\omega_\ell^n = \delta_{n,\ell}, \quad n \in \{0,1,2,\ldots,\ell\}.$$

Observe that these expressions above conform a system of linear equations whose unknown are the coefficients a_j, $j \in \{0,1,\ldots,\ell\}$. Its corresponding matrix equation is $\mathbf{M}\mathbf{x} = \mathbf{b}$ with

$$\mathbf{M} = \begin{pmatrix} 1 & 1 & \cdots & 1 \\ \omega_0 & \omega_1 & \cdots & \omega_\ell \\ \vdots & \vdots & \ddots & \vdots \\ \omega_0^\ell & \omega_1^\ell & \cdots & \omega_\ell^\ell \end{pmatrix}, \quad \mathbf{x} = \begin{pmatrix} a_0 \\ a_1 \\ \vdots \\ a_\ell \end{pmatrix}, \quad \text{and} \quad \mathbf{b} = \begin{pmatrix} 0 \\ \vdots \\ 0 \\ 1 \end{pmatrix}.$$

Using Cramer's rule we obtain that for each $j \in \{0,1,\ldots,\ell\}$

$$a_j = \frac{(-1)^j}{\prod_{k\neq j}(\omega_j - \omega_k)}. \tag{3.10}$$

Observe that

$$P(\zeta) = \prod_{k=1}^{\ell}(\zeta - \omega_k) = \zeta^{\ell+1} - z\zeta^\ell + \lambda^\ell\mu \quad P(\omega_j) = 0.$$

Taking its derivative we obtain that

$$P'(\zeta) = (\ell+1)\zeta^\ell - z\ell\zeta^{\ell-1} = \frac{\ell}{\zeta}\left(\zeta^{\ell+1} - z\zeta^\ell\right) + \zeta^\ell.$$

Hence

$$P'(\omega_j) = \omega_j^\ell - \frac{\ell\lambda^\ell\mu}{\omega_j} = \frac{\omega_j^{\ell+1} - \ell\lambda^\ell\mu}{\omega_j} = \prod_{k\neq j}^{\ell}(\omega_j - \omega_k).$$

We deduce

$$a_j = \frac{(-1)^j\omega_j}{\omega_j^{\ell+1} - m\lambda^\ell\mu} = \left(\frac{(-1)^j}{1 - \frac{\ell\lambda^\ell\mu}{\omega_j^{\ell+1}}}\right)\frac{1}{\omega_j^\ell}, \qquad j \in \{0,1,\ldots,\ell\},$$

which is the denominator in (3.10). This implies that

$$T_n(z) = \sum_{j=0}^{\ell} \frac{(-1)^j\omega_j^n(z)}{1 - \frac{\ell\lambda^\ell\mu}{\omega_j^{\ell+1}(z)}}, \quad z \in \bigcap_{j=0}^{\ell}\Omega_j,$$

and

$$\frac{T_n(z)}{\omega_0^n(z)} = \frac{1}{1 - \frac{\ell\lambda^\ell\mu}{\omega_0^{\ell+1}(z)}} + \sum_{j=1}^{\ell} \frac{(-1)^j}{1 - \frac{\ell\lambda^\ell\mu}{\omega_j^{\ell+1}(z)}} \frac{\omega_j^n(z)}{\omega_0^n(z)}, \quad z \in \bigcap_{j=0}^{\ell}\Omega_j. \tag{3.11}$$

From the inequality (3.8), for each $j \in \{1, \dots, \ell\}$, we can extend the equality

$$(3.12) \qquad \lim_{n\to\infty} \left(\frac{\omega_j}{\omega_0}\right)^n = 0, \quad \text{uniformly on any compact set} \quad K \subset \Omega_0.$$

Combining (3.11) and (3.12) we arrive at (3.9). Note that $\dfrac{1}{1-\frac{\ell\lambda^\ell\mu}{\omega_0^{\ell+1}}} = \dfrac{\omega_0^{\ell+1}}{\omega_0^{\ell+1}-\ell\mu\lambda^\ell}$ never vanishes on Ω_0. We now observe that this function is also holomorphic on Ω_0. The denominator only becomes zero when $\omega_o^{\ell+1}(z) = \ell\mu\lambda^\ell$, and this equality occurs at the endpoints of S_0. Hence the limit (3.9) implies that the starlike set S_0 attracts the roots of the polynomials T_n. This completes the proof. □

Let us obtain integral expressions for the functions $\dfrac{1}{\omega_0^j}$, $j = 1, 2, \dots, \ell$.

LEMMA 3.4. *There exists a function $\widetilde{\rho}$ defined on S_0 such that*

$$(3.13) \qquad \widetilde{\rho}(x) = \exp\left(-\frac{2\pi i k}{\ell+1}\right)\widetilde{\rho}(|x|), \quad x \in \left(0, a\exp\frac{2\pi i k}{\ell+1}\right),$$

$k = 0, 1, \dots, \ell$, where $\widetilde{\rho}(x)$ is a weight in $x \in (0, a)$), that satisfies

$$(3.14) \qquad \frac{1}{\omega_0(z)} = \frac{1}{2\pi i}\int \widetilde{\rho}(x)\frac{dx}{z-x} = \frac{1}{2\pi i}\int \widetilde{\rho}(|x|)\frac{d|x|}{z-x}, \quad z \in \overline{\mathbb{C}} \setminus S_0.$$

PROOF. Let us denote

$$\omega_{0\pm}(x) = \lim_{z\to x \in \overset{\circ}{S}_{0\pm}} \omega_0(z) \quad \text{with} \quad \overset{\circ}{S}_0 = \bigcup_{k=0}^{\ell}\left(0, a\exp\frac{2\pi i k}{\ell+1}\right).$$

Consider an arbitrary point $x \in (0, a) \subset S_0$ in the star's arm corresponding to $k = 0$. Using (3.8) we observe that the following functions defined on $\overset{\circ}{S}_0$, are real valued and never change sign

$$\left[\frac{1}{\omega_{0+}(x)} - \frac{1}{\omega_{0-}(x)}\right] = \frac{1}{\omega_0(x)} + \frac{1}{\omega_1(x)} = \rho(x), \quad x \in (0, a).$$

By symmetry we obtain that

$$(3.15) \qquad \widetilde{\rho}(x) = \exp\left(-\frac{2\pi i k}{\ell+1}\right)\widetilde{\rho}(|x|) = \left[\frac{1}{\omega_{0+}(x)} - \frac{1}{\omega_{0-}(x)}\right],$$

for all

$$x \in \left(0, a\exp\frac{2\pi i k}{m+1}\right), \quad k \in \{0, 1, \dots, \ell\}.$$

This proves that condition (3.13) holds. Since $\omega_0 \in \mathcal{H}(\Omega_0)$ never vanishes in Ω_0 and taking into account (3.7) we have that

$$(3.16) \qquad \frac{1}{\omega_0(z)} \in \mathcal{H}(\Omega_0) \quad \text{and} \quad \frac{1}{\omega_0(z)} = \mathcal{O}\left(\frac{1}{z}\right) \quad \text{as} \quad z \to \infty.$$

Let us recall (3.15) and the Sokhotski–Plemelj formula (see [**16**, Chapter I, equality (4.8)]. According to [**15**] the function $1/\omega_0$ behaves properly at the endpoints to apply the Sokhotski–Plemelj formula) to obtain that

$$\frac{1}{\omega_0(z)} = \frac{1}{2\pi i}\int \widetilde{\rho}(x)\frac{dx}{z-x}.$$

We have already proved (3.14). □

We are ready to give integral expressions for the functions $\left(\frac{1}{\omega_0}\right)^j$, $j = 1, \ldots, \ell$.

PROPOSITION 3.5. *There exists a system of weights $r_j(x) = x^{j-1}\widetilde{\rho}_j(x)$, $j = 1, \ldots, \ell$, defined on S_0 that satisfy the following identities*

$$\frac{1}{\omega_0^j(z)} = \frac{1}{z^{j-1}} \int x^{j-1} \widetilde{\rho}_j(x)\, \widetilde{\rho}(x) \frac{dx}{z-x}, \quad j = 1, \ldots, m, \quad z \in \overline{\mathbb{C}} \setminus S_0, \tag{3.17}$$

where $\widetilde{\rho}_1 \equiv 1$ and $\widetilde{\rho}$ is defined in Lemma 3.4.

PROOF. The case $j = 1$ in (3.17) is already proved in Lemma 3.4. For each $j \in \{2, \ldots, \ell\}$, we have that

$$\frac{1}{\omega_{0+}^j(x)} - \frac{1}{\omega_{0-}^j(x)} = \widetilde{\rho}(x) \sum_{k=0}^{j-1} \frac{1}{\omega_{0+}^{j-1-k}(x)\, \omega_{0-}^k(x)}, \quad x \in S_0.$$

Observe that the following functions are weights

$$r_j(x) = x^{j-1}\widetilde{\rho}_j(x) = \sum_{k=0}^{j-1} \frac{x^{j-1}}{\omega_{0+}^{j-1-k}(x)\, \omega_{0-}^k(x)} \in \mathbb{R}_+, \quad x \in S_0, \quad j = 1, \ldots, \ell.$$

From (3.16) we obtain that

$$\frac{z^{j-1}}{\omega_0^j(z)} \in \mathcal{H}(\Omega_0) \quad \text{and} \quad \frac{z^{j-1}}{\omega_0^j(z)} = \mathcal{O}\left(\frac{1}{z}\right) \quad \text{as} \quad z \to \infty, \quad j = 2, \ldots, \ell. \tag{3.18}$$

The equalities in (3.17) are proved combining (3.18) and Sokhotski–Plemelj ([**16**, Chapter I, equality (4.8)]). The proof is completed. □

PROPOSITION 3.6. *Let $\widetilde{\rho}, \widetilde{\rho}_1, \ldots, \widetilde{\rho}_\ell$ be the system of functions in Proposition 3.5. Then*

$$0 = \int x^\nu \rho_j(x)\, \rho(x)\, dx, \quad \nu = 0, \ldots, j-2, \quad j = 2, \ldots, \ell.$$

PROOF. From construction we have that

$$\widetilde{\rho}_j(x) = \sum_{k=0}^{j-1} \frac{1}{\omega_{0_+}^{j-1-1-k}(x)\omega_{0-}^k(x)}, \quad x \in S_0.$$

This means that for each $x \in \mathbb{R} \cap S_0$ we have that for each $\nu \in \{0, \ldots, j-1\}$ and $k \in \{0, 1, \ldots, \ell\}$

$$x^\nu \exp \frac{2\pi i k \nu}{m+1} \widetilde{\rho}_j \left(x \exp \frac{2\pi i k}{m+1}\right) = \exp\left(\frac{2\pi i k(\nu - j + 1)}{m+1}\right) x^\nu \widetilde{\rho}_j(x),$$

and

$$\int_{S_0} x^\nu \widetilde{\rho}_j(x)\, \widetilde{\rho}(x)\, dx = \sum_{k=0}^{m} \left(\exp \frac{-2\pi i(j-1-\nu)}{m+1}\right)^k \int_{S_0 \cap \mathbb{R}} x^\nu \widetilde{\rho}_j(x)\, \widetilde{\rho}(x)\, dx.$$

The proof is completed after realization that

$$\sum_{k=0}^{m} \left(\exp \frac{-2\pi i(j-1-\nu)}{m+1}\right)^k = 0, \quad \nu \in \{0, \ldots, j-2\}.$$

□

4. Proofs of Theorem 1.1, Theorem 1.2, and Theorem 1.3

Fix $j \in \{0, 1, \dots, \ell\}$ and consider the second kind of polynomials $\{L_{n,j}\}_{n\in\mathbb{Z}_+}$ defined in (2.12).

LEMMA 4.1. *Consider the sequence of polynomials $\{L_{n,j}\}_{n\in\mathbb{Z}_+}$. The following equality holds uniformly for any compact set*

$$K \subset \Omega_0 \setminus \bigcup_{k=1}^{\ell-j} \left\{ (-\lambda)^{\ell} \mu^{\ell+1} \exp\left(\frac{2\,\pi\, i(1-j)k}{\ell-j+1}\right) - \frac{1}{\mu}\exp\left(\frac{2\,\pi\, i\, k}{\ell-j+1}\right) \right\}$$

$$\lim_{n\to\infty} \frac{L_{n,j}(z)}{\omega_0^{n-j}(z)} = \frac{1}{1-\frac{\ell\lambda^{\ell}\mu}{\omega_0^{\ell+1}(z)}} \, \frac{1-(-\mu\,\omega_0(z))^{\ell-j+1}}{1+\mu\,\omega_0(z)}. \tag{4.1}$$

Consequently for any open set O that contains

$$\left(\bigcup_{k=1}^{\ell-j} \left\{ (-\lambda)^{\ell} \mu^{\ell+1} \exp\left(\frac{2\,\pi\, i(1-j)k}{\ell-j+1}\right) - \frac{1}{\mu}\exp\left(\frac{2\,\pi\, i\, k}{\ell-j+1}\right) \right\} \right) \bigcup S_0$$

there is a number $N \in \mathbb{Z}_+$ such that for all $n \geq N$ the polynomial $L_{n,j}$ has its roots in O. Even more, fixed $k \in \{1, \dots, \ell - j\}$, there is a domain O_k that contains the point

$$z_{k,j} = (-\lambda)^{\ell} \mu^{\ell+1} \exp\left(\frac{2\,\pi\, i(1-j)k}{\ell-j+1}\right) - \frac{1}{\mu}\exp\left(\frac{2\,\pi\, i\, k}{\ell-j+1}\right),$$

where for any open $A \subset O_k$ with $z_k \in A$, there is a large enough $N \in \mathbb{N}$ such that each polynomial $L_{n,j}$, $n \geq N$ vanishes in A once.

PROOF. Let us write its nth element as linear combination of T_k, $k \leq n$. Observe that any arbitrary sequence of polynomials $\{H_n\}_{n\in\mathbb{Z}_+}$ whose elements have form

$$H_n = T_{n-j} + b_{\ell-j-1}T_{n-j-1} + \cdots + b_0 T_{n-\ell}, \quad b_k \in \mathbb{R}, \quad k \in \{0, 1, \dots, \ell-j-1\},$$

satisfy the same recurrence relation as the elements of $\{L_{n,j}\}_{n\in\mathbb{Z}_+}$ hold in (2.16) of Lemma 2.7. If we requiere of H_n to satisfies the $L_{n,j}$'s initial conditions in (2.17), we obtain that $L_{n,j} = H_n$, $n \in \mathbb{Z}_+$. Let us find the proper coefficients $b_k \in \mathbb{R}$, $k = 0, 1, \dots, \ell - j$, to have:

$$L_{n,j}(x) = T_{n-j}(x) + \sum_{k=1}^{\ell-j} b_{\ell-j-k} T_{n-j-k}(x) = (x-\mu)^{n-j}, \quad n \in \{j, \dots, \ell\}.$$

Since the coefficients b_j are constant we fix $x = 0$ to obtain the following linear system of equations

$$T_{n-j}(0) + \sum_{k=1}^{n-j} b_{\ell-j-k} T_{n-j-k}(0) = (-\mu)^{n-j}, \quad n = j, \dots, \ell.$$

Taking into account that $T_k = x^k$, $j \in \{0, 1, \dots, \ell - j\}$, we have that

$$b_{\ell-n} = (-\mu)^{n-j}, \quad n = 0, \dots, \ell - 1.$$

Then

$$L_{n,j}(x) = T_{n-j}(x) + \sum_{j=1}^{\ell-j} (-\mu)^k \, T_{n-j-k}.$$

Substituting (3.11), we have for any compact set $K \subset D_0$ with

$$\lim_{n\to\infty} \frac{L_{n,j}(z)}{\omega_0^{n-j}(z)} = \lim_{n\to\infty} \left(\frac{T_{n-j}(z)}{\omega_0^{n-j}(z)} + \sum_{k=1}^{\ell-j} (-\mu\,\omega_0(z))^k \frac{T_{n-j-k}(z)}{\omega_0^{n-j-k}(z)} \right)$$

$$\lim_{n\to\infty} \frac{L_{n,j}(z)}{\omega_0^{n-j}(z)} = \frac{1}{1-\frac{\ell\lambda^\ell\mu}{\omega_0^{\ell+1}(z)}} \left(1 + \sum_{k=1}^{\ell-j} (-\mu\,\omega_0(z))^k \right)$$

$$= \frac{1}{1-\frac{\ell\lambda^\ell\mu}{\omega_0^{\ell+1}(\zeta)}} \frac{1-(-\mu\,\omega_0(z))^{\ell-j+1}}{1+\mu\,\omega_0(z)}.$$

This implies that given an arbitrary $\varepsilon > 0$ there is a large enough $N \in \mathbb{N}$ such that for all $n \geq N$, the polynomial $L_{n,j}$ has $n-\ell+j$ zeros $\zeta_{n,j,k}$, $k = n-\ell+j, \dots, n$ whose distance to S_0 is smaller than ε, (simbollicaly: $\operatorname{dist}(z_{n,j,k}, S_0) < \varepsilon$), and its $z_{n,j,k}$ $k \in \{1, \dots, \ell - j\}$ remainder roots satisfy that

$$\left| \mu\,\omega_0(z_{n,j,k}) + \exp\left(\frac{2\pi k\, i}{\ell - j + 1}\right) \right| < \varepsilon, \quad k = 1, \dots, \ell - j.$$

We complete the proof after combining the relations above and the algebraic equation in (3.5) with ω_0 being a solution function. □

4.1. Proof of of Theorem 1.1. Observe that Lemma 4.1 in the case $j = 0$ is reduced to

$$\lim_{n\to\infty} \frac{L_n(z)}{\omega_0^n(z)} = \lim_{n\to\infty} \frac{L_{n,0}(z)}{\omega_0^n(z)} = \frac{1}{1-\frac{\ell\lambda^\ell\mu}{\omega_0^{\ell+1}(z)}} \frac{1-(-\mu\,\omega_0(z))^{\ell+1}}{1+\mu\,\omega_0(z)}.$$

holds uniformly on every compact set

$$K \subset \Omega_0 \setminus \bigcup_{k=1}^{\ell} \left\{ \left[(-\lambda)^\ell \mu^{\ell+1} - \frac{1}{\mu} \right] \exp\left(\frac{2\,\pi\, i\, k}{\ell+1} \right) \right\},$$

and the set

$$\left(\bigcup_{k=1}^{\ell} \left\{ \left[(-\lambda)^\ell \mu^{\ell+1} - \frac{1}{\mu} \right] \exp\left(\frac{2\,\pi\, i\, k}{\ell+1} \right) \right\} \right) \bigcup S_0$$

attracts the L_n's roots such that for each $k \in \{1, \dots, \ell\}$ an only zero goes to

$$z_k = z_{k,0} = \left[(-\lambda)^\ell \mu^{\ell+1} - \frac{1}{\mu} \right] \exp\left(\frac{2\,\pi\, i\, k}{\ell+1} \right). \tag{4.2}$$

Then considering the relation in (2.1) and Remark 3.2 we complete the proof. □.

4.2. Proof of Theorem 1.2. First we study the Hermite-Padé approximants as in (2.18).

LEMMA 4.2. *Fix $j \in \{1, 2, \dots, \ell\}$. Given an arbitrary compact set*

$$K \subset \Omega_0 \setminus \bigcup_{k=1}^{\ell} \left\{ \left[(-\lambda)^\ell \mu^{\ell+1} - \frac{1}{\mu} \right] \exp\left(\frac{2\,\pi\, i\, k}{\ell+1} \right) \right\},$$

the following equality holds uniformly

$$\lim_{n\to\infty} \frac{L_{n,j}(z)}{L_n(z)} = \frac{1}{\omega_0^j(z)} + \frac{(-\mu)^j \left[1 + (-\lambda)^\ell \ell \mu^{\ell+2}\right]}{\ell+1} \sum_{k=1}^{\ell} \frac{1 - \exp\frac{2\pi i k j}{\ell+1}}{\exp\frac{2\pi i k(2j-1)}{\ell+1}} \frac{1}{z - z_k}, \tag{4.3}$$

where z_k are given in (4.2)*. This have an integral expression*

$$\lim_{n\to\infty}\frac{L_{n,j}(z)}{L_n(z)}=\int\frac{d\widetilde{\sigma}_j(x)}{z-x}$$

$$(4.4)\quad =\int\left(\widetilde{\rho}_j(x)\,\widetilde{\rho}(x)+\frac{(-\mu)^j\left[1+(-\lambda)^\ell\ell\mu^{\ell+2}\right]}{\ell+1}\sum_{k=1}^{\ell}\frac{1-\exp\frac{2\pi ikj}{\ell+1}}{\exp\frac{2\pi ik(2j-1)}{\ell+1}}\delta_{z_k}\right)\frac{dx}{z-x}.$$

PROOF. From (4.1) in Lemma 4.1, and using the method of partial fraction decomposition, we observe that

$$\lim_{n\to\infty}\frac{L_{n,j}(z)}{L_n(z)}=\lim_{n\to\infty}\frac{L_{n,j}(z)}{\omega_0^{n-j}(z)}\frac{\omega_0^n}{L_n(z)}\frac{1}{\omega_0^j(z)}=\frac{1}{\omega_0^j(z)}\frac{1-(-\mu\,\omega_0(z))^{\ell-j+1}}{1-(-\mu\,\omega_0(z))^{\ell+1}}$$

$$=\frac{1}{\omega_0^j(z)}+\frac{(-\mu)^j}{\ell+1}\sum_{k=1}^{\ell}\frac{1-\exp\frac{2\pi ikj}{\ell+1}}{\exp\frac{2\pi ik(2j-1)}{\ell+1}}\frac{1}{\exp\frac{2\pi ik}{\ell+1}+\mu\,\omega_0(z)}.$$

The summation conforms a rational functions whose poles are simple and lie where $\omega_0(z_k)=-\dfrac{1}{\mu}\exp\dfrac{2\pi ik}{\ell+1}$, $k=1,2,\ldots,\ell$, which are (4.2).

$$\frac{(-\mu)^j}{\ell+1}\sum_{k=1}^{\ell}\frac{1-\exp\frac{2\pi ikj}{\ell+1}}{\exp\frac{2\pi ik(2j-1)}{\ell+1}}\frac{1}{\exp\frac{2\pi ik}{\ell+1}+\mu\,\omega_0(z)}=\sum_{k=1}^{\ell}\frac{\lambda_k}{z-z_k},$$

where

$$\lambda_k=\frac{(-\mu)^j}{\ell+1}\frac{1-\exp\frac{2\pi ikj}{\ell+1}}{\exp\frac{2\pi ik(2j-1)}{\ell+1}}\frac{1}{\mu\,\omega_0'(z_k)},\qquad k-1,2,\ldots,\ell.$$

We use the equation (3.5) to arrive at (4.3). The integral expression (4.4) is deduced immediately from the combination of (4.3) and (3.17) in Proposition 3.5. This completes the proof. □

PROPOSITION 4.3. *Set the resolvent functions f_j, $j=1,\ldots,\ell$ as in* (2.9)*. Then for each $j\in\{1,\ldots,\ell\}$*

$$(4.5)\qquad f_j(z)=\int\frac{d\widetilde{\sigma}_j(x)}{z-x},\qquad j=1,2,\ldots,\ell$$

on

$$\Omega_0\setminus\bigcup_{k=1}^{\ell}\left\{\left[(-\lambda)^\ell\mu^{\ell+1}-\frac{1}{\mu}\right]\exp\left(\frac{2\,\pi\,i\,k}{\ell+1}\right)\right\}.$$

PROOF. From (4.4) we need to show that the quotients $L_{n,j}/L_n$ converge uniformly to f_j. Let us recall Lemma 2.6

$$(4.6)\qquad f_j(z)-\frac{L_{n,j}(z)}{L_n(z)}=\mathcal{O}\left(\frac{1}{z^{n+\tau(n,j)}}\right)\quad\text{as}\quad z\to\infty,$$

with

$$\tau(n,j)=\begin{cases}\kappa+\ell+j-s+1 & \text{if}\quad j\in\{1,\ldots,\ell\},\\[2ex] \kappa+j-s+1 & \text{if}\quad j\in\{s+1,\ldots,\ell\}.\end{cases}$$

According to Remark 4.1, given an arbitrary open set O containing

$$\Omega_0 \setminus \bigcup_{k=1}^{\ell} \left\{ \left[(-\lambda)^{\ell} \mu^{\ell+1} - \frac{1}{\mu} \right] \exp\left(\frac{2\pi i k}{\ell+1} \right) \right\}$$

there exists $N \in \mathbb{Z}_+$ such that for each $n \geq N$

$$f_j(z) - \frac{L_{n,j}(z)}{L_n(z)} \in \mathcal{H}\left(\overline{\mathbb{C}} \setminus O\right)\right). \tag{4.7}$$

Combining (4.6) and (4.7) we obtain

$$\left(f_j(\cdot, \mathbf{L}) - \frac{L_{n,j}}{L_n} \right) \omega_0^{n+\tau(n,j)-\ell} \prod_{k=1}^{\ell} (z - z_k) \in \mathcal{H}\left(\overline{\mathbb{C}} \setminus O\right).$$

Set $\alpha < 1$ such that the level curves

$$\gamma_{0,\alpha} = \left\{ z \in \mathbb{C} \setminus O : \left| \frac{1}{\omega_0(z)} \right| = \alpha \right\} \subset \mathbb{C} \setminus S_0,$$

and

$$\gamma_{k,\alpha} = \{|z - z_k| = \alpha\} \subset \mathbb{C} \setminus \{z_k\}, \quad k = 1, 2, \ldots, \ell.$$

Set a compact set $K \subset \overline{\mathbb{C}} \setminus O$ lying in the unbounded connected open region whose boundary is $\bigcup_{k=0}^{\ell} \gamma_{k\alpha}$. Since $f_j(\cdot, \mathbf{L})$ is analytic on K and Lemma 4.2, then there exist two constant $C > 0$ and $C' > 0$ such that $|f_j(z, \mathbf{L})| \leq C$ and $|L_{n,j}(z)/L_n(z)| < C'$, $z \in \mathcal{K}$. Using the maximum principle for $1/\omega_0$ we obtain that

$$\left\| \left(f_j(z, \mathbf{T}) - \frac{L_{n,j}(z)}{L_n(z)} \right) \right\|_K \leq (C + C')\, \alpha^{n+\tau(n,j)-2\ell}.$$

Hence

$$\limsup_{n\to\infty} \left\| \left(f_j(z) - \frac{L_{n,j}(z)}{L_n(z)} \right) \right\|_{\mathcal{K}}^{1/n+\tau(n,j)-2\ell} = \alpha < 1.$$

This proves the equality (4.5) in $\overline{\mathbb{C}} \setminus O$. After taking $N \to \infty$ the proof is completed. □

PROPOSITION 4.4. *For each $j \in \{1, 2, \ldots, \ell\}$ the moments in* (2.11) *have integral forms*

$$c_{\nu,j} = (\mathbf{L})^{\nu} \mathbf{e}_{j-1} \cdot \mathbf{e}_0 = \int x^{\nu} d\tilde{\sigma}_j(x), \quad \nu \in \mathbb{Z}_+.$$

PROOF. Fix $j \in \{1, \ldots, \ell\}$. In fact using Neumann's series for operator inversion to $(z\mathbb{I} - \mathbf{L})^{-1}$ in the weak ℓ_2 topology, we have the equality

$$f_j(z) = \sum_{\nu=0}^{\infty} \frac{(\mathbf{L})^{\nu} \mathbf{e}_{j-1} \cdot \mathbf{e}_0}{z^{\nu+1}} = \sum_{\nu=0}^{\infty} \frac{c_{\nu,j}}{z^{\nu+1}}$$

in any compact set contained in the widest open disk of $\overline{\mathbb{C}}$ centered at $z = \infty$ contained in

$$\Omega_0 \setminus \bigcup_{k=1}^{\ell} \left\{ \left[(-\lambda)^{\ell} \mu^{\ell+1} - \frac{1}{\mu} \right] \exp\left(\frac{2\pi i k}{\ell+1} \right) \right\}.$$

Hence

$$f_j(z) = \sum_{\nu=0}^{\infty} \frac{c_{\nu,j}}{z^{\nu+1}} = \sum_{\nu=0}^{\infty} \frac{1}{z^{\nu+1}} \int x^{\nu}\, d\widetilde{\sigma}_j(x)$$

uniformly. This completes the proof. □

In ℓ_2 we consider its regular inner product. This means that given two sequences $\mathbf{a} = \{a_j \in \mathbb{R}\}_{j\in\mathbb{Z}_+}$ and $\mathbf{b} = \{b_j \in \mathbb{R}\}_{j\in\mathbb{Z}_+}$ of ℓ^2

$$\mathbf{a}\cdot\mathbf{b} = \sum_{j=0}^{\infty} a_j b_j = \mathbf{a}^{\top}\mathbf{b}.$$

We think the sequences as infinite column vectors. Set an arbitrary vector polynomial $\mathbf{p} = (p_0, \dots, p_{\ell-1})$, let us denote for every system $\mathbf{d} = (\mathbf{d}_0, \dots, \mathbf{d}_{\ell-1})$ with $\mathbf{d}_j \in \ell_2$, $j = 0, \dots, \ell-1$ the "dot" product

$$\mathbf{p}(\mathbf{L})\cdot\mathbf{d} = \sum_{j=0}^{\ell-1} p_j(\mathbf{L})\mathbf{d}_j.$$

Observe that this vector gives a vector. Let $\{\mathbf{u}_m = (u_{m,0}, \dots, u_{m,\ell-1})\}_{m\in\mathbb{Z}_+}$ be sequence of vector polynomials which is dual with respect to $\{L_n\}_{n\in\mathbb{Z}_+}$.

$$\mathbf{u}_m = \begin{cases} \mathbf{q}_m & \text{if} \quad m \in \{-1, 0, 1, \dots, \ell-1\}, \\ \dfrac{1}{\lambda^m}\mathbf{q}_m(z-\lambda-\mu) & \text{if} \quad m \geq \ell, \end{cases} \tag{4.8}$$

whose elements satisfy the recurrence relation

$$\begin{gathered} \mathbf{u}_{-1} = (0, \dots, 0)\,,\ \mathbf{u}_0 = (1, 0, \dots, 0)\,, \dots, \mathbf{u}_{\ell-1} = (0, \dots, 0, 1)\,, \\ \lambda^{\ell}\mu\,\mathbf{u}_m(z) = (z-\mu)\mathbf{u}_{m-\ell}(z) - \mathbf{u}_{m-\ell-1}(z), \quad m \in \{\ell, \dots, 2\ell-1\}, \\ \lambda^{\ell}\mu\,\mathbf{u}_m(z) = z\mathbf{u}_{m-\ell}(z) - \mathbf{u}_{m-\ell-1}(z), \quad m \geq 2\ell. \end{gathered} \tag{4.9}$$

LEMMA 4.5. *Let $\mathbf{E} = (\mathbf{e}_0, \dots, \mathbf{e}_{\ell-1})$ denote. Then*

$$\mathbf{e}_m = \mathbf{u}_m(\mathbf{L})\cdot\mathbf{E}, \quad m \in \mathbb{Z}_+ = \{0, 1, 2, \dots\}.$$

PROOF. We observe that for each $m \in \{0, 1, \dots, \ell-1\}$

$$\mathbf{u}_m(\mathbf{L})\cdot\mathbf{E} = (\overbrace{0, \dots, 0, 1}^{m+1}, 0, \dots, 0)\cdot(\mathbf{e}_0, \dots, \mathbf{e}_{\ell-1}) = \mathbf{e}_m.$$

Let us consider $2\ell - 1 \geq m \geq \ell - 1$. For each $k = 0, \dots, m-1$, let us assume that $\mathbf{u}_k(\mathbf{L})\cdot\mathbf{E} = \mathbf{e}_k$. By the recurrence relation (4.9), we have that

$$\begin{aligned} \mathbf{u}_m(\mathbf{L})\cdot\mathbf{E} &= \frac{1}{\mu\lambda^{\ell}}\left[(\mathbf{L}-\mu)\,\mathbf{u}_{m-\ell}(\mathbf{L})\cdot\mathbf{E} - \mathbf{u}_{m-\ell-1}(\mathbf{L})\cdot\mathbf{E}\right] \\ &= \frac{1}{\mu\lambda^{\ell}}\left[\mathbf{L}\mathbf{u}_{m-\ell}(\mathbf{L})\cdot\mathbf{E} - \mu\mathbf{u}_{m-\ell}(\mathbf{L})\cdot\mathbf{E} - \mathbf{u}_{m-\ell-1}(\mathbf{L})\cdot\mathbf{E}\right] \\ &= \frac{1}{\mu\lambda^{\ell}}\left[\mathbf{L}\mathbf{e}_{m-\ell} - \mu\mathbf{e}_{m-\ell} - \mathbf{e}_{m-\ell-1}\right] \\ &= \frac{1}{\mu\lambda^{\ell}}\left[\mu\lambda^{\ell}\mathbf{e}_m + \mu\mathbf{e}_{m-\ell} + \mathbf{e}_{m-\ell-1} - \mu\mathbf{e}_{m-\ell} - \mathbf{e}_{m-\ell-1}\right] = \mathbf{e}_m. \end{aligned}$$

We now consider $m \geq 2\ell - 1$

$$\mathbf{u}_m(\mathbf{L}) \cdot \mathbf{E} = \frac{1}{\mu\lambda^\ell}\left[\mathbf{L}\mathbf{u}_{m-\ell}(\mathbf{L}) \cdot \mathbf{E} - \mathbf{u}_{m-\ell-1}(\mathbf{L}) \cdot \mathbf{E}\right]$$

$$= \frac{1}{\mu\lambda^\ell}\left[\mathbf{L}\mathbf{e}_{m-\ell} - \mathbf{e}_{m-\ell-1}\right] = \frac{1}{\mu\lambda^\ell}\left[\mu\lambda^\ell\mathbf{e}_m + \mathbf{e}_{m-\ell-1} - \mathbf{e}_{m-\ell-1}\right] = \mathbf{e}_m.$$

The proof is completed by induction. □

PROPOSITION 4.6. *Set* $\mathbf{E} = (\mathbf{e}_0, \ldots, \mathbf{e}_{\ell-1})$. *Given an order pair* $(n, m) \in \mathbb{Z}_+^2$, *the polynomial* L_n *in* (2.2) *and* $\mathbf{u}_m$ *in* (4.9) *satisfy the orthogonality relations:*

$$(\mathbf{u}_m(\mathbf{L}) \cdot \mathbf{E}) \cdot L_n(\mathbf{L}^\top)\mathbf{e}_0 = \delta_{n,m}. \tag{4.10}$$

PROOF. From Lemma 2.1 and Lemma 4.5 we have immediately

$$(\mathbf{u}_m(\mathbf{L}) \cdot \mathbf{E}) \cdot L_n(\mathbf{L}^\top)\mathbf{e}_0 = \mathbf{e}_m \cdot \mathbf{e}_n = \delta_{r,n}.$$

This proves (4.10). □

Combining the identity (4.10) in Propositions 4.6 and 4.4 we obtain that

$$\delta_{n,m} = \int L_n(x) \sum_{j=0}^{\ell-1} u_{m,j}(x)\, d\widetilde{\sigma}_j(x).$$

The equality (1.10) can be deduced using the change of variable $(z - \mu - \lambda) \to z$ in the orthogonality relations above, and considering the equalities (2.1) and (4.8) This completes the proof of Theorem 1.2. □

4.3. Proof of Theorem 1.3. We use the orthogonality relations in (1.10). Let us start by showing that the functions $P_{n,m}$, $(n, m) \in \mathbb{Z}_+^2$ in (1.12) are entries of a solution of (1.2). The integral expression in (1.12) is reduced to (1.10) when $t = 0$, hence the initial condition of the problem (1.2) holds. When we take the derivative $\mathbf{P}' = [P'_{n,m}]$ we obtain:

$$P'_{n,m}(t) = \int e^{xt} x\, Q_n(x) \sum_{j=0}^{\ell-1} q_{j,m}(x)\, d\sigma_j(x).$$

Using the recurrence relations that define Q_n (1.4) with initial conditions (1.6) in Section 1, we arrive at

$$P'_{n,m}(t) = \begin{cases} -\lambda P_{n,m}(t) + \lambda P_{n+1,m}(t), & (n, m) \in \{0, 1, \ldots, \ell - 1\} \times \mathbb{Z}_+ = \mathbb{F}, \\ \\ -(\lambda + \mu)P_{n,m}(t) + \lambda P_{n+1,m}(t) + \mu P_{n-\ell,m}(t), & (n, \ell) \in \mathbb{Z}_+^2 \setminus \mathbb{F}. \end{cases}$$

This proves that $P_{n,m}$, $(n, m) \in \mathbb{Z}_+^2$ as in (1.12) are entries of a matrix $\mathbf{P}$ which satisfies (1.2).

We observe that $P_{n,m}$, $(n, m) \in \mathbb{Z}_+^2$ in (1.12) have analytic continuations to the whole complex plane $\mathbb{C}$, this implies they conform the unique transition probability function which is solution of the initial value problem (1.2). We now show that $P_{n,m}$ has persistent states when there is a value of $k \in \{1, \ldots \ell\}$ such that

$$\left[(-\lambda)^\ell \mu^{\ell+1} - \frac{1}{\mu}\right] \cos\left(\frac{2\,\pi\, k}{\ell + 1}\right) > \mu + \lambda.$$

This condition implies that there is at least a Riemann delta measure δ_{z_k} conforming the measures σ_j, $j = 1, \ldots, \ell$ which is supported on a point z_k with positive real part. By symmetry if z_k is complex it must be another $k' \in \{0, 1, \ldots, \ell\}$ such that the point $z_{k'}$ is complex conjugated with respect to z_k. For each $n \in \{0, \ldots, \ell-1\}$ we have that $Q_n(z_k) = \left(\frac{z_k + \lambda}{\lambda}\right)^n$ and $\mathbf{q}_n$ is the constant vector whose nth component is one and the others are zero. Then considering the kth term in the integrand we have that

$$\int_0^\infty P_{n,n}(t)dt \geq \left| \Re \frac{(-\mu)^j \left[1 + (-\lambda)^\ell \ell \mu^{\ell+2}\right]}{\ell+1} \frac{1 - \exp \frac{2\pi i k j}{\ell+1}}{\exp \frac{2\pi i k(2j-1)}{\ell+1}} \left(\frac{z_k + \lambda}{\lambda}\right)^n \int_0^\infty e^{z_k t} dt \right|.$$

The integral diverges. The factor $1 - \exp \frac{2\pi i k j}{\ell+1}$ only vanishes when $\ell + 1$ divides the product $k \cdot j$. However the singularity at z_k is contained in any measure σ_j, $j = \{1, 2, \ldots, \ell\}$. This completes the proof. □

References

[1] Hessah Alqahtani and Lothar Reichel, *Multiple orthogonal polynomials applied to matrix function evaluation*, BIT **58** (2018), no. 4, 835–849, DOI 10.1007/s10543-018-0709-x. MR3882973

[2] Carlos Álvarez-Fernández, Ulises Fidalgo, and Manuel Mañas, *The multicomponent 2D Toda hierarchy: generalized matrix orthogonal polynomials, multiple orthogonal polynomials and Riemann-Hilbert problems*, Inverse Problems **26** (2010), no. 5, 055009, 15, DOI 10.1088/0266-5611/26/5/055009. MR2647151

[3] Carlos Álvarez-Fernández, Ulises Fidalgo Prieto, and Manuel Mañas, *Multiple orthogonal polynomials of mixed type: Gauss-Borel factorization and the multi-component 2D Toda hierarchy*, Adv. Math. **227** (2011), no. 4, 1451–1525, DOI 10.1016/j.aim.2011.03.008. MR2799802

[4] William J. Anderson, *Continuous-time Markov chains*, Springer Series in Statistics: Probability and its Applications, Springer-Verlag, New York, 1991. An applications-oriented approach, DOI 10.1007/978-1-4612-3038-0. MR1118840

[5] A. I. Aptekarev, V. A. Kalyagin, and E. B. Saff, *Higher-order three-term recurrences and asymptotics of multiple orthogonal polynomials*, Constr. Approx. **30** (2009), no. 2, 175–223, DOI 10.1007/s00365-008-9032-0. MR2519660

[6] A. Aptekarev, V. Kaliaguine, and J. Van Iseghem, *The genetic sums' representation for the moments of a system of Stieltjes functions and its application*, Constr. Approx. **16** (2000), no. 4, 487–524, DOI 10.1007/s003650010004. MR1771693

[7] G. Avondo Bodino and F. Brambilla, *Teoria delle Code (Theory of Queues)*, Instituto Editoriale Cisalpino, Milan, Italy, (1959)

[8] Norman T. J. Bailey, *On queueing processes with bulk service*, J. Roy. Statist. Soc. Ser. B **16** (1954), 80–87. MR63595

[9] T. S. Chihara, *An introduction to orthogonal polynomials*, Mathematics and its Applications, Vol. 13, Gordon and Breach Science Publishers, New York-London-Paris, 1978. MR481884

[10] Theodore S. Chihara and Mourad E. H. Ismail, *Orthogonal polynomials suggested by a queueing model*, Adv. in Appl. Math. **3** (1982), no. 4, 441–462, DOI 10.1016/S0196-8858(82)80017-1. MR682630

[11] Lotfi Tadj, Gautam Choudhury, and Chakib Tadj, *A bulk quorum queueing system with a random setup time under N-policy and with Bernoulli vacation schedule*, Stochastics **78** (2006), no. 1, 1–11, DOI 10.1080/17442500500397574. MR2219709

[12] Steven Delvaux and Abey López, *High-order three-term recursions, Riemann-Hilbert minors and Nikishin systems on star-like sets*, Constr. Approx. **37** (2013), no. 3, 383–453, DOI 10.1007/s00365-013-9184-4. MR3054610

[13] William Feller, *An Introduction to Probability Theory and Its Applications. Vol. I*, John Wiley & Sons, Inc., New York, 1950. MR38583

[14] U. Fidalgo, S. Medina Peralta, and J. Mínguez Ceniceros, *Mixed type multiple orthogonal polynomials: perfectness and interlacing properties of zeros*, Linear Algebra Appl. **438** (2013), no. 3, 1229–1239, DOI 10.1016/j.laa.2012.08.034. MR2997807

[15] Otto Forster, *Lectures on Riemann surfaces*, Graduate Texts in Mathematics, vol. 81, Springer-Verlag, New York-Berlin, 1981. Translated from the German by Bruce Gilligan. MR648106

[16] F. D. Gakhov, *Boundary value problems*, Pergamon Press, Oxford; Addison-Wesley Publishing Co., Inc., Reading, Mass.-London, 1966. Translation edited by I. N. Sneddon. MR198152

[17] Mourad E. H. Ismail, *A queueing model and a set of orthogonal polynomials*, J. Math. Anal. Appl. **108** (1985), no. 2, 575–594, DOI 10.1016/0022-247X(85)90046-0. MR793667

[18] V. A. Kalyagin, *Hermite-Padé approximants and spectral analysis of nonsymmetric operators* (Russian, with Russian summary), Mat. Sb. **185** (1994), no. 6, 79–100, DOI 10.1070/SM1995v082n01ABEH003558; English transl., Russian Acad. Sci. Sb. Math. **82** (1995), no. 1, 199–216. MR1280397

[19] S. Karlin and J. L. McGregor, *The differential equations of birth-and-death processes, and the Stieltjes moment problem*, Trans. Amer. Math. Soc. **85** (1957), 489–546, DOI 10.2307/1992942. MR91566

[20] David G. Kendall, *Stochastic processes occurring in the theory of queues and their analysis by the method of the imbedded Markov chain*, Ann. Math. Statistics **24** (1953), 338–354, DOI 10.1214/aoms/1177728975. MR56231

[21] Abey López-García and Guillermo López Lagomasino, *Nikishin systems on star-like sets: ratio asymptotics of the associated multiple orthogonal polynomials*, J. Approx. Theory **225** (2018), 1–40, DOI 10.1016/j.jat.2017.10.002. MR3733249

[22] Abey López-García and Guillermo López Lagomasino, *Nikishin systems on star-like sets: ratio asymptotics of the associated multiple orthogonal polynomials, II*, J. Approx. Theory **250** (2020), 105320, 28, DOI 10.1016/j.jat.2019.105320. MR4034652

[23] A. Lopes-Garsia and È. Min′ya-Dias, *Nikishin systems on star-like sets: algebraic properties and weak asymptotics of the associated multiple orthogonal polynomials* (Russian, with Russian summary), Mat. Sb. **209** (2018), no. 7, 139–177, DOI 10.4213/sm8768; English transl., Sb. Math. **209** (2018), no. 7, 1051–1088. MR3833531

[24] J. Medhi, *Stochastic models in queueing theory*, 2nd ed., Academic Press, Amsterdam, 2003. MR1991930

[25] E. M. Nikishin and V. N. Sorokin, *Rational approximations and orthogonality*, Translations of Mathematical Monographs, vol. 92, American Mathematical Society, Providence, RI, 1991. Translated from the Russian by Ralph P. Boas, DOI 10.1090/mmono/092. MR1130396

[26] Allen C. Pipkin, *A course on integral equations*, Texts in Applied Mathematics, vol. 9, Springer-Verlag, New York, 1991, DOI 10.1007/978-1-4612-4446-2. MR1125074

[27] W. Sternberg, *Über systeme unendlich vieler geowöhnlicher linearer Differentialgleichungen*, Heidelberg Akademie Sitzungsberichte Vol. **10**, 1–21 (1920).

[28] Herbert Stahl and Vilmos Totik, *General orthogonal polynomials*, Encyclopedia of Mathematics and its Applications, vol. 43, Cambridge University Press, Cambridge, 1992, DOI 10.1017/CBO9780511759420. MR1163828

[29] Richard S. Varga, *Gershgorin and his circles, 2nd ed.*, Classics in Applied Mathematics, vol. 36, Society for Industrial and Applied Mathematics (SIAM), Philadelphia, PA, 2004.

Science and Mathematics Division, Lorain County Community College, Physical Sciences Building, 1005 N Abbe Rd, Elyria, Ohio 44035

Email address: `ufidalgo@loraincccc.edu`

Contemporary Mathematics
Volume **822**, 2025
https://doi.org/10.1090/conm/822/16475

Bordered and framed Toeplitz and Hankel determinants with applications in integrable probability

Roozbeh Gharakhloo and Karl Liechty

ABSTRACT. Bordered and framed Toeplitz/Hankel determinants have the same structure as Toeplitz/Hankel determinants except in a small number of matrix rows and/or columns. We review these structured determinants and their connections to orthogonal polynomials, collecting well-known and perhaps less well-known results. We present some applications for these structured determinants to ensembles of non-intersecting paths and the six-vertex model, with an eye towards asymptotic analysis. We also prove some asymptotic formulas for the probability of non-intersection for an ensemble of continuous time random walks for certain choices of starting and ending points as the number of random walkers tends to infinity.

1. Introduction

1.1. Definitions and motivation. For $\phi \in L^1(\mathbb{T})$ denote the $n \times n$ (pure) Toeplitz determinant by

$$D_n[\phi] = \det_{0\le j,k\le n-1}\{\phi_{j-k}\}, \tag{1.1}$$

where

$$\phi_j = \int_{\mathbb{T}} \phi(\zeta)\zeta^{-j}\frac{\mathrm{d}\zeta}{2\pi \mathrm{i}\zeta}, \qquad j \in \mathbb{Z}, \tag{1.2}$$

is the j th Fourier coefficient of ϕ, and $\mathbb{T}$ denotes the unit circle oriented in the counterclockwise direction. More generally we can consider Toeplitz matrices with respect to a measure on $\mathbb{T}$. Indeed for a finite measure φ on $\mathbb{T}$, define φ_j as

$$\varphi_j = \int_{\mathbb{T}} \zeta^{-j}\mathrm{d}\varphi(\zeta), \qquad j \in \mathbb{Z}, \tag{1.3}$$

and the corresponding $n \times n$ Toeplitz determinant as

$$D_n[\varphi] = \det_{0\le j,k\le n-1}\{\varphi_{j-k}\}. \tag{1.4}$$

For a finite measure μ on $\mathbb{R}$ such that all moments exist, i.e.,

$$\mu_k := \int_{\mathbb{R}} x^k \mathrm{d}\mu(x) < \infty, \quad k = 0,1,2,3,\dots,$$

2020 *Mathematics Subject Classification.* Primary 15B05, 30E15, 30E25, 60K35.

©2025 American Mathematical Society

the Hankel matrix of size $n \times n$ corresponding to the measure μ is

$$\mathbf{H}_n[\mu] = \begin{pmatrix} \mu_0 & \mu_1 & \cdots & \mu_{n-1} \\ \mu_1 & \mu_2 & \cdots & \mu_n \\ \vdots & \vdots & \ddots & \vdots \\ \mu_{n-1} & \mu_n & \cdots & \mu_{2n-2} \end{pmatrix}.$$

We will denote the determinant of this matrix as

$$H_n[\mu] := \det \mathbf{H}_n[\mu].$$

Toeplitz and Hankel determinants appear in many different contexts in mathematical physics and random matrix theory. Here we list a few such applications. Toeplitz determinants are central objects in characterization of

- Diagonal and horizontal two point correlations in the 2D Ising model [**71**],
- Spectral analysis of large Toeplitz matrices [**41**],
- Gap probabilities in the circular unitary ensemble in random matrix theory [**33, 72**],
- The distribution of the length of the longest increasing subsequence of random permutations [**6**];

while Hankel determinants play fundamental roles in characterization of

- Gap probabilities in various random matrix ensembles, like the Gaussian unitary ensemble, Laguerre unitary ensemble, and Jacobi unitary ensemble [**31, 34, 72**],
- Correlations of the characteristic polynomials in the Gaussian Unitary Ensemble which are related to the mean values of the zeta-function on the critical line [**65**],
- Topological expansions in random matrix models and graph enumeration problems [**17, 22**],
- The partition function of the six-vertex model [**21**],

and much more. Well-known examples of structural deformations of Toeplitz and Hankel determinants are Toeplitz+Hankel [**9, 11–14, 53**], slant-Toeplitz determinants [**54, 56–60**], and Muttalib-Borodin determinants [**27, 32, 35, 48**].

Often in applications involving Toeplitz and Hankel determinants, the problem may be perturbed in such a way that the Toeplitz/Hankel matrix is altered in only a few rows or columns [**4, 10, 52**]. As a motivating example, consider the enumeration of non-intersecting random walks. That is, we consider n independent simple random walks on $\mathbb{Z}$, $X_1(t), X_2(t), \dots, X_n(t)$. At each discrete time step, the simple random walk either increases or decreases by 1, each with equal probability. We additionally restrict the random walks in the following way.

- At time $t = 0$ the walkers begin at the points $X_j(0) = 2x_j$ with $x_1 < x_2 < \cdots < x_n$ and $x_j \in \mathbb{Z}$ for all j.
- At the terminal time $2T \in 2\mathbb{N}$ the walkers end at the points $X_j(2T) = 2y_j$ with $y_1 < y_2 < \cdots < y_n$ and $y_j \in \mathbb{Z}$ for all j.
- The paths do not intersect, i.e., $X_j(t) \neq X_k(t)$ for $j \neq k$ and $t = 0, 1, \dots, 2T$.

The Lindström–Gessel–Viennot (LGV) formula [**51, 70**] states that the number of such collections of paths is given as a determinant

$$\det \left[\#(2x_j, 2y_k; 2T)\right]_{j,k=1}^{n},$$

where $\#(2x_j, 2y_k; 2T)$ is the number of paths beginning at $2x_j$ and ending at $2y_k$ after $2T$ time-steps. Since this is a binomial coefficient, the number of collections of non-intersecting paths beginning at $2x_1 < \cdots < 2x_n$ and ending at $2y_1 < \cdots < 2y_n$ is

$$\det\left[\binom{2T}{T+y_k-x_j}\right]_{j,k=1}^{n}. \tag{1.5}$$

The matrix can be written as

$$\begin{gathered}\binom{2T}{T+y_k-x_j}_{j,k=1}^{n} = \left(\int_{\mathbb{T}} \phi(\zeta)\zeta^{-(y_k-x_j)}\frac{\mathrm{d}\zeta}{2\pi\mathrm{i}\zeta}\right)_{j,k=1}^{n}, \\ \phi(\zeta) = \zeta^{-T}(\zeta+1)^{2T} = (\zeta+1/\zeta+2)^{T},\end{gathered} \tag{1.6}$$

i.e., $\phi(\zeta)$ is the generating function for the number of paths of a random walker with fixed displacement after $2T$ time steps. If the $x_j = y_j$ are equally spaced then (1.5) is a Toeplitz determinant. If most of the starting/ending points are equally spaced but a few are not, then the Toeplitz matrix may be altered in a few rows/columns. For example if $x_j = j$ for $j = 1, \dots, n$, $y_k = k$ for $k = 1, 2, \dots, n-m$, and $y_k > n-m$ are arbitrary for $k = n-m+1, \dots, n$, we get the a determinant of the form

$$\det\begin{pmatrix} \phi_0 & \phi_1 & \cdots & \phi_{n-m-1} & \phi_{y_{n-m+1}-1} & \phi_{y_{n-m+2}-1} & \cdots & \phi_{y_n-1} \\ \phi_{-1} & \phi_0 & \cdots & \phi_{n-m-2} & \phi_{y_{n-m+1}-2} & \phi_{y_{n-m+2}-2} & \cdots & \phi_{y_n-2} \\ \vdots & \vdots & \vdots & \vdots & \vdots & \cdots & \cdots & \vdots \\ \phi_{-n+1} & \phi_{-n+2} & \cdots & \phi_{-m} & \phi_{y_{n-m+1}-n} & \phi_{y_{n-m+2}-n} & \cdots & \phi_{y_n-n} \end{pmatrix}, \tag{1.7}$$

which has a Toeplitz structure in the first $n-m$ columns, and the remaining m columns consist of descending Fourier coefficients. More generally we can consider the top m starting points to be arbitrary as well. That is, if we take $x_j = y_j = j$ for $j = 1, 2, \dots, n-m$ and $x_j > n-m$ and $y_j > n-m$ to be arbitrary for $j = n-m+1, \dots, n$, then (1.5) becomes

$$\det\begin{pmatrix} \phi_0 & \phi_1 & \cdots & \phi_{n-m-1} & \phi_{y_{n-m+1}-1} & \phi_{y_{n-m+2}-1} & \cdots & \phi_{y_n-1} \\ \phi_{-1} & \phi_0 & \cdots & \phi_{n-m-2} & \phi_{y_{n-m+1}-2} & \phi_{y_{n-m+2}-2} & \cdots & \phi_{y_n-2} \\ \vdots & \vdots & \vdots & \vdots & \vdots & \cdots & \cdots & \vdots \\ \phi_{1-(n-m)} & \phi_{2-(n-m)} & \cdots & \phi_0 & \phi_{y_{n-m+1}-(n-m)} & \phi_{y_{n-m+2}-(n-m)} & \cdots & \phi_{y_n-(n-m)} \\ \phi_{1-x_{n-m+1}} & \phi_{2-x_{n-m+1}} & \cdots & \phi_{n-m-x_{n-m+1}} & \phi_{y_{n-m+1}-x_{n-m+1}} & \phi_{y_{n-m+2}-x_{n-m+1}} & \cdots & \phi_{y_n-x_{n-m+1}} \\ \vdots & \vdots & \vdots & \vdots & \vdots & \cdots & \cdots & \vdots \\ \phi_{1-x_n} & \phi_{2-x_n} & \cdots & \phi_{n-m-x_n} & \phi_{y_{n-m+1}-x_n} & \phi_{y_{n-m+2}-x_n} & \cdots & \phi_{y_n-x_n} \end{pmatrix}, \tag{1.8}$$

which has a Toeplitz structure in the top-left $(n-m)\times(n-m)$ block, descending Fourier coefficients in the first $n-m$ entries of each of the last m columns, and ascending Fourier coefficients in each of the first $n-m$ entries of the last m rows. The bottom-right block of size $m \times m$ has no particular structure.

REMARK 1.1. While it is not the topic of the current paper, we mention that the above application can also give rise to the so-called *slant*-Toeplitz structures. Indeed, let p and q be relatively prime positive integers and consider $x_j = pj$ and $y_k = qk$. Then, in view of (1.6), we have

$$\binom{2T}{T+y_k-x_j}_{j,k=1}^{n} = \left(\int_{\mathbb{T}} \phi(\zeta)\zeta^{-(qk-pj)}\frac{\mathrm{d}\zeta}{2\pi\mathrm{i}\zeta}\right)_{j,k=1}^{n}, \tag{1.9}$$

which is a slant-Toeplitz matrix. For example, the $2j-k$ and $j-2k$ matrices (with offset 0) look like:

$$\begin{pmatrix} \phi_0 & \phi_{-1} & \phi_{-2} & \cdots & \phi_{-n+1} \\ \phi_2 & \phi_1 & \phi_0 & \cdots & \phi_{-n+3} \\ \phi_4 & \phi_3 & \phi_2 & \cdots & \phi_{-n+5} \\ \vdots & \vdots & \ddots & \vdots & \vdots \\ \phi_{2n-2} & \phi_{2n-3} & \phi_{2n-4} & \cdots & \phi_{n-1} \end{pmatrix}$$

and

$$\begin{pmatrix} \phi_0 & \phi_{-2} & \phi_{-4} & \cdots & \phi_{-2n+2} \\ \phi_1 & \phi_{-1} & \phi_{-3} & \cdots & \phi_{-2n+3} \\ \phi_2 & \phi_0 & \phi_{-2} & \cdots & \phi_{-2n+4} \\ \vdots & \vdots & \ddots & \vdots & \vdots \\ \phi_{n-1} & \phi_{n-3} & \phi_{n-5} & \cdots & \phi_{-n+1} \end{pmatrix}.$$

The slant-Toeplitz operators have been studied in a series of works by Mark C. Ho **[56–60]** and the underlying system of orthogonal polynomials in the cases $(p,q) = (1,2)$ and $(p,q) = (2,1)$ have been studied in **[54]**.

We generalize (1.7) and (1.8) as follows.

DEFINITION 1.2. Fix $\phi \in L^1(\mathbb{T})$ as well as $\vec{\boldsymbol{\psi}}_m = (\psi^{(1)}, \dots, \psi^{(m)})$, a vector of m functions in $L^1(\mathbb{T})$. For $n > m$ we define the *(multi-)bordered Toeplitz determinant* of size $n \times n$ to be

$$(1.10)\quad D_n^B[\phi; \vec{\boldsymbol{\psi}}_m] := \det \begin{pmatrix} \phi_0 & \phi_1 & \cdots & \phi_{n-m-1} & \psi^{(1)}_{n-1} & \cdots & \psi^{(m)}_{n-1} \\ \phi_{-1} & \phi_0 & \cdots & \phi_{n-m-2} & \psi^{(1)}_{n-2} & \cdots & \psi^{(m)}_{n-2} \\ \vdots & \vdots & \vdots & \vdots & \vdots & \cdots & \vdots \\ \phi_{-n+1} & \phi_{-n+2} & \cdots & \phi_{-m} & \psi^{(1)}_0 & \cdots & \psi^{(m)}_0 \end{pmatrix}.$$

More generally, we will use the same definition and notation when the functions ϕ and $\psi^{(j)}$ are replaced with measures and the Fourier coefficients are replaced by (1.3).

DEFINITION 1.3. Fix $\phi \in L^1(\mathbb{T})$ as well as $\vec{\boldsymbol{\psi}}_m = (\psi^{(1)}, \dots, \psi^{(m)})$ and $\vec{\boldsymbol{\eta}}_m = (\eta^{(1)}, \dots, \eta^{(m)})$, two vectors of m functions in $L^1(\mathbb{T})$, as well as an $m \times m$ matrix $A = (a_{jk})_{j,k=1}^m$. For $n > m$ we define the *(multi-)framed Toeplitz determinant* of size $n \times n$ to be

$$(1.11)\quad D_n^F[\phi; \vec{\boldsymbol{\psi}}_m; \vec{\boldsymbol{\eta}}_m; A]$$
$$:= \det \begin{pmatrix} \phi_0 & \phi_1 & \cdots & \phi_{n-m-1} & \psi^{(1)}_{n-m-1} & \cdots & \psi^{(m)}_{n-m-1} \\ \phi_{-1} & \phi_0 & \cdots & \phi_{n-m-2} & \psi^{(1)}_{n-m-2} & \cdots & \psi^{(m)}_{n-m-2} \\ \vdots & \vdots & \vdots & \vdots & \vdots & \cdots & \vdots \\ \phi_{-(n-m-1)} & \phi_{-(n-m-1)+1} & \cdots & \phi_0 & \psi^{(1)}_0 & \cdots & \psi^{(m)}_0 \\ \eta^{(1)}_0 & \eta^{(1)}_1 & \cdots & \eta^{(1)}_{n-m-1} & a_{11} & \cdots & a_{1m} \\ \vdots & \vdots & \vdots & \vdots & \vdots & \cdots & \vdots \\ \eta^{(m)}_0 & \eta^{(m)}_1 & \cdots & \eta^{(m)}_{n-m-1} & a_{m1} & \cdots & a_{mm} \end{pmatrix}.$$

As with Definition 1.2, we will use the same definition and notation when the functions ϕ, $\psi^{(j)}$, and $\eta^{(j)}$ are replaced with measures and the Fourier coefficients are replaced by (1.3).

In this notation, (1.7) becomes the bordered Toeplitz determinant

$$D_n^B[\phi;\vec{\boldsymbol{\psi}}_m], \quad \psi^{(j)}(\zeta)=\phi(\zeta)\zeta^{-(y_{n-m+j}-n)},$$

and (1.8) becomes the multi framed Toeplitz determinant

$$D_n^F[\phi;\vec{\boldsymbol{\psi}}_m;\vec{\boldsymbol{\eta}}_m;A], \quad \psi^{(j)}(\zeta)=\phi(\zeta)\zeta^{-(y_{n-m+j}-n)},$$
$$\eta^{(j)}(\zeta)=\phi(\zeta)\zeta^{x_{n-m+j}-1}, \quad a_{jk}=\phi_{y_{n-m+k}-x_{n-m+j}}.$$

REMARK 1.4. A slightly different multi-framed Toeplitz determinant was defined in [**52**, Equation (1.6)]. In the current work we will deal exclusively with the multi-framed Toeplitz determinants as defined above, and will refer to them simply as framed Toeplitz determinants in the sequel.

REMARK 1.5. In the definitions of the bordered and framed Toeplitz determinants we have chosen the Fourier indices of $\psi^{(j)}$ to decrease from top to bottom in the last m columns, and in the framed Toeplitz determinant we have chosen the Fourier indices of $\eta^{(j)}$ to increase from left to right in the last m rows. It is worthwhile mentioning that the different ways to place the Fourier coefficients in the rows/columns could affect the leading order term in the large size asymptotics of the determinants. Instances of this phenomenon are shown in Theorems 1.10 and 1.11 of [**52**]. The other choices of positioning the Fourier coefficients in some of the columns (and/or some of the rows in the case of framed determinants) can be related to the form (1.10) (or to the form (1.11)) by simple manipulations. For example, for fixed $1\le i,\ell\le n$, the multi-framed determinant in which

the indices of $\psi^{(i)}$ increase from top to bottom,
the indices of $\eta^{(\ell)}$ decrease from left to right,
the indices of $\psi^{(j)}$ decrease from top to bottom, $j=1,\cdots,m$, $j\neq i$, and
the indices of $\eta^{(k)}$ increase from left to right, $k=1,\cdots,m$, $k\neq\ell$,

is given by $D_n^F[\phi;\vec{\boldsymbol{\psi}}_{*m};\vec{\boldsymbol{\eta}}_{*m};A]$ as defined in (1.11), where $\vec{\boldsymbol{\psi}}_{*m}=(\psi^{(1)},\dots,\psi^{(i-1)}, z^{n-m-1}\tilde{\psi}^{(i)},\psi^{(i+1)}\dots,\psi^{(m)})$, $\vec{\boldsymbol{\eta}}_{*m}=(\eta^{(1)},\dots,\eta^{(\ell-1)},z^{n-m-1}\tilde{\eta}^{(\ell)},\eta^{(\ell+1)}\dots,\psi^{(m)})$, and $\tilde{f}$ denotes the function $z\mapsto f(z^{-1})$.

By analogy we also introduce bordered and framed Hankel determinants.

DEFINITION 1.6. Let μ be a measure on $\mathbb{R}$ with finite moments, and $\vec{\boldsymbol{\nu}}_m=(\nu^{(1)},\nu^{(2)},\dots,\nu^{(m)})$ a vector of m measures on $\mathbb{R}$, each with finite moments. The *bordered Hankel determinant* of size $n\times n$ is

(1.12)
$$H_n^B[\mu;\vec{\boldsymbol{\nu}}_m]:=\det\begin{pmatrix} \mu_0 & \mu_1 & \cdots & \mu_{n-m-1} & \nu_0^{(1)} & \nu_0^{(2)} & \cdots & \nu_0^{(m)} \\ \mu_1 & \mu_2 & \cdots & \mu_{n-m} & \nu_1^{(1)} & \nu_1^{(2)} & \cdots & \nu_1^{(m)} \\ \mu_2 & \mu_3 & \cdots & \mu_{n-m+1} & \nu_2^{(1)} & \nu_2^{(2)} & \cdots & \nu_2^{(m)} \\ \vdots & \vdots & \ddots & \vdots & \vdots & \ddots & & \vdots \\ \mu_{n-1} & \mu_n & \cdots & \mu_{2n-m-2} & \nu_{n-1}^{(1)} & \nu_{n-1}^{(2)} & \cdots & \nu_{n-1}^{(m)} \end{pmatrix},$$

where μ_k is the kth moment of the measure μ, and $\nu_k^{(j)}$ is the kth moment of the measure $\nu^{(j)}$.

DEFINITION 1.7. Let μ be a measure on $\mathbb{R}$ with finite moments, $\vec{\boldsymbol{\nu}}_m = (\nu^{(1)}, \nu^{(2)}, \ldots, \nu^{(m)})$ and $\vec{\boldsymbol{\eta}}_m = (\eta^{(1)}, \eta^{(2)}, \ldots, \eta^{(m)})$ two vectors of m measures on $\mathbb{R}$, each with finite moments, and $A = (a_{jk})_{j,k=1}^m$ be a matrix of size $m \times m$. The *framed Hankel determinant* of size $n \times n$ is

$$H_n^F[\mu; \vec{\boldsymbol{\nu}}_m, \vec{\boldsymbol{\eta}}_m; A] \tag{1.13}$$
$$:= \det \begin{pmatrix} \mu_0 & \mu_1 & \cdots & \mu_{n-m-1} & \nu_0^{(1)} & \cdots & \nu_0^{(m)} \\ \mu_1 & \mu_2 & \cdots & \mu_{n-m} & \nu_1^{(1)} & \cdots & \nu_1^{(m)} \\ \vdots & \vdots & \ddots & \vdots & \vdots & \ddots & \vdots \\ \mu_{n-m-1} & \mu_{n-m} & \cdots & \mu_{2(n-m-1)} & \nu_{n-m-1}^{(1)} & \cdots & \nu_{n-m-1}^{(m)} \\ \eta_0^{(1)} & \eta_1^{(1)} & \cdots & \eta_{n-m-1}^{(1)} & a_{11} & \cdots & a_{1m} \\ \vdots & \vdots & \ddots & \vdots & \vdots & \ddots & \vdots \\ \eta_0^{(m)} & \eta_1^{(m)} & \cdots & \eta_{n-m-1}^{(m)} & a_{m1} & \cdots & a_{mm} \end{pmatrix},$$

where μ_k is the kth moment of the measure μ, $\nu_k^{(j)}$ is the kth moment of the measure $\nu^{(j)}$, and $\eta_k^{(j)}$ is the kth moment of the measure $\eta^{(j)}$.

REMARK 1.8. One could generalize the framed determinants so that the top-, bottom-, left-, and right-most m rows/columns differ from the Toeplitz/Hankel structure. In the Hankel case we could consider determinants of the form

$$\det \begin{pmatrix} a_{11} & \cdots & a_{1m} & \sigma_0^{(1)} & \sigma_1^{(1)} & \cdots & \sigma_{n-2m-1}^{(1)} & b_{11} & \cdots & b_{1m} \\ \vdots & \ddots & \vdots & \vdots & \vdots & \ddots & \vdots & \vdots & \ddots & \vdots \\ a_{m1} & \cdots & a_{mm} & \sigma_0^{(m)} & \sigma_1^{(m)} & \cdots & \sigma_{n-2m-1}^{(m)} & b_{m1} & \cdots & b_{mm} \\ \lambda_0^{(1)} & \cdots & \lambda_0^{(m)} & \mu_0 & \mu_1 & \cdots & \mu_{n-2m-1} & \nu_0^{(1)} & \cdots & \nu_0^{(m)} \\ \lambda_1^{(1)} & \cdots & \lambda_1^{(m)} & \mu_1 & \mu_2 & \cdots & \mu_{n-2m} & \nu_1^{(1)} & \cdots & \nu_1^{(m)} \\ \vdots & \ddots & \vdots & \vdots & \vdots & \ddots & \vdots & \vdots & \ddots & \vdots \\ \lambda_{n-2m-1}^{(1)} & \cdots & \lambda_{n-2m-1}^{(m)} & \mu_{n-2m-1} & \mu_{n-2m} & \cdots & \mu_{2(n-2m-1)} & \nu_{n-2m-1}^{(1)} & \cdots & \nu_{n-2m-1}^{(m)} \\ d_{11} & \cdots & d_{1m} & \eta_0^{(1)} & \eta_1^{(1)} & \cdots & \eta_{n-2m-1}^{(1)} & c_{11} & \cdots & c_{1m} \\ \vdots & \ddots & \vdots & \vdots & \vdots & \ddots & \vdots & \vdots & \ddots & \vdots \\ d_{m1} & \cdots & d_{mm} & \eta_0^{(m)} & \eta_1^{(m)} & \cdots & \eta_{n-2m-1}^{(m)} & c_{m1} & \cdots & c_{mm} \end{pmatrix}. \tag{1.14}$$

It is clear that the above determinant reduces to the one of the form (1.13) by row and column operations, so we restrict ourselves to matrices of the form (1.13) and (1.11).

The first major appearance of bordered Toeplitz determinants (1.10) in the mathematical physics literature was in the work **[4]** of Au-Yang and Perk in 1987 where it was established that the *next-to-diagonal* two-point correlation function in the 2D Ising model is given by

$$\langle \sigma_{0,0}\sigma_{n-1,n} \rangle = D_n^B[\widehat{\phi}; \widehat{\psi}], \tag{1.15}$$

with

$$\widehat{\phi}(z) = \sqrt{\frac{1-k^{-1}z^{-1}}{1-k^{-1}z}} \quad \text{and} \quad \widehat{\psi}(z) = \frac{C_v z\widehat{\phi}(z) + C_h}{S_v(z-c_*)}, \tag{1.16}$$

where k, C_v, C_h, S_v, and c_* are all physical parameters of the model and $\sigma_{i,j} \in \{-1, 1\}$ represents the magnetic dipole moment of the atomic spins at position (i, j).

The framed Toeplitz determinants have also already appeared in the calculations of entanglement entropy for disjoint subsystems in the XX spin chain [**30**, **61**] (also see the introduction of [**52**]).

In this paper we collect and review results which relate bordered and framed Toeplitz/Hankel determinants to systems of orthogonal polynomials. The orthogonal polynomial formulas for these structured determinants enables one to study their large-size asymptotics, since for the orthogonal polynomials, one has the associated Riemann–Hilbert problem first discovered by A. S. Fokas, A. R. Its, and A. V. Kitaev in 1992 [**47**]. More precisely, the asymptotic analysis of the Riemann–Hilbert problem via the Deift–Zhou method [**43**] allows for a detailed and rigorous asymptotic analysis, which in turn, allows for a successful asymptotic analysis for the framed and bordered Toeplitz/Hankel determinants. In the subsections below, we first give a brief review of the well known OPRL[1] and BOPUC[2] characterizations of the (pure) Hankel and Toeplitz determinants, respectively, before presenting orthogonal polynomial characterizations for the bordered and framed Toeplitz/Hankel determinants.

1.2. Toeplitz/Hankel determinants and orthogonal polynomials. Let us start with the OPRL characterization for Hankel determinants. Let $\mathcal{J} \subset \mathbb{R}$ and μ be a measure with finite moments on $\mathcal{J}$. Consider the associated set of monic orthogonal polynomials $\{P_n(z)\}_{n=0}^{\infty}$, $\deg P_n(z) = n$, satisfying the orthogonality conditions

$$\int_{\mathcal{J}} P_n(x)x^k \mathrm{d}\mu(x) = h_n\delta_{nk}, \qquad k = 0, 1, \cdots, n, \tag{1.17}$$

which is equivalent to

$$\int_{\mathcal{J}} P_n(x)P_k(x)\mathrm{d}\mu(x) = h_n\delta_{nk}, \qquad k, n \in \mathbb{Z}_{\geq 0}. \tag{1.18}$$

If $\det \boldsymbol{H}_n[\mu] \neq 0$, the polynomial $P_n(z)$ has the following determinantal representation

$$P_n(x) = \frac{1}{\det \boldsymbol{H}_n[\mu]} \det \begin{pmatrix} \mu_0 & \mu_1 & \cdots & \mu_{n-1} & \mu_n \\ \mu_1 & \mu_2 & \cdots & \mu_n & \mu_{n+1} \\ \vdots & \vdots & \iddots & \vdots & \vdots \\ \mu_{n-1} & \mu_n & \cdots & \mu_{2n-2} & \mu_{2n-1} \\ 1 & x & \cdots & x^{n-1} & x^n \end{pmatrix}. \tag{1.19}$$

Indeed the right-hand-side of (1.19) is a polynomial in x of degree n and it is easy to see that the integral of x^k times the right-hand-side of (1.19) vanishes for $k = 0, 1, 2, \ldots, n-1$ since two rows of the matrix coincide. Thus it is a constant times $P_n(x)$ and the leading coefficient is seen to be one by expanding the determinant in the last column. The formula (1.19) yields

$$h_n = \frac{\det \boldsymbol{H}_{n+1}[\mu]}{\det \boldsymbol{H}_n[\mu]}, \qquad \text{and thus} \qquad \det \boldsymbol{H}_n[\mu] = \prod_{j=0}^{n-1} h_j. \tag{1.20}$$

If the measure μ has a Hölder continuous density $\mathrm{d}\mu(x) = w(x)\,\mathrm{d}x$, then it is due to Fokas, Its, and Kitaev [**47**] that the following matrix-valued function which

[1] Orthogonal polynomials on the real line.

[2] Bi-orthogonal polynomials on the unit circle.

is built from the orthogonal polynomials and the Cauchy transforms of the weight w multiplied by the orthogonal polynomials

$$Y(z;n) = \begin{pmatrix} P_n(z) & \dfrac{1}{2\pi \mathrm{i}} \displaystyle\int_{\mathcal{J}} \frac{P_n(x)w(x)}{x-z} \mathrm{d}x \\ -\dfrac{2\pi \mathrm{i}}{h_{n-1}} P_{n-1}(z) & -\dfrac{1}{h_{n-1}} \displaystyle\int_{\mathcal{J}} \frac{P_{n-1}(x)w(x)}{x-z} \mathrm{d}x \end{pmatrix}, \qquad n \geq 1, \tag{1.21}$$

satisfies the following Riemann–Hilbert problem (RHP):

RH-Y1 $Y(\cdot;n) : \mathbb{C} \setminus \mathcal{J} \to \mathbb{C}^{2\times 2}$ is analytic.

RH-Y2 The limits of $Y(z;n)$ as z tends to $x \in \mathcal{J}$ from the upper and lower half plane exist, which are denoted by $Y_{\pm}(x;n)$ respectively. Moreover, the functions $x \mapsto Y_{\pm}(x;n)$ are continuous on $\mathcal{J}$ and are related by the *jump condition*

$$Y_+(x;n) = Y_-(x;n) \begin{pmatrix} 1 & w(x) \\ 0 & 1 \end{pmatrix}, \qquad x \in \mathcal{J}. \tag{1.22}$$

RH-Y3 As $z \to \infty$,

$$Y(z;n) = \big(I + O(z^{-1})\big) z^{n\sigma_3}, \tag{1.23}$$

where $\sigma_3 = \begin{pmatrix} 1 & 0 \\ 0 & -1 \end{pmatrix}$ is the third Pauli matrix.

A successful large-n asymptotic analysis of this RHP yields the asymptotics of the norms h_n's, and thus that of $\det \boldsymbol{H}_n[w]$ via (1.20) (see e.g. **[31, 34, 65, 66]** for this analysis for various choices of w). For discrete measures μ, there is a similar characterization in terms of a discrete Riemann–Hilbert problem **[5, 20, 21]**.

Now, let us briefly discuss the BOPUC characterization of Toeplitz determinants. Given a finite measure φ of $\mathbb{T}$, one can consider the associated sets of bi-orthonormal polynomials $\{Q_n(z)\}_{n=0}^{\infty}$ and $\{\widehat{Q}_n(z)\}_{n=0}^{\infty}$, where $Q_n(z) = \kappa_n z^n + l_n z^{n-1} + \cdots$, and $\widehat{Q}_n(z) = \widehat{\kappa}_n z^n + \widehat{l}_n z^{n-1} + \cdots$ satisfy the bi-orthonormality conditions

$$\int_{\mathbb{T}} Q_n(z) \widehat{Q}_k(z^{-1}) \mathrm{d}\varphi(z) = \delta_{nk}. \tag{1.24}$$

Similar to (1.19), if $D_n[\varphi], D_{n+1}[\varphi] \neq 0$, these polynomials have determinantal representations given by

$$Q_n(z) = \frac{1}{\sqrt{D_n[\varphi] D_{n+1}[\varphi]}} \det \begin{pmatrix} \varphi_0 & \varphi_{-1} & \cdots & \varphi_{-n} \\ \varphi_1 & \varphi_0 & \cdots & \varphi_{-n+1} \\ \vdots & \vdots & \ddots & \vdots \\ \varphi_{n-1} & \varphi_{n-2} & \cdots & \varphi_{-1} \\ 1 & z & \cdots & z^n \end{pmatrix}, \tag{1.25}$$

and

$$\widehat{Q}_n(z) = \frac{1}{\sqrt{D_n[\varphi] D_{n+1}[\varphi]}} \det \begin{pmatrix} \varphi_0 & \varphi_{-1} & \cdots & \varphi_{-n+1} & 1 \\ \varphi_1 & \varphi_0 & \cdots & \varphi_{-n+2} & z \\ \vdots & \vdots & \ddots & \vdots & \vdots \\ \varphi_n & \varphi_{n-1} & \cdots & \varphi_1 & z^n \end{pmatrix}, \tag{1.26}$$

from which it is clear that

$$\kappa_n = \widehat{\kappa}_n = \sqrt{\frac{\det T_n[\varphi]}{\det T_{n+1}[\varphi]}}. \tag{1.27}$$

Now let us consider the case $\mathrm{d}\varphi(z) = \frac{\phi(z)\,\mathrm{d}z}{2\pi \mathrm{i} z}$ for a Hölder continuous function $\phi(z)$, and define the 2×2 matrix-valued function

$$X(z;n) := \begin{pmatrix} \kappa_n^{-1} Q_n(z) & \kappa_n^{-1} \displaystyle\int_{\mathbb{T}} \frac{Q_n(\zeta)}{(\zeta - z)} \frac{\phi(\zeta)\mathrm{d}\zeta}{2\pi \mathrm{i}\zeta^n} \\ -\kappa_{n-1} z^{n-1} \widehat{Q}_{n-1}(z^{-1}) & -\kappa_{n-1} \displaystyle\int_{\mathbb{T}} \frac{\widehat{Q}_{n-1}(\zeta^{-1})}{(\zeta - z)} \frac{\phi(\zeta)\mathrm{d}\zeta}{2\pi \mathrm{i}\zeta} \end{pmatrix}. \tag{1.28}$$

It is due to J. Baik, P. Deift and K. Johansson [**6**] (inspired by the RHP for OPRL [**47**]) that the function X defined above satisfies the following associated Riemann–Hilbert problem:

RH-X1 $X : \mathbb{C} \setminus \mathbb{T} \to \mathbb{C}^{2\times 2}$ is analytic.

RH-X2 The limits of $X(\xi;n)$ as ξ tends to $z \in \mathbb{T}$ from the inside and outside of the unit circle exist, which are denoted by $X_\pm(z;n)$ respectively. Moreover, the functions $z \mapsto X_\pm(z;n)$ are continuous on $\mathbb{T}$ and are related by the *jump condition*

$$X_+(z;n) = X_-(z;n) \begin{pmatrix} 1 & z^{-n}\phi(z) \\ 0 & 1 \end{pmatrix}, \qquad z \in \mathbb{T}. \tag{1.29}$$

RH-X3 As $z \to \infty$

$$X(z;n) = \left(I + \frac{\overset{\infty}{X}_1(n)}{z} + \frac{\overset{\infty}{X}_2(n)}{z^2} + O(z^{-3}) \right) z^{n\sigma_3}. \tag{1.30}$$

For the purpose of calculations in Section 7, we present the asymptotic analysis of this RHP in Appendix A for Szegő-type symbols[3]. For such asymptotic analysis in the case of Fisher–Hartwig symbols we refer the interested reader to [**15, 45**].

1.3. Bordered Toeplitz/Hankel determinants and orthogonal polynomials. We start with the extension of (1.19) to several variables. Let $H_n^B[\mu; z_1, \dots, z_m]$ be the determinant

(1.31)

$$H_n^B[\mu; z_1, \dots, z_m] := \det \begin{pmatrix} \mu_0 & \mu_1 & \cdots & \mu_{n-m-1} & 1 & 1 & \cdots & 1 \\ \mu_1 & \mu_2 & \cdots & \mu_{n-m} & z_1 & z_2 & \cdots & z_m \\ \mu_2 & \mu_3 & \cdots & \mu_{n-m+1} & z_1^2 & z_2^2 & \cdots & z_m^2 \\ \vdots & \vdots & \ddots & \vdots & \vdots & \vdots & \ddots & \vdots \\ \mu_{n-1} & \mu_n & \cdots & \mu_{2n-m-2} & z_1^{n-1} & z_2^{n-1} & \cdots & z_m^{n-1} \end{pmatrix}.$$

This determinant may also be expressed in terms of orthogonal polynomials.

[3]We will occasionally refer to a symbol as *Szegő-type*, if a) it is smooth and nonzero on the unit circle, b) has no winding number, and c) admits an analytic continuation in a neighborhood of the unit circle. For a Szegő-type symbol $\phi(z)$, we refer to the piecewise analytic function $\alpha(z)$ satisfying the jump condition $\alpha_+(z) = \alpha_-(z)\phi(z)$ and the large-z asymptotic condition $\alpha(z) = 1 + O(1/z)$, as the *Szegő function* associated with ϕ (see e.g. (7.7) and (7.8)).

PROPOSITION 1.9. *The determinant $H_n[\mu; z_1, \dots, z_m]$ is given as*

$$H_n^B[\mu; z_1, \dots, z_m] = H_{n-m}[\mu] \det \begin{pmatrix} P_{n-m}(z_1) & \cdots & P_{n-m}(z_m) \\ \vdots & \ddots & \vdots \\ P_{n-1}(z_1) & \cdots & P_{n-1}(z_m) \end{pmatrix}. \tag{1.32}$$

The above proposition is well known in the literature (see, e.g., [**36**, Equation (5.5)]) though we were unable to find a proof written explicitly anywhere. The formula (1.32) is rather similar to a formula of Christoffel ([**73**, Theorem 2.5], and Proposition 1.9 can be proved based on that formula). In this paper we give a different proof based on the Dodgson Condensation identity. This proof is presented in Section 5. Using an identical argument we can prove the analogous result for the system of bi-orthogonal polynomials on the unit circle.

PROPOSITION 1.10. *Let $T_n[\varphi; z_1, \cdots, z_m]$ be the following $n \times n$ multi-bordered Toeplitz determinant*

$$\begin{aligned} T_n[\varphi; z_1, \cdots, z_m] \\ := \det \begin{pmatrix} \varphi_0 & \varphi_1 & \cdots & \varphi_{n-m-1} & 1 & 1 & \cdots & 1 \\ \varphi_{-1} & \varphi_0 & \cdots & \varphi_{n-m-2} & z_1 & z_2 & \cdots & z_m \\ \vdots & \vdots & \ddots & \vdots & \vdots & \vdots & \ddots & \vdots \\ \varphi_{-n+1} & \varphi_{-n+2} & \cdots & \varphi_{-m} & z_1^{n-1} & z_2^{n-1} & \cdots & z_m^{n-1} \end{pmatrix}. \end{aligned} \tag{1.33}$$

It holds that

$$T_n[\varphi; z_1, \dots, z_m] = \sqrt{D_{n-m}[\varphi] D_{n-m+1}[\varphi]} \det \begin{pmatrix} Q_{n-m}(z_1) & \cdots & Q_{n-m}(z_m) \\ \vdots & \ddots & \vdots \\ Q_{n-1}(z_1) & \cdots & Q_{n-1}(z_m) \end{pmatrix}, \tag{1.34}$$

and

$$T_n[\tilde{\varphi}; z_1, \dots, z_m] = \sqrt{D_{n-m}[\varphi] D_{n-m+1}[\varphi]} \det \begin{pmatrix} \widehat{Q}_{n-m}(z_1) & \cdots & \widehat{Q}_{n-m}(z_m) \\ \vdots & \ddots & \vdots \\ \widehat{Q}_{n-1}(z_1) & \cdots & \widehat{Q}_{n-1}(z_m) \end{pmatrix}, \tag{1.35}$$

where $Q_k(z)$ and $\widehat{Q}_k(z)$ are the biorthogonal polynomials on the unit circle given by (1.25) and (1.26), and the measures φ and $\tilde{\varphi}$ are related by $d\tilde{\varphi}(z) := d\varphi(z^{-1})$.

Using Propositions 1.9 and 1.10, it is simple to give a similar formula for any bordered Hankel or Toeplitz determinant as a (multiple) integral.

COROLLARY 1.11. *The bordered determinants $H_n^B[\mu; \vec{\boldsymbol{\nu}}_m]$ and $D_n^B[\phi; \vec{\boldsymbol{\psi}}_m]$ are given by*

$$\begin{aligned} H_n^B[\mu; \boldsymbol{\nu}_m] \\ = H_{n-m}[\mu] \int_{\mathbb{R}} \cdots \int_{\mathbb{R}} \det \begin{pmatrix} P_{n-m}(z_1) & \cdots & P_{n-m}(z_m) \\ \vdots & \ddots & \vdots \\ P_{n-1}(z_1) & \cdots & P_{n-1}(z_m) \end{pmatrix} d\nu^{(1)}(z_1) \cdots d\nu^{(m)}(z_m), \end{aligned} \tag{1.36}$$

(1.37)
$$D_n^B[\varphi;\vec{\psi}_m] = \sqrt{D_{n-m}[\varphi]D_{n-m+1}[\varphi]}\int_{\mathbb{T}}\cdots\int_{\mathbb{T}}\det\begin{pmatrix} Q_{n-m}(z_1) & \cdots & Q_{n-m}(z_m)\\ \vdots & \ddots & \vdots \\ Q_{n-1}(z_1) & \cdots & Q_{n-1}(z_m)\end{pmatrix} \times\prod_{j=1}^{m} z_j^{-n+1}d\psi^{(j)}(z_j).$$

PROOF. By multilinearity of the determinant,

(1.38)
$$H_n^B[\mu;\boldsymbol{\nu}_m] = \det\begin{pmatrix} \mu_0 & \mu_1 & \cdots & \mu_{n-m-1} & \int_{\mathbb{R}} d\nu^{(1)}(z) & \int_{\mathbb{R}} d\nu^{(2)}(z) & \cdots & \int_{\mathbb{R}} d\nu^{(m)}(z)\\ \mu_1 & \mu_2 & \cdots & \mu_{n-m} & \int_{\mathbb{R}} z d\nu^{(1)}(z) & \int_{\mathbb{R}} z d\nu^{(2)}(z) & \cdots & \int_{\mathbb{R}} z d\nu^{(m)}(z)\\ \mu_2 & \mu_3 & \cdots & \mu_{n-m+1} & \int_{\mathbb{R}} z^2 d\nu^{(1)}(z) & \int_{\mathbb{R}} z^2 d\nu^{(2)}(z) & \cdots & \int_{\mathbb{R}} z^2 d\nu^{(m)}(z)\\ \vdots & \vdots & \ddots & \vdots & \vdots & & \ddots & \vdots \\ \mu_{n-1} & \mu_n & \cdots & \mu_{2n-m-2} & \int_{\mathbb{R}} z^{n-1} d\nu^{(1)}(z) & \int_{\mathbb{R}} z^{n-1} d\nu^{(2)}(z) & \cdots & \int_{\mathbb{R}} z^{n-1} d\nu^{(m)}(z)\end{pmatrix}$$

$$= \int_{\mathbb{R}}\cdots\int_{\mathbb{R}}\begin{pmatrix} \mu_0 & \mu_1 & \cdots & \mu_{n-m-1} & 1 & 1 & \cdots & 1\\ \mu_1 & \mu_2 & \cdots & \mu_{n-m} & z_1 & z_2 & \cdots & z_m\\ \mu_2 & \mu_3 & \cdots & \mu_{n-m+1} & z_1^2 & z_2^2 & \cdots & z_m^2\\ \vdots & \vdots & \ddots & \vdots & \vdots & \vdots & \ddots & \vdots\\ \mu_{n-1} & \mu_n & \cdots & \mu_{2n-m-2} & z_1^{n-1} & z_2^{n-1} & \cdots & z_m^{n-1}\end{pmatrix} d\nu^{(1)}(z_1)d\nu^{(2)}(z_2)\cdots d\nu^{(m)}(z_m),$$

which is (1.36) using (1.32). Equation (1.37) follows in a similar way. □

REMARK 1.12. The proof of Corollary 1.11 is strictly algebraic and makes no assumptions on the border measures $\nu^{(j)}$ and $\psi^{(j)}$. The corollary holds quite broadly, including for signed and complex measures $\nu^{(j)}$ and $\psi^{(j)}$. These will appear in some of the examples in subsequent sections.

1.4. Framed Toeplitz/Hankel determinants and orthogonal polynomials. It was shown in **[52]** (based on an idea introduced in **[54]**) that the single-framed Toeplitz determinants are closely connected to the reproducing kernel of BOPUC. In this section we extend those results to multi-framed Toeplitz determinants and present their analogues for (multi-)framed Hankel determinants as well. The formulas established in this section for an m-framed Toeplitz/Hankel determinant involve an $m \times m$ determinant using the reproducing kernel of the OPRL or BOPUC. For $m = 1$ such formulas were written in Theorem 1.9 of **[52]**, and a similar approach establishes analogous formulas for the single-framed Hankel determinants. However for $m \geq 2$, unlike what we show in this work, no determinantal formulas were shown in **[52]**. In fact, for $m > 2$, there exists an alternative (and perhaps less efficient) approach to write an m-framed Toeplitz/Hankel determinant in terms of the reproducing kernels of orthogonal polynomials. This alternative representation is achieved by expressing the m-framed determinant in terms of several single-framed determinants, utilizing successive applications of proper Dodgson Condensation identities (see Section 5.1). Consequently, as a result of Theorem 1.9 in **[52]**, the m-framed determinant can be expressed in terms of the reproducing kernels of orthogonal polynomials. This idea is displayed in section 3.3 of **[52]** and will not be pursued in this section.

The reproducing kernel for the OPRL (1.17) is defined as

$$K_n(x,y)=\sum_{k=0}^{n}\frac{P_k(x)P_k(y)}{h_k}. \tag{1.39}$$

For an $m\times m$ matrix $A=(a_{jk})_{j,k=1}^m$, and variables $x_1,\dots,x_m$, and $y_1,\dots,y_m$, define the determinant $H_n^F(\mu;x_1,\dots,x_m;y_1,\dots,y_m;A)$ as

$$H_n^F[\mu;x_1,\dots,x_m;y_1,\dots,y_m;A]$$
$$:=\det\begin{pmatrix}\mu_0 & \mu_1 & \cdots & \mu_{n-m-1} & 1 & \cdots & 1\\ \mu_1 & \mu_2 & \cdots & \mu_{n-m} & x_1 & \cdots & x_m\\ \mu_2 & \mu_3 & \cdots & \mu_{n-m+1} & x_1^2 & \cdots & x_m^2\\ \vdots & \vdots & \ddots & \vdots & \vdots & \ddots & \vdots\\ \mu_{n-m-1} & \mu_{n-m} & \cdots & \mu_{2(n-m-1)} & x_1^{n-m-1} & \cdots & x_m^{n-m-1}\\ 1 & y_1 & \cdots & y_1^{n-m-1} & a_{11} & \cdots & a_{1m}\\ \vdots & \vdots & \ddots & \vdots & \vdots & \ddots & \vdots\\ 1 & y_m & \cdots & y_m^{n-m-1} & a_{m1} & \cdots & a_{mm}\end{pmatrix}.$$

PROPOSITION 1.13. *Let $A=(a_{jk})_{j,k=1}^m$. We have*

$$H_n^F[\mu;x_1,\dots,x_m;y_1,\dots,y_m;A]=H_{n-m}[\mu]\det\left[A-(K_{n-m-1}(x_k,y_j))_{j,k=1}^m\right], \tag{1.40}$$

where $K_n(x,y)$ is the reproducing kernel of the system of orthogonal polynomials on the real line given by (1.39).

The above result, which will be proved in Section 5.2, implies a general formula for framed Hankel determinants.

COROLLARY 1.14. *Let $A=(a_{jk})_{j,k=1}^m$. We have*

$$\begin{aligned}H_n^F(\mu;\nu^{(1)},\dots,\nu^{(m)};\eta^{(1)},\dots,\eta^{(m)};A)=\\ H_{n-m}[\mu]\det\left[A-\left(\int_{\mathbb{R}}\int_{\mathbb{R}}K_{n-m-1}(x_k,y_j)d\nu^{(k)}(x_k)d\eta^{(j)}(y_j)\right)_{j,k=1}^m\right],\end{aligned} \tag{1.41}$$

where $K_n(x,y)$ is the reproducing kernel of the system of orthogonal polynomials on the real line given by (1.39).

PROOF. The proof is nearly identical to that of (1.36). □

REMARK 1.15. If the powers of x and/or y are reversed in order, similar results follow as corollaries. For $m=1$ we record the rest of the four different orders of the powers of x and y below:

$$\begin{aligned}\det\begin{pmatrix}\mu_0 & \mu_1 & \cdots & \mu_{n-2} & x^{n-2}\\ \mu_1 & \mu_2 & \cdots & \mu_{n-1} & x^{n-3}\\ \mu_2 & \mu_3 & \cdots & \mu_n & x^{n-4}\\ \vdots & \vdots & \ddots & \vdots & \vdots\\ \mu_{n-2} & \mu_{n-1} & \cdots & \mu_{2n-4} & 1\\ 1 & y & \cdots & y^{n-2} & a\end{pmatrix} &= H_n^F[\mu;1/x,y;a/x^{n-2}]x^{n-2}\\ &= H_{n-1}[\mu](a-x^{n-2}K_{n-2}(1/x,y)),\end{aligned} \tag{1.42}$$

$$(1.43)\quad \det\begin{pmatrix} \mu_0 & \mu_1 & \cdots & \mu_{n-2} & 1 \\ \mu_1 & \mu_2 & \cdots & \mu_{n-1} & x^2 \\ \mu_2 & \mu_3 & \cdots & \mu_n & x^3 \\ \vdots & \vdots & \ddots & \vdots & \vdots \\ \mu_{n-2} & \mu_{n-1} & \cdots & \mu_{2n-4} & x^{n-2} \\ y^{n-2} & y^{n-3} & \cdots & 1 & a \end{pmatrix} = H_n^F[\mu; x, 1/y; a/y^{n-2}]y^{n-2}$$

$$= H_{n-1}[\mu](a - y^{n-2}K_{n-2}(x, 1/y)),$$

(1.44)

$$\det\begin{pmatrix} \mu_0 & \mu_1 & \cdots & \mu_{n-2} & x^{n-2} \\ \mu_1 & \mu_2 & \cdots & \mu_{n-1} & x^{n-3} \\ \mu_2 & \mu_3 & \cdots & \mu_n & x^{n-4} \\ \vdots & \vdots & \ddots & \vdots & \vdots \\ \mu_{n-2} & \mu_{n-1} & \cdots & \mu_{2n-4} & 1 \\ y^{n-2} & y^{n-3} & \cdots & 1 & a \end{pmatrix} = H_n^F[\mu; 1/x, 1/y; a/(xy)^{n-2}](xy)^{n-2}$$

$$= H_{n-1}[\mu](a - (xy)^{n-2}K_{n-2}(1/x, 1/y)).$$

Analogously in the Toeplitz case, define the determinant

$$D_n^F[\varphi; x_1, \dots, x_m; y_1, \dots, y_m; A]$$

$$= \begin{pmatrix} \varphi_0 & \varphi_{-1} & \cdots & \varphi_{-n+m+1} & 1 & \cdots & 1 \\ \varphi_1 & \varphi_0 & \cdots & \varphi_{-n+m+2} & x_1 & \cdots & x_m \\ \vdots & \vdots & \ddots & \vdots & \vdots & \ddots & \vdots \\ \varphi_{n-m-1} & \varphi_{n-m-2} & \cdots & \varphi_0 & x_1^{n-m-1} & \cdots & x_m^{n-m-1} \\ 1 & y_1 & \cdots & y_1^{n-m-1} & a_{11} & \cdots & a_{1m} \\ \vdots & \vdots & \ddots & \vdots & \vdots & \ddots & \vdots \\ 1 & y_m & \cdots & y_m^{n-m-1} & a_{m1} & \cdots & a_{mm} \end{pmatrix}.$$

The following results can be proven almost identically as Proposition 1.13, and respectively generalize Theorems 3.5 and 1.9 of [**52**].

PROPOSITION 1.16. *Let $A = (a_{jk})_{j,k=1}^m$. We have*

(1.45)

$$D_n^F[\varphi; x_1, \dots, x_m; y_1, \dots, y_m; A] = D_{n-m}[\varphi] \det\left[A - (\mathcal{K}_{n-m-1}(x_k, y_j))_{j,k=1}^m\right],$$

where

$$(1.46)\qquad \mathcal{K}_n(x, y) := \sum_{j=0}^{n} Q_j(y)\widehat{Q}_j(x),$$

is the reproducing kernel of the system of biorthogonal polynomials on the unit circle:

$$(1.47)\qquad \int_{\mathbb{T}} Q_j(\zeta)\widehat{Q}_k(\zeta^{-1})\,d\varphi(\zeta) = \delta_{jk}, \qquad j, k \in \mathbb{N} \cup \{0\}.$$

COROLLARY 1.17. *Let $A = (a_{jk})_{j,k=1}^m$. We have*

(1.48)
$$\begin{aligned} &D_n^F[\phi;\psi^{(1)},\dots,\psi^{(m)};\eta^{(1)},\dots,\eta^{(m)};A] \\ &= D_{n-m}[\phi]\det\left[A - \left(\int_{\mathbb{T}}\int_{\mathbb{T}} \mathcal{K}_{n-m-1}(x_k^{-1},y_j^{-1})\,d\psi^{(k)}(x_k)\,d\eta^{(j)}(y_j)\right)_{j,k=1}^m\right], \end{aligned}$$

where $\mathcal{K}_n(x,y)$ is the reproducing kernel of the system of biorthogonal polynomials on the unit circle given by (1.46).

REMARK 1.18. As with Corollary 1.11, Corollaries 1.14 and 1.17 hold when the border measures $\nu^{(j)}$, $\eta^{(j)}$, and $\psi^{(j)}$ are signed or complex measures.

The formulas of this subsection and the previous one for bordered and framed Toeplitz/Hankel determinants pave the way for their large-size asymptotic analysis, due to the Riemann-Hilbert characterizations for OPRL and BOPUC. Using such orthogonal polynomial representations, a few strong Szegő limit theorems of the form

$$G^n[\phi]E[\phi]\left(H[\phi;\boldsymbol{\psi};\boldsymbol{\eta}] + O(e^{-cn})\right), \qquad c>0, \tag{1.49}$$

were proven for certain bordered/framed Toeplitz determinants with Szegő-type symbols ϕ in **[10]** and **[52]**. In Section 7, we present a concise overview of the strong Szegö limit theorem, accompanied by precise definitions of $G[\phi]$ and $E[\phi]$ as given in equation (7.2)[4]. Although the primary focus of this paper is not the computation of asymptotics using these orthogonal polynomial characterizations, we include some calculations in Section 7 to underscore their relevance.

1.5. Applications. In Sections 2, 3, and 4 below, we present several examples of bordered and framed Toeplitz/Hankel determinants which appear in various integrable models in probability.

- Section 2 considers the maximal height of an ensemble of non-intersecting Brownian excursions, previously studied in **[49, 67]** in the case of all particles beginning and ending at the origin, in which case the distribution of the maximal height is given by a Hankel determinant with a discrete weight. In Section 2 we consider the case in which a few of the starting/ending points may be non-zero, in which case the distribution of the maximal height is described by a bordered or framed Hankel determinant of a discrete weight. Using Corollaries 1.11 and 1.14 we describe the corresponding formulation in terms of orthogonal polynomials and give some idea of how the asymptotic analysis would proceed and describe expected asymptotic results.
- Section 3 considers the width of several ensembles of non-intersecting paths: Brownian bridges, discrete time random walks, and continuous time random walks. Each of these ensembles was previously studied in **[8]**, in which the authors showed that for the simplest initial conditions

[4]For a more comprehensive exploration of this theorem, we direct readers to **[46]**. The expression for $H[\phi;\boldsymbol{\psi};\boldsymbol{\eta}]$ (see (1.49)) varies depending on whether bordered or framed Toeplitz determinants are under consideration. We refer the interested reader to **[10]** and **[52]** for detailed descriptions of H in various settings.

on paths, the distribution of the width is the ratio of two Hankel determinants for non-intersecting Brownian bridges, and the ratio of two Toeplitz determinants for the non-intersecting discrete or continuous time random walks. We once again consider the case in which a few particles have starting/ending positions different from those of [**8**], and formulate the distribution in terms of bordered and framed Toeplitz/Hankel determinants, and consequently in terms of orthogonal polynomials using Corollaries 1.11 and 1.14.

- Section 4 considers the six-vertex model with domain wall boundary conditions (DWBC), one of the most important models in two-dimensional statistical physics. The partition function of the model may be formulated as a Hankel determinant [**21, 79**], and a certain variation known as the *partially inhomogeneous partition function* may be expressed as a bordered or framed Hankel determinant. The partially inhomogeneous partition function has a probabilistic interpretation as the generating function for certain statistics near the boundary of a state of the model [**24, 36–38, 55**], and for the bordered version of the Hankel determinant, an asymptotic analysis was recently performed in [**55**]. We briefly review this result and present the framed version as well.

The formulas derived in Sections 2, 3, and 4 all invite detailed asymptotic analysis in upcoming works. We do not give a complete asymptotic analysis of the examples mentioned above, but we do give some partial results and analysis to highlight the capacity of the orthogonal polynomial characterizations. We choose to study the asymptotics of the probability of non-intersection for an ensemble of continuous time random walks as the number of random walkers approach infinity. This is the quantity which appears in the denominator of (3.23) and (3.24). See Section 3.2 for the definition of a continuous time random walk as the difference of two independent Poisson processes of rate $\lambda = 1/2$. We prove

THEOREM 1.19. *Consider $n+2$ independent continuous time random walks $X_1(t), \dots, X_{n+2}(t)$ which begin at locations $x_1 < x_2 < \cdots < x_{n+2}$ at time $t=0$. For a given $y_1 < y_2 < \cdots < y_{n+2}$, let $\mathbb{P}_{n+2}(x_1, \dots, x_{n+2}; y_1, \dots, y_{n+2}; T)$ be the probability that $X_1(t), \dots, X_{n+2}(t)$ satisfy $X_j(T) = y_j$ for $j = 1, \dots, n+2$, and $X_1(t), \dots, X_{n+2}(t)$ do not intersect for $0 \le t \le T$. Assume $x_j = j$ for $1 \le j \le n+2$, $y_k = k$ for $1 \le k \le n$, while both y_{n+1} and y_{n+2} are arbitrary except for the restriction $n < y_{n+1} < y_{n+2}$. If T, $\beta_2 := y_{n+2} - n$, and $\beta_1 := y_{n+1} - n$ are bounded as $n \to \infty$, then $\mathbb{P}_{n+2}(x_1, \dots, x_{n+2}; y_1, \dots, y_{n+2}; T)$ as $n \to \infty$ has the asymptotics*

$$\mathbb{P}_{n+2}(x_1, \dots, x_{n+2}; y_1, \dots, y_{n+2}; T) = \frac{e^{-(n+2)T} \left(\frac{T}{2}\right)^{\beta_1+\beta_2-3} e^{\frac{T^2}{4}}}{(\beta_1-1)!(\beta_2-1)!} (\beta_2 - \beta_1) \left(1 + O\left(\frac{1}{n^2}\right)\right). \tag{1.50}$$

THEOREM 1.20. *Let $\mathbb{P}_{n+1}(x_1, \dots, x_{n+1}; y_1, \dots, y_{n+1}; T)$ be as defined in Theorem 1.19. Suppose that $x_j = j$ for $1 \le j \le n$, $y_k = k$ for $1 \le k \le n$, while both x_{n+1} and y_{n+1} are arbitrary. If T, $\beta_1 := y_{n+1} - n$, and $\gamma_1 := x_{n+1} - n$ are bounded as $n \to \infty$, then $\mathbb{P}_{n+1}(x_1, \dots, x_{n+1}; y_1, \dots, y_{n+1}; T)$ as $n \to \infty$ has the asymptotics*

$$e^{-(n+1)T} e^{\frac{T^2}{4}} \left[I_{y_{n+1}-x_{n+1}}(T) + \left(\frac{T}{2}\right)^{\gamma_1+\beta_1-2} \sum_{k=0}^{k_0} \frac{\left(\frac{T}{2}\right)^{-2k}}{(\gamma_1-1-k)!(\beta_1-1-k)!} \right] (1+O(e^{-cn})), \tag{1.51}$$

where $I_\nu(\cdot)$ *is the modified Bessel function of order* ν, $k_0 := \min\{\gamma_1 - 1, \beta_1 - 1\}$, *and* c *is any positive constant.*

The proofs of Theorems 1.19 and 1.20 are presented in Section 7. These calculations are performed under the simplifying assumptions that the total time T and the length of the gaps between consecutive starting and ending points of the random walks are bounded. It is very interesting to study the situations in which these quantities grow at various rates with n, the number of random walks.

REMARK 1.21. It may be worth mentioning that when $x_j = j$ and $y_k = k$ for all $1 \leq j, k \leq n$, then one has to analyze the (pure) Toeplitz determinant with symbol $e^{\frac{T}{2}(z+z^{-1}-2)}$ and the strong Szegő theorem implies

$$\mathbb{P}_n(1,2,\cdots,n;1,2,\cdots,n;T) = e^{-nT} e^{\frac{T^2}{4}} \left(1 + O(e^{-cn})\right), \qquad c > 0. \tag{1.52}$$

Also, comparing with Theorem 1.19, if we have $n+1$ continuous time random walks where only the top path has a gapped ending point $y_{n+1} > n+1$, then one has to asymptotically analyze the associated single bordered Toeplitz determinant, and as shown in Theorem 7.2,

$$\begin{aligned}\mathbb{P}_{n+1}&(1,2,\cdots,n+1;1,2,\cdots,n,y_{n+1};T) \\ &= \frac{e^{-(n+1)T}\left(\frac{T}{2}\right)^{\beta_1-1} e^{\frac{T^2}{4}}}{(\beta_1-1)!}\left(1 + O(e^{-cn})\right), \qquad c > 0,\end{aligned}$$

where $\beta_1 := y_{n+1} - n$. Comparing this with (1.50) invites one to expect the asymptotic formula

$$\begin{aligned}\mathbb{P}_{n+m}&(1,\ldots,n+m;1,\ldots,n,y_{n+1},\ldots,y_{n+m};T) = \\ &\frac{e^{\frac{T^2}{4}} e^{-(n+m)T} \prod_{j=1}^m \left(\frac{T}{2}\right)^{\beta_j - j}}{\prod_{j=1}^m (\beta_j - 1)!} \prod_{1 \leq j < k \leq m} (\beta_k - \beta_j)\left(1 + o(1)\right),\end{aligned} \tag{1.53}$$

with $\beta_j := y_{n+j} - n$, although we do not propose it as a conjecture yet.

1.6. Outline for the rest of the paper. Sections 2, 3, and 4 deal with the applications described above. Propositions 1.9, 1.10, 1.13, and 1.16 are then proven in Section 5. The bordered and framed Hankel determinant formulas appearing in Sections 2 and 3 are proven in Section 6. Finally in Section 7 we prove the asymptotic results of Theorems 1.19 and 1.20.

2. Height of non-intersecting Brownian excursions with wanderers

Consider n Brownian excursions $X_1(t), \ldots, X_n(t)$, i.e., Brownian bridges on $\mathbb{R}_+$ with an absorbing wall at zero. For $0 \leq m < n$ and vectors $\vec{\alpha} = (\alpha_1, \ldots, \alpha_m)$, $\vec{\beta} = (\beta_1, \ldots, \beta_m)$, satisfying $\alpha_1 \geq \alpha_2 \geq \cdots \geq \alpha_m \geq 0$ and $\beta_1 \geq \beta_2 \geq \cdots \geq \beta_m \geq 0$, we will condition the Brownian excursions as follows (also see Figure 1):

- At time $t = 0$ we start the paths at $X_1(0) = X_2(0) = \cdots = X_{n-m}(0) = 0$, and $X_n(0) = \alpha_1, X_{n-1}(0) = \alpha_2, \cdots X_{n-m+1}(0) = \alpha_m$.
- At time $t = 1$ the paths end at the points $X_1(1) = X_2(1) = \cdots = X_{n-m}(1) = 0$, and $X_n(1) = \beta_1, X_{n-1}(1) = \beta_2, \ldots, X_{n-m+1}(1) = \beta_m$.
- The particles do not intersect at times $0 < t < 1$, and are ordered as $X_1(t) < X_2(t) < \cdots < X_n(t)$.

We denote the conditional probability described above as $\mathbb{P}_m^{\vec{\alpha},\vec{\beta}}$, and refer to the model as n *non-intersecting Brownian excursions with wanderers on the time interval* $[0,1]$. Our primary interest in this section is in the random variable

$$\mathcal{M}_n := \max_{t\in(0,1)} X_n(t).$$

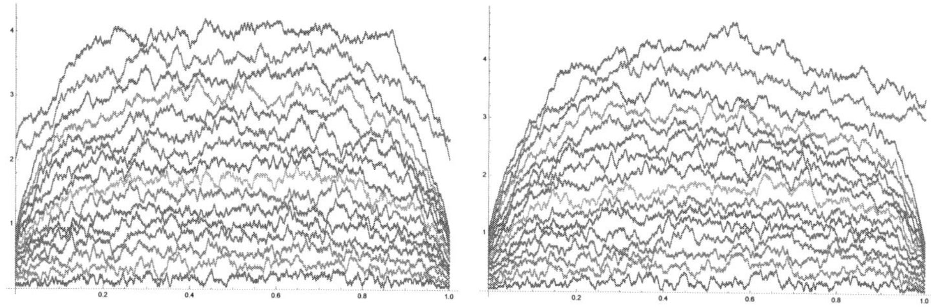

FIGURE 1. Two simulations of non-intersecting Brownian excursions with wanderers with $n = 16$ and $m = 2$. In the figure on the left, $\alpha_1, \alpha_2, \beta_1$, and β_2 are all positive and the distribution of $\mathcal{M}_n$ is described in (2.4) by a framed Hankel determinant. In the figure on the right, $\alpha_1 = \alpha_2 = 0$, and the distribution of $\mathcal{M}_n$ is described in (2.5) by a bordered Hankel determinant.

In the paper **[67]** (see also **[49]**) it was shown that the in the case $m = 0$, i.e., all Brownian particles begin and end at 0, distribution of the random variable $\mathcal{M}_n$ is expressed as a Hankel determinant with respect to a discrete measure. Here we describe the case $m > 0$, in which most of the particles begin and end at 0, but a few have positive starting/ending points. In this case the discrete Hankel determinant is replaced by a bordered or framed Hankel determinant. To describe these determinants introduce the discrete measure supported on perfect squares[5]

$$\mu = \sum_{h=1,4,9,16,\dots} h e^{-\frac{\pi^2 h}{2M^2}} \delta_h, \tag{2.1}$$

as well as the signed measure depending on a parameter α,

$$\nu(\alpha) = \frac{1}{\alpha} \sum_{h=1,4,9,16,\dots} \left(\sqrt{h} e^{-\frac{\pi^2 h}{2M^2}} \sin\left(\frac{\sqrt{h}\pi\alpha}{M}\right)\right) \delta_h, \tag{2.2}$$

and the constants depending on two parameters

$$\begin{aligned} s(\alpha,\beta) &= \frac{1}{\alpha\beta} \sum_{h=1,4,9,16,\dots} e^{-\frac{\pi^2 h}{2M^2}} \sin\left(\frac{\sqrt{h}\pi\alpha}{M}\right) \sin\left(\frac{\sqrt{h}\pi\beta}{M}\right) \\ &= \frac{1}{\alpha\beta} \sum_{h=1}^{\infty} e^{-\frac{\pi^2 h^2}{2M^2}} \sin\left(\frac{h\pi\alpha}{M}\right) \sin\left(\frac{h\pi\beta}{M}\right). \end{aligned} \tag{2.3}$$

[5]While it is perhaps more natural to replace $h \mapsto h^2$ in the definition of μ so that the measure is supported on the positive integers, we choose to write μ as supported on the perfect squares so that the relevant Hankel matrices are written with all moments of μ rather than just the even ones.

We have the following proposition describing the distribution of $\mathcal{M}_n$ in terms of a framed or bordered Hankel determinant.

PROPOSITION 2.1. *Fix integers n and m with $0 \le m < n$ as well as vectors $\vec{\alpha}$ and $\vec{\beta}$ satisfying $\alpha_1 > \alpha_2 > \cdots > \alpha_m > 0$ and $\beta_1 > \beta_2 > \cdots > \beta_m > 0$. The distribution of the maximum height $\mathcal{M}_n$ of $X_n(t)$ is given by the formula*

$$\mathbb{P}_m^{\vec{\alpha},\vec{\beta}}(\mathcal{M}_n < M) = \tag{2.4}$$

$$\frac{2^{n/2}\pi^{2(n-m)^2+n/2}\prod_{j=1}^m e^{\frac{1}{2}(\alpha_j^2+\beta_j^2)}}{M^{2(n-m)^2+n}\det\left[\frac{\sinh(\alpha_{m-j+1}\beta_{m-k+1})}{\alpha_{m-j+1}\beta_{m-k+1}}-\sum_{\ell=0}^{n-m-1}\frac{(\alpha_{m-j+1}\beta_{m-k+1})^{2\ell}}{(2\ell+1)!}\right]_{j,k=1}^m\prod_{j=1}^{n-m}(2j-1)!}$$
$$\times H_n^F(\mu;\nu(\beta_m),\nu(\beta_{m-1}),\dots,\nu(\beta_1);\nu(\alpha_m),\nu(\alpha_{m-1}),\dots,\nu(\alpha_1);S),$$

for all $M \ge \max\{\alpha_1,\beta_1\}$, where the framed Hankel determinant $H_n^F(\mu;\boldsymbol{\nu}_m;\boldsymbol{\eta}_m;S)$ is as defined in (1.13), the measures μ and $\nu(\alpha)$ are defined in (2.1) and (2.2), and the $m \times m$ matrix S is given by

$$S = \Big(s(\alpha_{m-j+1},\beta_{m-k+1})\Big)_{j,k=1}^m,$$

with $s(\alpha,\beta)$ defined in (2.3).

In the case all particles begin at zero, but the top m particles end at $0 < \beta_m < \beta_{m-1} < \cdots < \beta_1$, the distribution of the maximum height $\mathcal{M}_n$ is given by the formula

(2.5)

$$\mathbb{P}_m^{\vec{0},\vec{\beta}}(\mathcal{M}_n < M) = (-1)^{m(n-1)}\frac{2^{n/2}\pi^{n^2+(n-m)^2+n/2}}{M^{n^2+(n-m)^2+n}\Delta(\vec{\beta}^2)}\prod_{j=1}^m\frac{e^{\frac{1}{2}\beta_j^2}}{\beta_j^{2(n-m)}}\prod_{j=1}^{n-m}\left(\frac{1}{(2j-1)!}\right)$$
$$\times H_n^B[\mu;\nu(\beta_m),\dots,\nu(\beta_1)],$$

where $\Delta(\vec{\beta}^2) = \prod_{1\le j<k\le m}(\beta_k^2-\beta_j^2)$ and the bordered Hankel determinant $H_n^B[\mu;\vec{\boldsymbol{\nu}}_m]$ is as defined in (1.12).

REMARK 2.2. The above Proposition is stated for distinct $\alpha_1,\dots,\alpha_m$ and $\beta_1,\dots,\beta_m$. If any of the α_j's or β_j's coincide, then the framed or bordered Hankel determinants in (2.4) and (2.5) vanish, as do the $m \times m$ determinants in the denominators of (2.4) and (2.5). In this case those formulas may be understood using l'Hôpitals rule.

In the case $m = 0$, i.e., all starting and ending points of the non-intersecting Brownian excursions are at 0, it was proven in [**67**] that for all $x \in \mathbb{R}$,

$$\lim_{n\to\infty}\mathbb{P}_0(\mathcal{M}_n < \sqrt{2n} - 2^{-7/6}n^{1/6}x) = F_1(2^{2/3}x),$$

where $F_1(x)$ is the cumulative distribution function for the Tracy–Widom GOE (TW-GOE) distribution which describes the location of the largest eigenvalue in the Gaussian orthogonal ensemble of real symmetric random matrices [**76**]. This result is quite expected since the top path $X_n(t)$ converges to the Airy$_2$ process minus a parabola in the proper scaling limit[6], and it is well known that that the distribution of the maximum of the Airy$_2$ process minus a parabola is given by TW-GOE (up to a constant rescaling factor) [**40, 62**].

When $n-m$ starting and ending points are at zero, but m of them are positive so that $X_{n-j+1}(0) = \alpha_j$ and $X_{n-j+1}(1) = \beta_j$ for $j = 1, \dots, m$, then a different limiting process may emerge as the scaling limit of the top path. In particular, if m is fixed and α_j and β_j are scaled with n as

$$\alpha_j = \sqrt{2n} - (2n)^{1/6} a_j, \quad \beta_j = \sqrt{2n} - (2n)^{1/6} b_j, \tag{2.6}$$

then the top path $X_n(t)$ should converge[7] in a certain scaling limit as $n \to \infty$ to a perturbation of the Airy$_2$ process known as the *Airy process with wanderers*, again minus a parabola. The Airy process with wanderers depends on the parameters $a_1, \dots, a_m$ and $b_1, \dots, b_m$, and we denote it as $\mathcal{A}_m^{(\vec{a},\vec{b})}(t)$, where $\vec{a} = (a_1, \dots, a_m)$ and $\vec{b} = (b_1, \dots, b_m)$. This first appeared in the paper [**26**] in the context of last passage percolation with boundary sources, and then in [**2, 3**] as a scaling limit of non-intersecting paths. More recently in [**68**], several formulas were obtained for the distribution of the supremum of $\mathcal{A}_m^{(\vec{a},\vec{b})}(t) - t^2$ for $0 \le a_1 \le \cdots \le a_m$ and $0 \le b_1 \le \cdots \le b_m$[8]. We denote this distribution function as $F_m^{(\vec{a},\vec{b})}(x)$:

$$F_m^{(\vec{a},\vec{b})}(x) := \mathbb{P}\left(\sup_{t \in \mathbb{R}} \left(\mathcal{A}_m^{(\vec{a},\vec{b})}(t) - t^2 \right) < x \right). \tag{2.7}$$

In [**68**], $F_m^{(\vec{a},\vec{b})}(x)$ is expressed as a Fredholm determinant (Theorem 1) and in terms of a Lax pair for the Painlevé II equation (Theorem 2). Both formulas have well defined limits as the parameters a_j and/or b_j go to $+\infty$. In view of the scaling (2.6), this corresponds to pushing the starting and/or ending points of the non-intersecting paths to 0. In particular we have

$$\lim_{\substack{\vec{a} \to \vec{0} \\ \vec{\beta} \to \vec{0}}} F_m^{(\vec{a},\vec{b})}(x) =: F_m^{(\infty,\infty)}(x) = F_1(2^{2/3}x),$$

where $F_1(x)$ is the TW-GOE distribution function. We can also take the parameters a_j to $+\infty$ but leave $\vec{b}$ finite, which corresponds to the starting points of the non-intersecting paths going to 0 but the top m ending points still positive. We denote this limit as

$$F_m^{(\infty,\vec{b})}(x) := \lim_{\vec{\alpha} \to \vec{0}} F_m^{(\vec{a},\vec{b})}(x).$$

In light of the expected convergence of $X_n(t)$ to $\mathcal{A}_m^{(\vec{a},\vec{b})}(t)$ in the proper scaling limit, we therefore expect that for α_j and β_j scaled as in (2.6) with $a_j > 0$ and $b_j > 0$ for

[6] It is well known that the top path in n non-intersecting Brownian *bridges* conditioned to have all starting and ending points at zero converges in the proper scaling limit to the Airy$_2$ process minus a parabola. For non-intersecting Brownian excursions, the convergence of the top path to the Airy$_2$ process minus a parabola follows from combining a finite dimensional convergence implied by [**77**, Section 5.3] with the results of [**39**].

[7] The works [**2**] and [**3**] proved this convergence for non-intersecting Brownian bridges, not non-intersecting Brownian excursions.

[8] The condition that all a_j's and b_j's are positive is essential to get a nontrivial limit. Otherwise the probability converges to 0.

$j = 1, \ldots, m$,

$$\begin{aligned} &\lim_{n\to\infty} \mathbb{P}_m^{\vec{\alpha},\vec{\beta}}(\mathcal{M}_n < \sqrt{2n} - 2^{-7/6}n^{1/6}x) = F_m^{(\vec{a},\vec{b})}(x), \\ &\lim_{n\to\infty} \mathbb{P}_m^{\vec{0},\vec{\beta}}(\mathcal{M}_n < \sqrt{2n} - 2^{-7/6}n^{1/6}x) = F_m^{(\infty,\vec{b})}(x). \end{aligned} \tag{2.8}$$

Somewhat surprisingly, [**68**, Corollary 1] states that the two limiting functions described above are essentially the same. More precisely it states

$$F_m^{(\vec{a},\vec{b})}(x) = F_{2m}^{(\infty,\langle\vec{a},\vec{b}\rangle)}(x),$$

where $\langle\vec{a},\vec{b}\rangle$ is the reordering of the combined elements of $\vec{a}$ and $\vec{b}$ in non-decreasing order. Assuming (2.8), this would imply that $\mathbb{P}_m^{\vec{\alpha},\vec{\beta}}(\mathcal{M}_n < \sqrt{2n} - 2^{-7/6}n^{1/6}x)$, expressed in (2.4) in terms of a framed Hankel determinant, and and $\mathbb{P}_{2m}^{\vec{0},(\vec{\alpha},\vec{\beta})}(\mathcal{M}_n < \sqrt{2n} - 2^{-7/6}n^{1/6}x)$, expressed in (2.5) as a bordered Hankel determinant, have the same scaling limit as $n \to \infty$ when $\vec{\alpha}$ and $\vec{\beta}$ are scaled as (2.6) with $a_j, b_j > 0$, and $(\vec{\alpha},\vec{\beta})$ is the reordering of the combined elements of $\vec{a}$ and $\vec{b}$ in non-increasing order.

In order to evaluate the formulas (2.4) and (2.5) as $n \to \infty$, it is useful to use Corollaries 1.11 and 1.14 to express them in terms of orthogonal polynomials. Then one may hope to evaluate them asymptotically as $n \to \infty$ provided asymptotics of the orthogonal polynomials are known. To apply the Corollaries 1.11 and 1.14, we first introduce the system of monic orthogonal polynomials $\{P_k(x) = x^k + \cdots\}_{k=0}^{\infty}$ with respect to the discrete measure μ given in (2.1):

$$\sum_{x=1,4,9,16,\ldots} P_k(x)P_j(x)xe^{-\frac{\pi^2}{2M^2}x} = h_k\delta_{jk}, \tag{2.9}$$

or equivalently

$$\sum_{x=1}^{\infty} P_k(x^2)P_j(x^2)x^2e^{-\frac{\pi^2}{2M^2}x^2} = h_k\delta_{jk}. \tag{2.10}$$

Applying Corollary 1.17 to (2.4) gives

(2.11)

$$\mathbb{P}_m^{\vec{\alpha},\vec{\beta}}(\mathcal{M}_n < M) = \frac{2^{n/2}\pi^{2(n-m)^2+n/2}\prod_{j=1}^m e^{\frac{1}{2}(\alpha_j^2+\beta_j^2)}H_{n-m}[\mu]}{M^{2(n-m)^2+n}\det\left[\frac{\sinh(\alpha_{m-j+1}\beta_{m-k+1})}{\alpha_{m-j+1}\beta_{m-k+1}} - \sum_{\ell=0}^{n-m-1}\frac{(\alpha_{m-j+1}\beta_{m-k+1})^{2\ell}}{(2\ell+1)!}\right]_{j,k=1}^m \prod_{j=1}^{n-m}(2j-1)!}$$

$$\times \det\left[S - \left(\sum_{x,y=1}^{\infty} K_{n-m-1}(x^2,y^2)xye^{-\frac{\pi^2(x_j+y_k^2)}{2M^2}}\sin\left(\frac{x\pi\beta_k}{M}\right)\sin\left(\frac{x\pi\alpha_j}{M}\right)\right)_{j,k=1}^m\right],$$

where

$$K_{n-m-1}(x,y) = \sum_{k=0}^{n-m-1}\frac{P_k(x)P_k(y)}{h_k},$$

$$S = \left(\sum_{h=1}^{\infty}\frac{e^{-\frac{\pi^2h^2}{2M^2}}\sin\left(\frac{h\pi\alpha_{m-j+1}}{M}\right)\sin\left(\frac{h\pi\beta_{m-k+1}}{M}\right)}{\alpha_{m-j+1}\beta_{m-k+1}}\right)_{j,k=1}^m.$$

Applying Corollary 1.11 to (2.5) gives

$$(2.12)\quad \mathbb{P}_m^{\vec{0},\vec{\beta}}(\mathcal{M}_n < M) =$$

$$(-1)^{n(n-1)/2+(n-m)(n-m-1)/2} \frac{2^{n/2}\pi^{n^2+(n-m)^2+n/2}}{M^{n^2+(n-m)^2+n}\Delta(\vec{\beta}^2)} \prod_{j=1}^{m} \frac{e^{\frac{1}{2}\beta_j^2}}{\beta_j^{2(n-m)}} \prod_{j=1}^{n-m}\left(\frac{1}{(2j-1)!}\right)$$

$$\times H_{n-m}[\mu] \det\left[\sum_{x=1}^{\infty}\left(xe^{-\frac{\pi^2 x_j^2}{2M^2}}\sin\left(\frac{x\pi\beta_j}{M}\right)\right)P_{n-m+k-1}(x^2)\right]_{j,k=1}^{m}.$$

To evaluate (2.11) and (1.36) asymptotically as $n \to \infty$, one first needs large-degree asymptotic formulas for the orthogonal polynomials (2.10). In general this can be accomplished by the Riemann–Hilbert method [**5**,**21**,**42**]. Indeed, for the orthogonal polynomials (2.10), the relevant Riemann–Hilbert analysis was performed in [**67**], although explicit asymptotic formulas for the orthogonal polynomials do not appear there. Nonetheless, one can use the steepest descent analysis of the Riemann–Hilbert problem from [**67**] to extract asymptotic formulas for the orthogonal polynomials (2.10) and then insert the asymptotic formulas for the orthogonal polynomials into (2.11) and (1.36). This is the approach outlined in [**68**, Appendix A] to prove (2.8) with $m = 2$ and $\vec{\alpha} = \vec{0}$, though few details are given there. Here we repeat the remark from [**68**, Appendix A] that larger values of m require finer asymptotic formulas for the orthogonal polynomials, thus the analysis becomes increasingly difficult for larger values of m.

3. Width of non-intersecting processes

The paper [**8**] considers the width of various ensembles of non-intersecting paths: non-intersecting Brownian bridges, discrete random walks, and continuous time random walks. For the simplest initial conditions on the random paths, the distribution of the width of the ensemble is described as the ratio of two Hankel determinants for the non-intersecting Browinan bridges, and the ratio of two Toeplitz determinants for the non-intersecting random walks. In this section we describe how the initial conditions may be perturbed slightly so that the Toeplitz/Hankel determinants from [**8**] are replaced with bordered or framed Toeplitz/Hankel determinants.

3.1. Non-intersecting Brownian bridges with wanderers. Consider n independent Brownian motions $X_1(t), \dots, X_n(t)$. For integers $m_1 \geq 0$ and $m_2 \geq 0$ satisfying $m_1+m_2 = m < n$. define the vectors $\vec{\alpha} = (\alpha_1, \dots, \alpha_m)$, $\vec{\beta} = (\beta_1, \dots, \beta_m)$, satisfying $\alpha_1 \leq \alpha_2 \leq \cdots \leq \alpha_{m_1} < 0 < \alpha_{m_1+1} \leq \cdots \leq \alpha_{m_1+m_2}$ and $\beta_1 \leq \beta_2 \leq \cdots \leq \beta_{m_1} < 0 < \beta_{m_1+1} \leq \cdots \leq \beta_{m_1+m_2}$. Similar to the non-intersecting Brownian excursions in Section 2, we will condition the Brownian motions as follows (also see Figure 2):

- At time $t = 0$ we start the paths at $X_{m_1+1}(0) = X_{m_1+2}(0) = \cdots = X_{n-m_2}(0) = 0$, as well as $X_1(0) = \alpha_1, X_2(0) = \alpha_2, \dots, X_{m_1}(0) = \alpha_{m_1}$, and $X_{n-m_2+1}(0) = \alpha_{m_1+1}, X_{n-m_2+2}(0) = \alpha_{m_1+2}, \dots, X_n(0) = \alpha_{m_1+m_2}$.
- At time $t = 1$ the paths end at the points $X_{m_1+1}(1) = X_{m_1+2}(1) = \cdots = X_{n-m_2}(1) = 0$, as well as $X_1(1) = \beta_1, X_2(1) = \beta_2, \dots, X_{m_1}(1) = \beta_{m_1}$, and $X_{n-m_2+1}(1) = \beta_{m_1+1}, X_{n-m_2+2}(1) = \beta_{m_1+2}, \dots, X_n(1) = \beta_{m_1+m_2}$.

- The particles not to intersect at times $0 < t < 1$, and are ordered as $X_1(t) < X_2(t) < \cdots < X_n(t)$.

We refer to this model as *non-intersecting Brownian bridges with wanderers* and denote the conditional probability described above as $\mathbb{P}^{\vec{\alpha},\vec{\beta}}_{m_1,m_2}$.

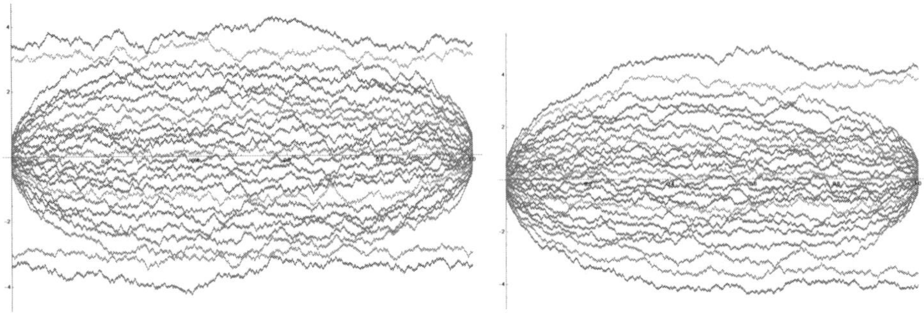

FIGURE 2. Two simulations of non-intersecting Brownian bridges with wanderers with $n = 24$ and $m_1 = m_2 = 2$. In the figure on the left, $\alpha_1, \alpha_2, \alpha_3, \alpha_4, \beta_1, \beta_2, \beta_3$, and β_4 are all non-zero and the distribution of W_n is described in (3.4) in terms of framed Hankel determinants. In the figure on the right, $\alpha_1 = \alpha_2 = \alpha_3 = \alpha_4 = 0$, and the distribution of W_n is described in (3.6) in terms of bordered Hankel determinants.

We define the width of the ensemble as

$$\mathcal{W}_n := \max_{0\leq t\leq 1}\{X_n(t) - X_1(t)\}.$$

For the case $m_1 = m_2 = 0$, in which all particles begin and end at 0, the distribution of $\mathcal{W}_n$ was presented in **[8]** as the integral of the ratio of two Hankel determinants with respect to Gaussian (or Gaussian-like) measures, one continuous and one discrete. Here we present the analogous formulas in terms of bordered and/or framed Hankel determinants when m_1 or m_2 is nonzero. In order to present the formulas we first define the continuous measure μ and discrete measure $\mu^{M,\tau}$ as

$$\text{(3.1)} \qquad d\mu(x) = e^{-x^2/2}dx, \quad \mu^{M,\tau} = \frac{2\pi}{M}\sum_{x\in\mathcal{D}_{M,\tau}} e^{-x^2/2}\delta_x,$$

where

$$\text{(3.2)} \qquad \mathcal{D}_{M,\tau} = \left\{\frac{2\pi(k-\tau)}{M} \; : \; k \in \mathbb{Z}\right\}.$$

We also introduce the complex measures μ_β and $\mu_\beta^{M,\tau}$ as

$$\text{(3.3)} \qquad d\mu_\beta(x) = e^{-x^2/2-\mathrm{i}\beta x}dy, \quad \mu_\beta^{M,\tau} = \frac{2\pi}{M}\sum_{x\in\mathcal{D}_{M,\tau}} e^{-x^2/2-\mathrm{i}\beta x}\delta_x,$$

We then have the following proposition, which relies on the Karlin–McGregor formula **[64]** for non-intersecting processes. We present a proof in Section 6.

PROPOSITION 3.1. *Fix integers n, m_1, and m_2 with $m_1 \geq 0$, $m_2 \geq 0$, and $m = m_1 + m_2 < n$ as well as vectors $\vec{\alpha} = (\alpha_1, \dots, \alpha_m)$ and $\vec{\beta} = (\beta_1, \dots, \beta_m)$ satisfying $\alpha_1 < \alpha_2 < \cdots < \alpha_{m_1} < 0 < \alpha_{m_1+1} < \cdots < \alpha_{m_1+m_2}$ and $\beta_1 < \beta_2 < \cdots < \beta_{m_1} < 0 < \beta_{m_1+1} < \cdots < \beta_{m_1+m_2}$. The distribution of the width of the non-intersecting Brownian bridges with wanderers is given by the formula*

$$(3.4)\quad \mathbb{P}^{\alpha,\beta}_{m_1,m_2}(\mathcal{W}_n < M) = \int_0^1 \frac{H_n^F[\mu^{M,\tau}; \mu^{M,\tau}_{\beta_1}, \mu^{M,\tau}_{\beta_2}, \dots, \mu^{M,\tau}_{\beta_m}; \mu^{M,\tau}_{(-\alpha_1)}, \dots, \mu^{M,\tau}_{(-\alpha_m)}; A^{M,\tau}]}{H_n^F[\mu; \mu_{\beta_1}\mu_{\beta_2}, \dots, \mu_{\beta_m}; \mu_{(-\alpha_1)}, \dots, \mu_{(-\alpha_m)}; A]}\, d\tau,$$

for all $M \geq \max\{\alpha_m - \alpha_1, \beta_m - \beta_1\}$, where the framed Hankel determinant $H_n^F(\mu; \vec{\nu}_m; \vec{\eta}_m; A)$ is as defined in (1.13), and the $m \times m$ matrix A is given by

$$(3.5)\quad \begin{aligned} A &= \left(\int_{\mathbb{R}} e^{-y^2/2+iy(\alpha_k-\beta_j)}\, dy\right)_{j,k=1}^m = (2\pi)^{m/2}\left(e^{-(\alpha_k-\beta_j)^2/2}\right)_{j,k=1}^m \\ A^{M,\tau} &= \left(\frac{2\pi}{M}\sum_{y\in\mathcal{D}_{M,\tau}} e^{-y^2/2+iy(\alpha_k-\beta_j)}\right)_{j,k=1}^m . \end{aligned}$$

In the case all particles begin at zero, but the bottom m_1 particles end at $\beta_1 < \beta_2 < \cdots < \beta_{m_1} < 0$ and the top m_2 particles end at $0 < \beta_{m_1+1} < \cdots < \beta_m$, the distribution of the $\mathcal{W}_n$ is given by the formula

$$(3.6)\quad \mathbb{P}^{\vec{0},\vec{\beta}}_{m_1,m_2}(\mathcal{W}_n < M) = \int_0^1 \frac{H_n^B[\mu^{M,\tau}; \mu^{M,\tau}_{\beta_1}, \dots, \mu^{M,\tau}_{\beta_m}]}{H_n^B[\mu; \mu_{\beta_1}, \dots, \mu_{\beta_m}]}\, d\tau,$$

where the bordered Hankel determinant $H_n^B[\mu; \vec{\nu}_m]$ is as defined in (1.12).

REMARK 3.2. As in Remark 2.2 following Proposition 2.1, the above formulas are also valid when two or more of the α_j's or β_j's coincide, after applyng l'Hôpital's rule.

In the paper [**8**], it was shown that for the case $m_1 = m_2 = 0$, so that all Brownian particles begin and end at 0, for any $x \in \mathbb{R}$,

$$(3.7)\quad \lim_{n\to\infty} \mathbb{P}_0\left(\mathcal{W}_n \leq 2\sqrt{n} + 2^{-2/3}n^{-1/6}x\right) = F_2(x),$$

where $F_2(x)$ is the Tracy–Widom GUE (TW-GUE) distribution function which describes the location of the largest eigenvalue in the Gaussian unitary ensemble of complex Hermitian random matrices [**75**]. Since the top and bottom curves of the ensemble each converge in the proper scaling limit to the Airy$_2$ process minus a parabola, the result (3.7) has the interpretation that [**8**, Equation (19)]

$$\mathbb{P}\left(2^{-1/3}\sup_{t\in\mathbb{R}}\left(\mathcal{A}^{(1)}(t) + \mathcal{A}^{(2)}(t) - 2t^2\right) \leq x\right) = F_2(x),$$

where $\mathcal{A}^{(2)}$ and $\mathcal{A}^{(2)}$ are two independent copies of the Airy$_2$ process. In other words the supremum of the sum of two independent Airy$_2$ processes minus a parabola is given by the Tracy–Widom GUE distribution. In the setting of non-intersecting Brownian motions with wanderers we expect a similar result for Airy processes with wanderers. Namely, for fixed m_1 and m_2 if we take the scalings

$$(3.8)\quad M = 2\sqrt{n} + 2^{-2/3}n^{-1/6}x,$$

and

(3.9)
$$\alpha_j = -2\sqrt{n} + 2^{2/3}n^{1/6}a_j, \quad \beta_j = -2\sqrt{n} + 2^{2/3}n^{1/6}b_j, \quad j = 1, \ldots, m_1,$$
(3.10)
$$\alpha_j = 2\sqrt{n} - 2^{2/3}n^{1/6}a_j, \quad \beta_j = 2\sqrt{n} - 2^{2/3}n^{1/6}b_j, \quad j = m_1+1, \ldots, m_1+m_2,$$

with all a_j's and b_j's positive, then we expect (3.4) to converge to the distribution of the supremum of the sum of to Airy processes with wanderers minus a parabola. That is, we expect

$$\lim_{n\to\infty} \mathbb{P}^{\alpha,\beta}_{m_1,m_2}(\mathcal{W}_n < M) = \mathbb{P}\left(2^{-1/3}\sup_{t\in\mathbb{R}}\left(\mathcal{A}^{(\vec{a}_{m_1},\vec{b}_{m_1})}(t) + \mathcal{A}^{(\vec{a}_{m_2},\vec{b}_{m_2})}(t) - 2t^2\right) \le x\right),$$

where $\vec{a}_{m_1} = (a_1, \ldots, a_{m_1})$, $\vec{a}_{m_2} = (a_{m_1+1}, \ldots, a_{m_1+m_2})$, $\vec{b}_{m_1} = (b_1, \ldots, b_{m_1})$, $\vec{b}_{m_2} = (b_{m_1+1}, \ldots, b_{m_1+m_2})$, and the Airy processes with wanderers $\mathcal{A}^{(\vec{a}_{m_1},\vec{b}_{m_1})}$ and $\mathcal{A}^{(\vec{a}_{m_2},\vec{b}_{m_2})}(t)$ are assumed to be independent. Similarly we expect (3.6) to converge to the distribution of

$$2^{-1/3}\sup_{t\in\mathbb{R}}\left(\mathcal{A}(t) + \mathcal{A}^{(\vec{a}_m,\vec{b}_m)}(t) - 2t^2\right),$$

where $\mathcal{A}$ is the usual Airy$_2$ process, and is independent from the Airy process with wanderers $\mathcal{A}^{(\vec{a}_m,\vec{b}_m)}$.

With an eye towards asymptotic analysis, we can write (3.4) and (3.6) in terms of orthogonal polynomials using Corollaries 1.11 and 1.17. Introduce the systems of monic orthogonal polynomials $\{P_k(x)\}_{k=0}^\infty$, $P_k(x) = x^k + \cdots$, and $\{P_k^{M,\tau}(x)\}_{k=0}^\infty$, $P_k(x) = x^k + \cdots$, via the orthogonality conditions
(3.11)
$$\int_{\mathbb{R}} P_j(x)P_k(x)e^{-x^2/2}\mathrm{d}x = h_k\delta_{jk}, \quad \frac{2\pi}{M}\sum_{x\in\mathcal{D}_{M,\tau}} P_j^{M,\tau}(x)P_k^{M,\tau}(x)e^{-x^2/2} = h_k^{M,\tau}\delta_{jk}.$$

Then using Corollary 1.17 in (3.4) gives (with $m = m_1 + m_2$),

$$\text{(3.12)}\quad \mathbb{P}^{\vec{\alpha},\vec{\beta}}_{m_1,m_2}(\mathcal{W}_n < M) = \int_0^1 \frac{H_{n-m}[\mu^{M,\tau}]}{H_{n-m}[\mu]}$$
$$\times \frac{\det\left[A^{M,\tau} - \left(\left(\frac{2\pi}{M}\right)^2 \sum_{y_j\in\mathcal{D}_{M,\tau}}\sum_{x_k\in\mathcal{D}_{M,\tau}} K^{M,\tau}_{n-m-1}(x_k,y_j)e^{-x_k^2/2+\mathrm{i}x_k\alpha_k - y_j^2/2 - \mathrm{i}y_j\beta_j}\right)^m_{j,k=1}\right]}{\det\left[A - \left(\int_{\mathbb{R}}\int_{\mathbb{R}} K_{n-m-1}(x_k,y_j)e^{-x_k^2/2+\mathrm{i}x_k\alpha_k - y_j^2/2 - \mathrm{i}y_j\beta_j}\,\mathrm{d}x_k\,\mathrm{d}y_j\right)^m_{j,k=1}\right]}\mathrm{d}\tau,$$

where

$$K_{n-m-1}(x,y) = \sum_{k=0}^{n-m-1}\frac{P_k(x)P_k(y)}{h_k}, \quad K^{M,\tau}_{n-m-1}(x,y) = \sum_{k=0}^{n-m-1}\frac{P_k^{M,\tau}(x)P_k^{M,\tau}(y)}{h_k^{M,\tau}},$$

and using Corollary 1.11 in (3.6) gives (with $m = m_1 + m_2$),

$$\text{(3.13)}\quad \mathbb{P}^{\vec{0},\beta}_{m_1,m_2}(W_n < M)\int_0^1 \frac{H_{n-m}[\mu^{M,\tau}]}{H_{n-m}[\mu]} \times \frac{\left(\frac{2\pi}{M}\right)^m \displaystyle\sum_{x_1,x_2,\dots,x_m\in\mathcal{D}_{M,\tau}} \det\begin{pmatrix} P^{M,\tau}_{n-m}(x_1) & \cdots & P^{M,\tau}_{n-m}(x_m)\\ \vdots & \ddots & \vdots\\ P^{M,\tau}_{n-1}(x_1) & \cdots & P^{M,\tau}_{n-1}(x_m)\end{pmatrix}\prod_{j=1}^m e^{-x_j^2/2-\mathrm{i}\beta_j}}{\displaystyle\int_{\mathbb{R}}\cdots\int_{\mathbb{R}} \det\begin{pmatrix} P_{n-m}(x_1) & \cdots & P_{n-m}(x_m)\\ \vdots & \ddots & \vdots\\ P_{n-1}(x_1) & \cdots & P_{n-1}(x_m)\end{pmatrix}\prod_{j=1}^m \left(e^{-x_j^2/2-\mathrm{i}\beta_j}\mathrm{d}x_j\right)}\mathrm{d}\tau.$$

It was shown in [**8**] that in the scaling (3.8),

$$\lim_{n\to\infty}\frac{H_{n-m}[\mu^{M,\tau}]}{H_{n-m}[\mu]} = F_2(x),$$

where this limit is independent of τ. To evaluate the rest of (3.12), (3.13) as $n\to\infty$, one needs to know the large degree behavior of the orthogonal polynomials $P_k(z)$ and $P_j^{M,\tau}(z)$, preferably throughout the complex plane. The polynomials $\{P_j(x)\}_{j=0}^{\infty}$ are nothing but the Hermite polynomials, for which asymptotics are well known. For the polynomials $\{P_j^{M,\tau}(z)\}_{j=0}^{\infty}$, asymptotic formulas may be obtained by the Riemann–Hilbert method [**5**, **20**, **21**]. For this system of orthogonal polynomials with the scaling (3.8), the nonlinear steepest analysis of the corresponding discrete Riemann–Hilbert problem was carried out in [**67**] (see also [**69**]). Although asymptotic formulas for $P_n^{M,\tau}(z)$ do not appear directly in that paper, they follow from the analysis which appears there. If one has asymptotic formulas for $P_n^{M,\tau}(z)$ as $n\to\infty$, they could be inserted into (3.12), (3.13). Then, at least for small values of m, the sums appearing in the numerators of (3.12), (3.13) could be evaluated by first replacing the sum by a complex integral using the residue theorem, and then evaluating the integral using the method of steepest descent. This approach was used in [**69**, Section 5] to asymptotically calculate a sum very similar to the one in the numerator of (3.12).

3.2. Discrete and continuous time random walks. In the paper [**8**], the authors also consider the width of the non-intersecting random walks described in the introduction. That is, we consider n independent simple random walks on $\mathbb{Z}$, $X_1(t), X_2(t), \dots, X_n(t)$ which are conditioned to begin at the points $X_j(0) = 2x_j$ and end at the points $X_j(2T) = 2y_j$ after a fixed time $2T\in 2\mathbb{N}$, with $x_1 < x_2 < \cdots < x_n$ and $y_1 < y_2 < \cdots < y_n$. The *width* of this ensemble of non-intersecting paths as

$$W_n(2T) = \max_{0\le t\le 2T}\{X_n(t) - X_1(t)\}.$$

As discussed in the introduction, the probability that the paths begin at the points $2x_1,\dots,2x_n$ and end at the points $2y_1,\dots,2y_n$ at time $2T$ without intersecting for $0\le t\le 2T$ is

$$\text{(3.14)}\qquad \det\left[\phi_{y_k-x_j}\right]_{j,k=1}^n,\quad \phi(\zeta) = \zeta^{-T}\left(\frac{\zeta+1}{2}\right)^{2T},$$

where the extra factor of 1/2 in $\phi(\zeta)$ as compared to (1.6) is due to the fact that (3.14) gives a probability whereas (1.6) is counting paths.

A variation of the above set-up is to consider continuous time random walks. Let $N_1(t)$, $N_2(t)$ be independent Poisson processes with rate $\lambda = 1/2$, and let $X(t) = N_1(t) - N_2(t)$ be their difference. The probability generating function for the transition probabilities for $N_1(t)$ (and $N_2(t)$) is

$$p_{N_1}(z;t) = \sum_{k=0}^{\infty} \mathbb{P}(N_1(t) = k) z^k = \sum_{k=0}^{\infty} e^{-\frac{t}{2}} \frac{1}{k!} \left(\frac{t}{2}\right)^k z^k = e^{\frac{t}{2}(z-1)}.$$

The probability generating function for $X(t) = N_1(t) - N_2(t)$ is therefore

$$p_X(z;t) = p_{N_1}(z;t) p_{N_2}(1/z;t) = e^{\frac{t}{2}(z+1/z-2)}. \tag{3.15}$$

We refer to the process $X(t) = N_1(t) - N_2(t)$ as a continuous time random walk; it increases or decreases by 1 after exponentially distributed waiting times. We can consider n continuous time random walks $X_1(t), \dots, X_n(t)$ beginning at locations $x_1 < x_2 < \cdots < x_n$ at time $t = 0$ and conditioned not to intersect for $0 \le t \le T$. Then, similar to (3.14), by the Karlin–McGregor/LGV formula, the transition probability for the vector $(X_1(t), \dots, X_n(t))$ to be at the location $(y_1, \dots, y_n)$ at time t is

$$e^{-nT} \det\left(\phi_{y_k - x_j}\right), \qquad \phi(\zeta) = e^{\frac{T}{2}(\zeta + 1/\zeta)}. \tag{3.16}$$

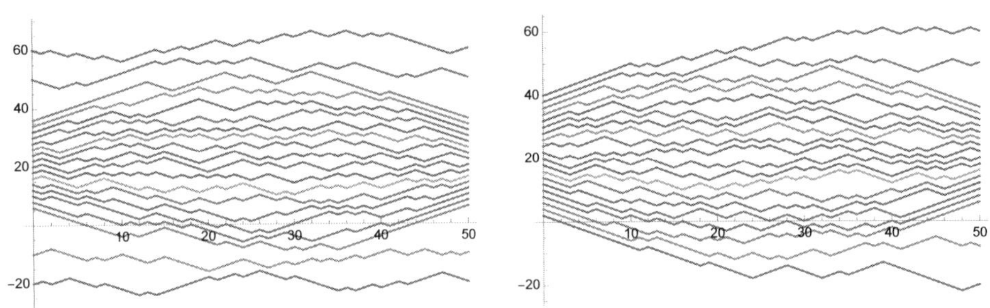

FIGURE 3. Two simulations of non-intersecting discrete time random walks with wanderers with $n = 20$, $2T = 50$, and $m_1 = m_2 = 2$. In the figure on the left, the starting and ending points of the walkers are equispaced except for the lowest and highest pair, and the distribution of W_n is described in (3.21) in terms of framed Hankel determinants. In the figure on the right, the starting points are all equispaced but the ending points are not and the distribution of W_n is described in (3.23) in terms of bordered Hankel determinants.

The width of non-intersecting ensembles of both discrete and continuous random walks were discussed in [**8**], where it was shown that they can be expressed in terms of a discrete analogue of determinants described above. For a positive integer $M \in \mathbb{Z}_{>0}$ and a point $s \in \mathbb{T}$, define the discrete set $\mathcal{D}_{M,s}$ as[9]

$$\mathcal{D}_{M,s} = \{z \in \mathbb{T} \,:\, z^M = s\},$$

[9]Note that the set $\mathcal{D}_{M,s}$ defined here is different from the set $\mathcal{D}_{M,\tau}$ defined in (3.2), though they play similar roles.

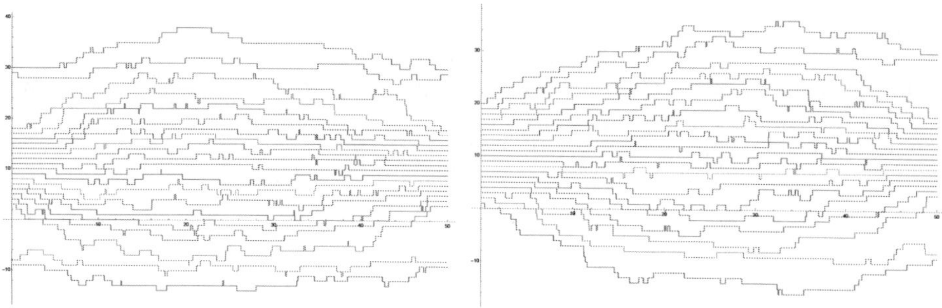

FIGURE 4. Two simulations of non-intersecting continuous time random walks with wanderers with $n = 20$, $T = 50$, and $m_1 = m_2 = 2$. In the figure on the left, the starting and ending points of the walkers are equispaced except for the lowest and highest pair, and the distribution of W_n is described in (3.22) in terms of framed Hankel determinants. In the figure on the right, the starting points are all equispaced but the ending points are not and the distribution of W_n is described in (3.24) in terms of bordered Hankel determinants.

and for a function $f \in L^1(\mathbb{T})$ define the discrete measure

$$f^{M,s} := \frac{1}{M} \sum_{z \in \mathcal{D}_{M,s}} f(z)\delta_z. \tag{3.17}$$

Note that as $M \to \infty$, the discrete measure $f^{M,s}$ converges to $\frac{f(z)\mathrm{d}z}{2\pi \mathrm{i} z}$.

PROPOSITION 3.3. *Let $x_1 < x_2 < \cdots < x_n$ and $y_1 < y_2 < \cdots < y_n$ be integers, and consider the ensemble of n non-intersecting discrete time random walks beginning at points $X_j(0) = 2x_j$ for $j = 1, \ldots, n$, conditioned to end at points $X_k(2T) = 2y_k$ for $j = 1, \ldots, n$. The width of this ensemble of non-intersecting paths is given as*

$$\begin{gathered} \mathbb{P}(W_n(2T) < 2M) = \frac{1}{\det\left(\phi_{y_k - x_j}\right)_{j,k=1}^n} \int_{|s|=1} \det\left(\phi^{M,s}_{y_k - x_j}\right)_{j,k=1}^n \frac{ds}{2\pi i s}, \\ \phi(\zeta) = \zeta^{-T}(\zeta + 1)^{2T}. \end{gathered} \tag{3.18}$$

Similarly, in the ensemble of n non-intersecting continuous time random walks beginning at points $X_j(0) = x_j$ for $j = 1, \ldots, n$ and conditioned to end at points $X_k(T) = y_k$ for $k = 1, \ldots, n$, the width of the ensemble is given as

$$\begin{gathered} \mathbb{P}(W_n(T) < M) = \frac{1}{\det\left(\phi_{y_k - x_j}\right)_{j,k=1}^n} \int_{|s|=1} \det\left(\phi^{M,s}_{y_k - x_j}\right)_{j,k=1}^n \frac{ds}{2\pi i s}, \\ \phi(\zeta) = e^{\frac{T}{2}(\zeta + 1/\zeta)}. \end{gathered} \tag{3.19}$$

PROOF. For the pure Toeplitz case $x_j = y_j = j, j = 1, 2, \ldots, n$, the formulas (3.18) and (3.19) are exactly [**8**, Proposition 4.2] and [**8**, Proposition 4.1], respectively. The proofs there do not depend on the Toeplitz structure of the determinant, and our proof proceeds in exactly the same way. $\square$

Similar to the previous section, we can choose the starting/ending points so that the determinants in (3.18) and (3.19) have bordered or framed structure. In addition to the symbol ϕ and its discretization $\phi^{M,s}$, define the symbols $\psi[\beta]$ and $\eta[\alpha]$ as

$$\psi[\beta](\zeta) := \phi(\zeta)\zeta^{n-m_2-\beta}, \quad \eta[\alpha](\zeta) := \phi(\zeta)\zeta^{-(m_1+1-\alpha)}, \tag{3.20}$$

and their discretizations

$$\psi[\beta]^{M,s} := \sum_{z\in\mathcal{D}_{M,s}} \psi[\beta](z)\delta_z, \quad \eta[\alpha]^{M,s} := \sum_{z\in\mathcal{D}_{M,s}} \eta[\beta](z)\delta_z.$$

Also let A and $A^{M,s}$ be the $m\times m$ matrices

$$A = \left(\phi_{\beta_k-\alpha_j}\right)_{j,k=1}^m, \quad A^{M,s} = \left(\phi^{M,s}_{\beta_k-\alpha_j}\right)_{j,k=1}^m.$$

We then have the following propositions which follow immediately from Proposition 3.3 and elementary row operations.

PROPOSITION 3.4. *Fix non-negative integers m_1 and m_2 such that $m_1+m_2<n$, and integers $\alpha_1<\alpha_2<\cdots<\alpha_{m_1}<m_1<n-m_2<\alpha_{m_1+1}<\cdots<\alpha_{m_1+m_2}$ and $\beta_1<\beta_2<\cdots<\beta_{m_1}<m_1<n-m_2<\beta_{m_1+1}<\cdots<\beta_{m_1+m_2}$. Fix the numbers $x_1,\ldots,x_n$ and $y_1,\ldots,y_n$ as follows: $x_j=y_j=j$ for $j=m_1+1,\ldots,n-m_2$; $x_j=\alpha_j$ and $y_j=\beta_j$ for $j=1,\ldots,m_1$; and $x_{n-m_2+j}=\alpha_j$ and $y_{n-m_2+j}j=\beta_j$ for $j=m_1+1,\ldots,m_1+m_2$.*

In the ensemble of discrete time simple random walks with starting points $X_j(0)=2x_j$ and ending points $X_j(2T)=2y_j$, the distribution of the width (3.18) *has the form (with $m=m_1+m_2$)*

$$\mathbb{P}(W_n(2T)<2M) = \int_{|s|=1} \frac{D_n^F[\phi^{M,s};\psi[\beta_1]^{M,s},\ldots,\psi[\beta_m]^{M,s};\eta[\alpha_1]^{M,s},\ldots,\eta[\alpha_m]^{M,s};A^{M,s}]}{D_n^F[\phi;\psi[\beta_1],\ldots,\psi[\beta_m];\eta[\alpha_1],\ldots,\eta[\alpha_m];A]}\frac{ds}{2\pi i s}, \tag{3.21}$$

where ϕ is as in (3.18) *.*

Similarly in the ensemble of continuous time random walks with starting points $X_j(0)=x_j$ and ending points $X_j(T)=y_j$, the distribution of the width (3.19) *has the form (with $m=m_1+m_2$)*

$$\mathbb{P}(W_n(T)<M) = \int_{|s|=1} \frac{D_n^F[\phi^{M,s};\psi[\beta_1]^{M,s},\ldots,\psi[\beta_m]^{M,s};\eta[\alpha_1]^{M,s},\ldots,\eta[\alpha_m]^{M,s};A^{M,s}]}{D_n^F[\phi;\psi[\beta_1],\ldots,\psi[\beta_m];\eta[\alpha_1],\ldots,\eta[\alpha_m];A]}\frac{ds}{2\pi i s}, \tag{3.22}$$

where ϕ is as in (3.19).

If the starting points are all consecutive, the framed determinants above reduce to bordered ones.

PROPOSITION 3.5. *Fix non-negative integers m_1 and m_2 such that $m_1+m_2<n$, and integers $\beta_1<\beta_2<\cdots<\beta_{m_1}<m_1<n-m_2<\beta_{m_1+1}<\cdots<\beta_{m_1+m_2}$. Fix the numbers $x_1,\ldots,x_n$ and $y_1,\ldots,y_n$ as follows: $x_j=j$ for $j=1,\ldots,n$; and $y_j=j$ for $j=m_1+1,\ldots,n-m_2$; $y_j=\beta_j$ for $j=1,\ldots,m_1$; and $y_{n-m_2+j}=\beta_j$ for $j=m_1+1,\ldots,m_1+m_2$.*

In the ensemble of discrete time simple random walks with starting points $X_j(0) = 2x_j$ and ending points $X_j(2T) = 2y_j$, the distribution of the width (3.18) *has the form (with $m = m_1 + m_2$)*

$$\mathbb{P}(W_n(2T) < 2M) = \int_{|s|=1} \frac{D_n^B[\phi^{M,s}; \psi[\beta_1]^{M,s}, \dots, \psi[\beta_m]^{M,s}]}{D_n^B[\phi; \psi[\beta_1], \dots, \psi[\beta_m]]} \frac{ds}{2\pi is}, \tag{3.23}$$

where ϕ is as in (3.18) .

In the ensemble of continuous time simple random walks with starting points $X_j(0) = x_j$ and ending points $X_j(T) = y_j$, the distribution of the width (3.19) *has the form (with $m = m_1 + m_2$)*

$$\mathbb{P}(W_n(T) < M) = \int_{|s|=1} \frac{D_n^B[\phi^{M,s}; \psi[\beta_1]^{M,s}, \dots, \psi[\beta_m]^{M,s}]}{D_n^B[\phi; \psi[\beta_1], \dots, \psi[\beta_m]]} \frac{ds}{2\pi is}, \tag{3.24}$$

where ϕ is as in (3.19).

Two simulations, related respectively to (3.21) and (3.23), for the discrete random walks with $m_1 = m_2 = 2$ are displayed in Figure 3. Similarly for continuous time random walks, two simulations related respectively to (3.22) and (3.24) are shown in Figure 4.

As in the previous subsection on the width of non-intersecting Brownian bridges, we expect (3.21), (3.22), (3.23) and (3.24) to have scaling limits related to the supremum of the sum of two Airy processes with wanderers minus a parabola. We do not describe that limit in detail, but we do give the orthogonal polynomials which may be relevant for asymptotic analysis.

For ϕ given by either (3.18) or (3.19), introduce the bi-orthogonal system of polynomials $Q_k(z) = \kappa_k z^k + \cdots$, and $\widehat{Q}_n(z) = \widehat{\kappa}_k z^k + \cdots$ via the bi-orthonormality conditions

$$\int_{\mathbb{T}} Q_j(\zeta)\widehat{Q}_k(\zeta^{-1})\phi(\zeta)\frac{\mathrm{d}z}{2\pi\mathrm{i}\zeta} = \delta_{jk}, \qquad j,k \in \mathbb{Z}_{>0}, \tag{3.25}$$

as well as bi-orthogonal system of polynomials $Q_k^{M,s}(z) = \kappa_k^{M,s} z^k + \cdots$, and $\widehat{Q}_k^{M,s}(z) = \widehat{\kappa}_k^{M,s} z^k + \cdots$ defined via

$$\frac{1}{M}\sum_{z^M=s} Q_j^{M,s}(z)\widehat{Q}_k^{M,s}(z^{-1})\phi(z) = \delta_{jk}, \qquad j,k \in \mathbb{Z}_{\geq 0}. \tag{3.26}$$

Then applying Corollary 1.11 gives that the right-hand-side of (3.23) and (3.24) is,

$$\int_{|s|=1} \sqrt{\frac{D_{n-m}[\phi^{M,s}]D_{n-m+1}[\phi^{M,s}]}{D_{n-m}[\phi]D_{n-m+1}[\phi]}} \tag{3.27}$$
$$\times \frac{\displaystyle\sum_{\{z_1^M, z_2^M, \dots, z_m^M = s\}} \begin{pmatrix} Q_{n-m}^{M,s}(z_1) & \cdots & Q_{n-m}^{M,s}(z_m) \\ \vdots & \ddots & \vdots \\ Q_{n-1}^{M,s}(z_1) & \cdots & Q_{n-1}^{M,s}(z_m) \end{pmatrix} \prod_{j=1}^m z_j^{1-m_2-\beta_j}\phi_j(z_j)}{\displaystyle\int \cdots \int_{\mathbb{T}^m} \begin{pmatrix} Q_{n-m}(z_1) & \cdots & Q_{n-m}(z_m) \\ \vdots & \ddots & \vdots \\ Q_{n-1}(z_1) & \cdots & Q_{n-1}(z_m) \end{pmatrix} \prod_{j=1}^m z_j^{1-m_2-\beta_j}\frac{\phi_j(z_j)\mathrm{d}z_j}{2\pi\mathrm{i}z_j}} \frac{\mathrm{d}s}{2\pi\mathrm{i}s}.$$

Applying Corollary 1.17 gives that the right-hand-side of (3.21) and (3.22) is,

$$\int_{|s|=1}\frac{D_{n-m}[\phi^{M,s}]}{D_{n-m}[\phi]} \tag{3.28}$$
$$\times\frac{\det\left[A^{M,s}-\left(\sum_{z,w\in\mathcal{D}_{M,s}}\mathcal{K}^{M,s}_{n-m-1}(z^{-1},w^{-1})\phi(z)\phi(w)z^{n-m_2-\beta_k}w^{-(m_1+1-\alpha_j)}\right)^m_{j,k=1}\right]}{\det\left[A-\left(\int_{\mathbb{T}}\int_{\mathbb{T}}\mathcal{K}_{n-m-1}(z^{-1},w^{-1})\frac{\phi(z)\phi(w)z^{n-m_2-\beta_k}w^{-(m_1+1-\alpha_j)}}{(2\pi \mathrm{i})^2zw}\mathrm{d}z\mathrm{d}w\right)^m_{j,k=1}\right]}\frac{\mathrm{d}s}{2\pi \mathrm{i}s},$$

where

$$\mathcal{K}_{n-m-1}(z,w):=\sum_{j=0}^{n-m-1}Q_j(w)\widehat{Q}_j(z),$$
$$\mathcal{K}^{M,s}_{n-m-1}(z,e):=\sum_{j=0}^{n-m-1}Q^{M,s}_j(w)\widehat{Q}^{M,s}_j(z).$$

For ϕ given by (3.16), a steepest descent analysis of the Riemann–Hilbert problem for the orthogonal system (3.25) was carried out in **[6]**; for the discrete version of the same symbol, a steepest descent analysis of the discrete Riemann–Hilbert problem for the orthogonal system (3.25) was carried out in **[7]**. To evaluate (3.27), one could first extract asymptotics of $Q_n(z)$ from the analysis of **[6]**, **[7]** and insert those asymptotics into (3.27). We do not do this in the general case in this work, but we give partial results in some special cases. For the case $m=2$, the asymptotic result for the denominator of (3.27) is presented in Theorem 1.19. Also, for $m=1$ the asymptotic result for the denominator of (3.28) is given in Theorem 1.20. The asymptotic analysis leading to these results is presented in Section 7.

4. The six-vertex model with DWBC

The six-vertex model is stated on a square lattice of size $n\times n$, with arrows drawn on edges satisfying *ice rule*: at each vertex, exactly two adjacent arrows point inward and two point outward. There are six possible configurations of arrows at each vertex, and we label them as shown in Figure 5.

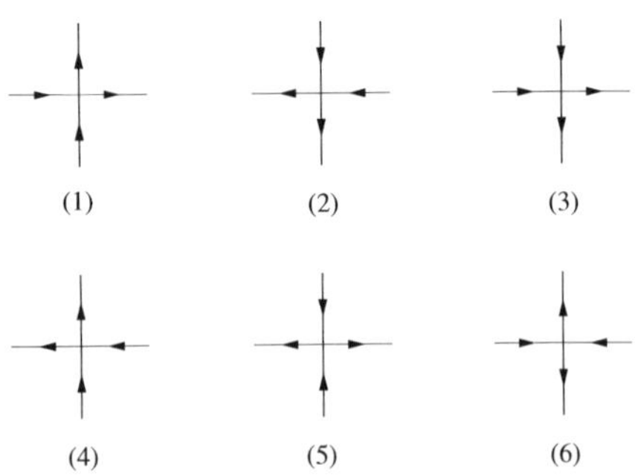

FIGURE 5. The six arrow configurations allowed at a vertex.

We can give a weight $w_i > 0$ to each vertex type $i = 1, 2, \dots, 6$, and define the weight of an arrow configuration σ as

$$w_n(\sigma) = \prod_{i=1}^{6} w_i^{N_i(\sigma)},$$

where $N_i(\sigma)$ is the number of vertices of type i in the configuration σ. The Gibbs measure is then given by the normalized weight of a configuration

$$\mu_n(\sigma) = \frac{w_n(\sigma)}{Z_n}, \quad Z_n = \sum_{\sigma} w(\sigma). \tag{4.1}$$

The quantity Z_n is the partition function for the model, and the sum defining it is over all allowable configurations σ on the $n \times n$ lattice, which may be restricted by some boundary conditions as well as by the ice-rule. For certain boundary conditions known as *Domain Wall Boundary Conditions* (DWBC), the partition function may be expressed as a Hankel determinant. In DWBC all arrows on the left and right boundary point outward, and all arrows on the top and bottom boundaries point inward. See Figure 6 for an example of an arrow configuration with DWBC on the 5×5 lattice. The determinantal formula for the partition function of the DWBC six-vertex model is known as the *Izergin–Korepin formula.*

For fixed boundary conditions like DWBC, the Gibbs measure of the six-vertex model is invariant under reversal of arrows (see e.g., [**21**, Chapter 5, Section 2]), so we may give the weight $a > 0$ to the vertices of type 1 and 2; the weight $b > 0$ to the vertices of type 3 and 4; and the weight $c > 0$ to the vertices of type 5 and 6. Then the phase diagram for the model is as follows.

- **Ferroelectric phase**. If $a > b + c$ (resp. $b > a + c$) then vertices of Type 1 and 2 (resp. Type 3 and 4) are dominant. In the ferroelectric phase it is typical to parametrize the weights as

$$\begin{aligned} a \equiv a(t,\gamma) = \sinh(t-\gamma), \quad & b \equiv b(t,\gamma) = \sinh(\gamma+t), \\ c \equiv c(\gamma) = \sinh(2\gamma), \quad & 0 < |\gamma| < t. \end{aligned} \tag{4.2}$$

- **Disordered phase**. If the quantities a, b, and c form a triangle, then there is no dominant weight. In the disordered phase it is typical to parametrize the weights as

$$\begin{aligned} a \equiv a(t,\gamma) = \sin(\gamma-t), \quad & b \equiv b(t,\gamma) = \sin(\gamma+t), \\ c \equiv c(\gamma) = \sin(2\gamma), \quad & |t| < \gamma < \pi/2. \end{aligned} \tag{4.3}$$

- **Antiferroelectric phase**. If $c > a + b$, then vertices of Type 5 and 6 are dominant. In the antiferroelectric phase it is typical to parametrize the weights as

$$a \equiv a(t,\gamma) = \sinh(\gamma-t), \quad b \equiv b(t,\gamma) = \sinh(\gamma+t), \quad c \equiv c(\gamma) = \sinh(2\gamma), \quad |t| < \gamma. \tag{4.4}$$

In order to describe the Izergin–Korepin formula in its most general form, it is convenient to introduce vertex weights which depend on the location of the vertex in the $n \times n$ lattice as well as the vertex type. To do so, fix $n = 1, 2, \dots,$ and $2n + 1$ complex parameters $\chi_1, \dots, \chi_n; \psi_1, \dots, \psi_n; \gamma$. For a configuration σ of the six-vertex model in an $n \times n$ square with DWBC and one of the parametrizations

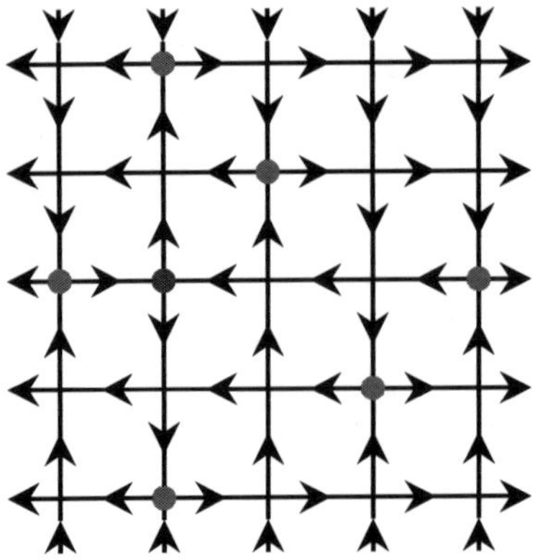

FIGURE 6. A six-vertex configuration on the 5×5 lattice with DWBC. Type 5 vertices are marked in red, and type-6 vertices are marked in blue. The domain wall boundary conditions and the ice rule ensure that there is exactly one type-5 vertex along each boundary, and no more than k type 5 vertices in the kth row from the right or left and the kth column from the top or bottom. Vertices of types 5 and 6 alternate along each row and column.

(4.2)–(4.4), we define the weight of the vertex at (x, y) to be

$$\omega(x, y; \sigma) = \begin{cases} a(\psi_y - \chi_x, \gamma), & \text{if } \sigma(x, y) \text{ is of Type 1 or 2,} \\ b(\psi_y - \chi_x, \gamma), & \text{if } \sigma(x, y) \text{ is of Type 3 or 4,} \\ c(\gamma), & \text{if } \sigma(x, y) \text{ is of Type 5 or 6.} \end{cases}$$

The inhomogeneous partition function is then defined as

$$Z_n(\chi_1, \ldots, \chi_n; \psi_1, \ldots, \psi_n; \gamma) = \sum_{\sigma} \prod_{x=1}^{n} \prod_{y=1}^{n} \omega(x, y; \sigma), \tag{4.5}$$

where the sum is over all configurations satisfying DWBC.

The Izergin–Korepin formula states that for any of the parametrizations (4.2)–(4.4),

$$\begin{aligned} Z_n(\chi_1, \ldots, \chi_n;\, &\psi_1, \ldots, \psi_n;\, \gamma) \\ &= \frac{\prod_{i,j=1}^{n} \big(a(\psi_j - \chi_i, \gamma) b(\psi_j - \chi_i, \gamma)\big)}{\prod_{i<j} \big(b(\chi_i - \chi_j, 0) b(\psi_i - \psi_j, 0)\big)} \det \left[\frac{c(\gamma)}{a(\psi_j - \chi_i, \gamma) b(\psi_j - \chi_i, \gamma)} \right]_{i,j=1}^{n}. \end{aligned} \tag{4.6}$$

The partition function for the homogeneous model is then obtained by taking all parameters $\chi_1, \ldots, \chi_n$ to be the same, and $\psi_1, \ldots, \psi_n$ to be the same. For convenience we can take $\chi_1, \ldots, \chi_n = 0$ and $\psi_1, \ldots, \psi_n = t$. The determinant vanishes when two of the ψ's or χ's coincide, as does the denominator of the prefactor, so one must be careful to evaluate the limit as $\chi_1, \ldots, \chi_n \to 0$ and $\psi_1, \ldots, \psi_n \to t$. To do so it is convenient to define the function

$$\varphi(t) = \frac{c(\gamma)}{a(t, \gamma) b(t, \gamma)},$$

so that the Izergin–Korepin determinant is

$$Z_n(\chi_1,\dots,\chi_n;\,\psi_1,\dots,\psi_n;\,\gamma) = \frac{\prod\limits_{i,j=1}^{n}\big(a(\psi_j-\chi_i,\gamma)b(\psi_j-\chi_i,\gamma)\big)}{\prod\limits_{i<j}\big(b(\chi_i-\chi_j,0)b(\psi_i-\psi_j,0)\big)}\det\left[\varphi(\psi_j-\chi_i)\right]_{i,j=1}^{n}.$$

We can take $\chi_i \to 0$ in sequential order row-by-row. By expanding $\varphi(\psi_j-\chi_i)$ as a Taylor series around ψ_j and using row operations we obtain

$$Z_n(0^n;\,\psi_1,\dots,\psi_n;\,\gamma) = \frac{\prod\limits_{j=1}^{n}\big(a(\psi_j,\gamma)b(\psi_j,\gamma)\big)^n}{\left(\prod_{j=0}^{n-1}j!\right)\prod\limits_{i<j}\big(b(\psi_i-\psi_j,0)\big)}\det\left[\varphi^{(i-1)}(\psi_j)\right]_{i,j=1}^{n}.$$

We can then take $\psi_j \to t$ in sequential order column-by-column, again expanding $\varphi^{(i-1)}(\psi_j)$ in a Taylor series around $\psi_j = t$ and using column operations to obtain

$$Z_n(0^n;\,t^n;\,\gamma) = \frac{\big(a(t,\gamma)b(t,\gamma)\big)^{n^2}}{\left(\prod_{j=0}^{n-1}j!\right)^2}\det\left[\varphi^{(i+j-2)}(t)\right]_{i,j=1}^{n}.$$

To recognize this as a Hankel determinant, one must recognize the function $\varphi(t)$ as the Laplace transform of some measure m:

$$\varphi(t) = \int_{\mathbb{R}} e^{tx}\mathrm{dm}(x),$$

so that

$$\varphi^{(i+j-2)}(t) = \int_{\mathbb{R}} x^{i+j-2}e^{tx}\mathrm{dm}(x) = \mu_{i+j-2}, \quad \mathrm{d}\mu(x) = e^{tx}\mathrm{dm}(x), \tag{4.7}$$

and the homogeneous partition function becomes

$$Z_n = Z_n(0^n;\,t^n;\,\gamma) = \frac{\big(a(t,\gamma)b(t,\gamma)\big)^{n^2}}{\left(\prod_{j=0}^{n-1}j!\right)^2}H_n[\mu].$$

This characterization as a Hankel determinant was exploited in a series of works to compute the large-n asymptotics of the partition function in various regions of the phase diagram [**16**, **18**, **19**, **21**, **23**, **79**].

Variations on the above formula can be obtained by taking a partially homogeneous limit of (4.6). Such partially inhomogeneous partition functions in fact function as generating functions for the location of the c-type vertices near the boundary of the $n \times n$ domain. For example, if we take $\chi_1,\dots,\chi_n \to 0$ and

$\psi_1, \ldots, \psi_{n-1} \to t$ as before, but leave ψ_n free, we obtain

$$(4.8) \quad Z_n(0^n; t^{n-1}, \psi_n; \gamma) = \frac{\big(a(t,\gamma)b(t,\gamma)\big)^{n^2-n}\big(a(\psi_n,\gamma)b(\psi_n,\gamma)\big)^n}{(n-1)!b(\psi_n - t, 0)^{n-1}\left(\prod_{j=0}^{n-2} j!\right)^2}$$
$$\times \det \begin{pmatrix} \varphi(t,\gamma) & \varphi'(t,\gamma) & \cdots & \varphi^{(n-2)}(t,\gamma) & \varphi(\psi_n,\gamma) \\ \varphi'(t,\gamma) & \varphi''(t,\gamma) & \cdots & \varphi^{(n-1)}(t,\gamma) & \varphi'(\psi_n,\gamma) \\ \vdots & \vdots & \ddots & \vdots & \vdots \\ \varphi^{(n-2)}(t,\gamma) & \varphi^{(n-1)}(t,\gamma) & \cdots & \varphi^{(2n-4)}(t,\gamma) & \varphi^{(n-2)}(\psi_n,\gamma) \\ \varphi^{(n-1)}(t,\gamma) & \varphi^{(n)}(t,\gamma) & \cdots & \varphi^{(2n-3)}(t,\gamma) & \varphi^{(n-1)}(\psi_n,\gamma) \end{pmatrix}.$$

If we write $\psi_n = t + \xi$ and use the Laplace transform (4.7), the above formula becomes

$$(4.9) \quad Z_n(0^n; t^{n-1}, \psi_n; \gamma) = \frac{\big(a(t,\gamma)b(t,\gamma)\big)^{n^2-n}\big(a(\psi_n,\gamma)b(\psi_n,\gamma)\big)^n}{(n-1)!b(\psi_n - t, 0)^{n-1}\left(\prod_{j=0}^{n-2} j!\right)^2}$$
$$\times \det \begin{pmatrix} \mu_0 & \mu_1 & \cdots & \mu_{n-2} & \nu_0 \\ \mu_1 & \mu_2 & \cdots & \mu_n & \nu_1 \\ \vdots & \vdots & \ddots & \vdots & \vdots \\ \mu_{n-2} & \mu_{n-1} & \cdots & \mu_{2n-4} & \nu_{n-2} \\ \mu_{n-1} & \mu_n & \cdots & \mu_{2n-3} & \nu_{n-1} \end{pmatrix}.$$

where μ is as in (4.7), and ν is defined as

$$\mathrm{d}\nu(x) = e^{(t+\xi)x}\mathrm{dm}(x) = e^{\xi x}\mathrm{d}\mu(x).$$

This is the bordered Hankel determinant $H_n^B[\mu;\nu]$ and Corollary 1.11 gives the formula

$$(4.10) \quad Z_n(0^n; t^{n-1}, \psi_n; \gamma) = \frac{\big(a(t,\gamma)b(t,\gamma)\big)^{n^2-n}\big(a(t+\xi,\gamma)b(t+\xi,\gamma)\big)^n}{(n-1)!b(\xi, 0)^{n-1}\left(\prod_{j=0}^{n-2} j!\right)^2}$$
$$\times H_{n-1}[\mu] \int_{\mathbb{R}} P_{n-1}(x) e^{\xi x}\mathrm{d}\mu(x),$$

where the system of monic polynomials $\{P_k(x)\}_{k=0}^{\infty}$ is defined by the orthogonality condition

$$(4.11) \qquad \int_{\mathbb{R}} P_j(x)P_k(x)\mathrm{d}\mu(x) = h_k\delta_{jk}.$$

This partially inhomogeneous partition function tracks the location of the c-type vertex in the last column of the six-vertex state, which we may denote Λ_n. This fact has been used throughout the past 20 years to investigate the distribution of Λ_n **[24]** and subsequently to investigate other correlation functions in the six-vertex model **[36–38]**. The formula (4.10) in particular was analyzed asymptotically in the recent work **[55]** in each of the phase regions to prove Gaussian fluctuations for Λ_n as $n \to \infty$ in the disordered and anti-ferroelectric phases, and convergence of Λ_n to a geometric random variable in the ferroelectric phase[10]. The orthogonal polynomials (4.11) depend on the phase region, and their asymptotics can be ascertained from the Riemann–Hilbert analysis presented in [**19**, **21**, **23**] (see also [**55**, Appendix A.2] for the orthogonal polynomials in the ferroelectric phase region).

One can also jointly track the location of the c-type vertex in the last row and last column by taking $\chi_1, \ldots, \chi_{n-1} \to 0$, $\psi_1, \ldots, \psi_{n-1} \to t$, $\psi_n = t + \xi_2$, and $\chi_n = -\xi_1$. Similar to (4.8), one finds

$$(4.12)\quad Z_n(0^{n-1}, -\xi_1; t^{n-1}, t+\xi_2; \gamma) = \frac{\left[a(t,\gamma)b(t,\gamma)\right]^{(n-1)^2}\left[a(t+\xi_1,\gamma)b(t+\xi_1,\gamma)a(t+\xi_2,\gamma)b(t+\xi_2,\gamma)\right]^{n-1}a(t+\xi_1+\xi_2,\gamma)b(t+\xi_1+\xi_2,\gamma)}{[b(\xi_1,0)b(\xi_2,0)]^{n-1}\left(\prod_{j=0}^{n-2} j!\right)^2} \times \det\begin{pmatrix} \mu_0 & \mu_1 & \cdots & \mu_{n-2} & \nu_0 \\ \mu_1 & \mu_2 & \cdots & \mu_{n-1} & \nu_1 \\ \vdots & \vdots & \ddots & \vdots & \vdots \\ \mu_{n-2} & \mu_{n-1} & \cdots & \mu_{2n-4} & \nu_{n-2} \\ \eta_0 & \eta_1 & \cdots & \eta_{n-2} & a \end{pmatrix},$$

where μ is as in (4.7) and

$$\mathrm{d}\eta(x) = e^{(t+\xi_1)x}\mathrm{d}m(x) = e^{\xi_1 x}\mathrm{d}\mu(x), \quad \mathrm{d}\nu(x) = e^{(t+\xi_2)x}\mathrm{d}m(x) = e^{\xi_2 x}\mathrm{d}\mu(x),$$

and the corner entry a is

$$a = \varphi(t+\xi_1+\xi_2) = \int_{\mathbb{R}} e^{(t+\xi_1+\xi_2)x}\mathrm{d}m(x).$$

This is a framed Hankel determinant, and we have

$$(4.13)\quad Z_n(0^{n-1}, -\xi_1; t^{n-1}, t+\xi_2; \gamma) = \frac{\left[a(t,\gamma)b(t,\gamma)\right]^{(n-1)^2}\left[a(t+\xi_1,\gamma)b(t+\xi_1,\gamma)a(t+\xi_2,\gamma)b(t+\xi_2,\gamma)\right]^{n-1}a(t+\xi_1+\xi_2,\gamma)b(t+\xi_1+\xi_2,\gamma)}{[b(\xi_1,0)b(\xi_2,0)]^{n-1}\left(\prod_{j=0}^{n-2} j!\right)^2} \times H_n^F[\mu; \nu, \eta; a].$$

[10]In fact [**55**] did much more. There it was shown that in the disordered and anti-ferroelectric phase regions the type-5 vertices near the boundary converge in a scaling limit to the GUE corners process from random matrix theory [**63**] and in the ferroelectric phase region they converge to the stochastic six-vertex model [**25**].

Corollary 1.17 then gives

$$(4.14)\quad Z_n(0^{n-1},-\xi_1;t^{n-1},t+\xi_2;\gamma)=$$

$$\frac{[a(t,\gamma)b(t,\gamma)]^{(n-1)^2}[a(t+\xi_1,\gamma)b(t+\xi_1,\gamma)a(t+\xi_2,\gamma)b(t+\xi_2,\gamma)]^{n-1}a(t+\xi_1+\xi_2,\gamma)b(t+\xi_1+\xi_2,\gamma)}{[b(\xi_1,0)b(\xi_2,0)]^{n-1}\left(\prod_{j=0}^{n-2}j!\right)^2}$$

$$\times H_{n-1}[\mu]\left(a-\int_{\mathbb{R}}\int_{\mathbb{R}}e^{\xi_1x+\xi_2y}K_{n-2}(x,y)\mathrm{d}\mu(x)\mathrm{d}\mu(y)\right)$$

$$=\frac{[a(t,\gamma)b(t,\gamma)]^{(n-1)^2}[a(t+\xi_1,\gamma)b(t+\xi_1,\gamma)a(t+\xi_2,\gamma)b(t+\xi_2,\gamma)]^{n-1}a(t+\xi_1+\xi_2,\gamma)b(t+\xi_1+\xi_2,\gamma)}{[b(\xi_1,0)b(\xi_2,0)]^{n-1}\left(\prod_{j=0}^{n-2}j!\right)^2}$$

$$\times H_{n-1}[\mu]\left(\int_{\mathbb{R}}e^{(\xi_1+\xi_2)x}d\mu(x)-\int_{\mathbb{R}}\int_{\mathbb{R}}e^{\xi_1x+\xi_2y}\sum_{k=0}^{n-2}\frac{P_k(x)P_k(y)}{h_k}\mathrm{d}\mu(x)\mathrm{d}\mu(y)\right).$$

It is an open and interesting question to evaluate (4.14) as $n\to\infty$ in order to obtain information on the joint distribution of the location of the c-type vertices in the first row and column.

5. Proofs of Propositions 1.9, 1.10, 1.13, and 1.16

5.1. Proof of Propositions 1.9 and 1.10. We prove Proposition 1.9 in detail. The proof of Proposition 1.10 is similar.

Before proceeding let us recall the *Dodgson Condensation Identity*[11] , which we occasionally abbreviate as DCI (see **[1, 29, 50, 54]** and references therein). Let $\mathcal{M}$ be an $n\times n$ matrix. By

$$\mathcal{M}\begin{Bmatrix} j_1 & j_2 & \cdots & j_\ell \\ k_1 & k_2 & \cdots & k_\ell \end{Bmatrix},$$

we mean the determinant of the $(n-\ell)\times(n-\ell)$ matrix obtained from $\mathcal{M}$ by removing the rows j_i and the columns k_i, $1\le i\le \ell$. Although the order of writing the row and column indices is immaterial for this definition, in this work we prefer to respect the order of indices, for example we prefer to write

$$\mathcal{M}\begin{Bmatrix} 3 & 5 \\ 1 & 4 \end{Bmatrix},$$

and not

$$\mathcal{M}\begin{Bmatrix} 5 & 3 \\ 1 & 4 \end{Bmatrix}\quad\text{or}\quad\mathcal{M}\begin{Bmatrix} 3 & 5 \\ 4 & 1 \end{Bmatrix}\quad\text{or}\quad\mathcal{M}\begin{Bmatrix} 5 & 3 \\ 4 & 1 \end{Bmatrix},$$

although all of these are the same determinant. Let $j_1<j_2$ and $k_1<k_2$. The Dodgson Condensation identity reads

$$(5.1)\qquad \mathcal{M}\cdot\mathcal{M}\begin{Bmatrix} j_1 & j_2 \\ k_1 & k_2 \end{Bmatrix}=\mathcal{M}\begin{Bmatrix} j_1 \\ k_1 \end{Bmatrix}\cdot\mathcal{M}\begin{Bmatrix} j_2 \\ k_2 \end{Bmatrix}-\mathcal{M}\begin{Bmatrix} j_1 \\ k_2 \end{Bmatrix}\cdot\mathcal{M}\begin{Bmatrix} j_2 \\ k_1 \end{Bmatrix}.$$

Now we prove Proposition 1.9. Let $n\in\mathbb{Z}_{\ge 2}$ be arbitrary. The case $m=n$ is obvious by row operations. For $m<n$, we prove this Proposition by induction on $m\in\mathbb{N}$, the number of borders. For $m=1$, the result obviously holds due to (1.19). To prepare for proving the inductive step, we introduce the notations

$$(5.2)\qquad \mathcal{M}_1\equiv H_n[\mu;z_1,\ldots,z_m],\quad\text{and}\quad \mathcal{M}_2\equiv H_n[\mu;z_1,\ldots,z_{m-1}],$$

[11] Also known as the *Desnanot–Jacobi* identity or the *Sylvester determinant* identity.

for the $n \times n$ multi-bordered Hankel determinants respectively with m and $m-1$ borders of monomials. Recalling (5.1) we use the following two Dodgson Condensation identities:

(5.3)

$$\mathcal{M}_1 \cdot \mathcal{M}_1 \begin{Bmatrix} 0 & n-1 \\ n-2 & n-1 \end{Bmatrix} = \mathcal{M}_1 \begin{Bmatrix} 0 \\ n-2 \end{Bmatrix} \cdot \mathcal{M}_1 \begin{Bmatrix} n-1 \\ n-1 \end{Bmatrix} - \mathcal{M}_1 \begin{Bmatrix} 0 \\ n-1 \end{Bmatrix} \cdot \mathcal{M}_1 \begin{Bmatrix} n-1 \\ n-2 \end{Bmatrix},$$

and

(5.4)

$$\mathcal{M}_2 \cdot \mathcal{M}_2 \begin{Bmatrix} 0 & n-1 \\ n-m & n-1 \end{Bmatrix} = \mathcal{M}_2 \begin{Bmatrix} 0 \\ n-m \end{Bmatrix} \cdot \mathcal{M}_2 \begin{Bmatrix} n-1 \\ n-1 \end{Bmatrix} - \mathcal{M}_2 \begin{Bmatrix} 0 \\ n-1 \end{Bmatrix} \cdot \mathcal{M}_2 \begin{Bmatrix} n-1 \\ n-m \end{Bmatrix}.$$

We study the determinants on the right hand side of these DCIs one by one. Notice that, by the induction hypothesis

$$\mathcal{M}_1 \begin{Bmatrix} n-1 \\ n-1 \end{Bmatrix} = H_{n-1}[\mu; z_1, \cdots, z_{m-1}] = \tag{5.5}$$

$$H_{n-m}[\mu] \det_{(m-1)\times(m-1)} \begin{pmatrix} P_{n-m}(z_1) & \cdots & P_{n-m}(z_{m-2}) & P_{n-m}(z_{m-1}) \\ \vdots & \ddots & \vdots & \vdots \\ P_{n-2}(z_1) & \cdots & P_{n-2}(z_{m-2}) & P_{n-2}(z_{m-1}) \end{pmatrix},$$

and

$$\mathcal{M}_1 \begin{Bmatrix} n-1 \\ n-2 \end{Bmatrix} = \tag{5.6}$$

$$H_{n-1}[\mu; z_1, \cdots, z_{m-2}, z_m]$$

$$= H_{n-m}[\mu] \det_{(m-1)\times(m-1)} \begin{pmatrix} P_{n-m}(z_1) & \cdots & P_{n-m}(z_{m-2}) & P_{n-m}(z_m) \\ \vdots & \ddots & \vdots & \vdots \\ P_{n-2}(z_1) & \cdots & P_{n-2}(z_{m-2}) & P_{n-2}(z_m) \end{pmatrix}.$$

The remaining two determinants on the right hand side of (5.3) are not directly expressible in terms of H_{n-1}, since due to the removal of the first row, their 11-entries are μ_1 as opposed to μ_0. Let us now turn our attention to (5.4). We can already observe that

$$\mathcal{M}_2 \begin{Bmatrix} 0 & n-1 \\ n-m & n-1 \end{Bmatrix} = \mathcal{M}_1 \begin{Bmatrix} 0 & n-1 \\ n-2 & n-1 \end{Bmatrix}, \tag{5.7}$$

$$\mathcal{M}_2 \begin{Bmatrix} 0 \\ n-m \end{Bmatrix} \equiv \mathcal{M}_2 \begin{Bmatrix} 0 \\ n-m \end{Bmatrix} [z_1, \cdots, z_{m-1}] = \mathcal{M}_1 \begin{Bmatrix} 0 \\ n-1 \end{Bmatrix}, \tag{5.8}$$

$$\mathcal{M}_2 \begin{Bmatrix} 0 \\ n-m \end{Bmatrix} [z_1, \cdots, z_{m-2}, z_m] = \mathcal{M}_1 \begin{Bmatrix} 0 \\ n-2 \end{Bmatrix}, \tag{5.9}$$

$$\mathcal{M}_2 \begin{Bmatrix} n-1 \\ n-m \end{Bmatrix} \equiv \mathcal{M}_2 \begin{Bmatrix} n-1 \\ n-m \end{Bmatrix} [z_1, \cdots, z_{m-1}] = \mathcal{M}_1 \begin{Bmatrix} n-1 \\ n-1 \end{Bmatrix}, \quad \text{and} \tag{5.10}$$

$$\mathcal{M}_2 \begin{Bmatrix} n-1 \\ n-m \end{Bmatrix} [z_1, \cdots, z_{m-2}, z_m] = \mathcal{M}_1 \begin{Bmatrix} n-1 \\ n-2 \end{Bmatrix}. \tag{5.11}$$

We can now express the two determinants $\mathcal{M}_1 \begin{Bmatrix} 0 \\ n-1 \end{Bmatrix}$ and $\mathcal{M}_1 \begin{Bmatrix} 0 \\ n-2 \end{Bmatrix}$ in terms of elements in the Dodgson Condensation identity (5.4). In fact, from (5.4) and we have

$$\mathcal{M}_1 \begin{Bmatrix} 0 \\ n-1 \end{Bmatrix} [z_1, \cdots, z_{m-1}] = \frac{1}{\mathcal{M}_2 \begin{Bmatrix} n-1 \\ n-1 \end{Bmatrix} [z_1, \cdots, z_{m-2}]} \times \left(\mathcal{M}_2[z_1, \cdots, z_{m-1}] \mathcal{M}_2 \begin{Bmatrix} 0 & n-1 \\ n-m & n-1 \end{Bmatrix} [z_1, \cdots, z_{m-2}] + \mathcal{M}_2 \begin{Bmatrix} 0 \\ n-1 \end{Bmatrix} [z_1, \cdots, z_{m-2}] \mathcal{M}_2 \begin{Bmatrix} n-1 \\ n-m \end{Bmatrix} [z_1, \cdots, z_{m-1}] \right). \tag{5.12}$$

In view of (5.9), for $\mathcal{M}_1 \begin{Bmatrix} 0 \\ n-2 \end{Bmatrix}$ we have the same expression given on the right hand side of the equation above, except that we only need to replace z_{m-1} by z_m. Indeed,

$$\mathcal{M}_1 \begin{Bmatrix} 0 \\ n-2 \end{Bmatrix} [z_1, \cdots, z_{m-2}, z_m] = \frac{1}{\mathcal{M}_2 \begin{Bmatrix} n-1 \\ n-1 \end{Bmatrix} [z_1, \cdots, z_{m-2}]} \times \left(\mathcal{M}_2[z_1, \cdots, z_{m-2}, z_m] \mathcal{M}_2 \begin{Bmatrix} 0 & n-1 \\ n-m & n-1 \end{Bmatrix} [z_1, \cdots, z_{m-2}] + \mathcal{M}_2 \begin{Bmatrix} 0 \\ n-1 \end{Bmatrix} [z_1, \cdots, z_{m-2}] \mathcal{M}_2 \begin{Bmatrix} n-1 \\ n-m \end{Bmatrix} [z_1, \cdots, z_{m-2}, z_m] \right). \tag{5.13}$$

Using (5.10) and (5.11), we can write these DCIs as

$$\mathcal{M}_1 \begin{Bmatrix} 0 \\ n-1 \end{Bmatrix} = \frac{1}{\mathcal{M}_2 \begin{Bmatrix} n-1 \\ n-1 \end{Bmatrix}} \times \left(\mathcal{M}_2 \cdot \mathcal{M}_2 \begin{Bmatrix} 0 & n-1 \\ n-m & n-1 \end{Bmatrix} + \mathcal{M}_2 \begin{Bmatrix} 0 \\ n-1 \end{Bmatrix} \cdot \mathcal{M}_1 \begin{Bmatrix} n-1 \\ n-1 \end{Bmatrix} \right), \tag{5.14}$$

and

$$\mathcal{M}_1 \begin{Bmatrix} 0 \\ n-2 \end{Bmatrix} = \frac{1}{\mathcal{M}_2 \begin{Bmatrix} n-1 \\ n-1 \end{Bmatrix}} \times \left(\mathcal{M}_2[z_1, \cdots, z_{m-2}, z_m] \mathcal{M}_2 \begin{Bmatrix} 0 & n-1 \\ n-m & n-1 \end{Bmatrix} + \mathcal{M}_2 \begin{Bmatrix} 0 \\ n-1 \end{Bmatrix} \mathcal{M}_1 \begin{Bmatrix} n-1 \\ n-2 \end{Bmatrix} \right). \tag{5.15}$$

Solving for $\mathcal{M}_1$ from (5.3) yields

$$\mathcal{M}_1 = \frac{\mathcal{M}_1 \begin{Bmatrix} 0 \\ n-2 \end{Bmatrix} \cdot \mathcal{M}_1 \begin{Bmatrix} n-1 \\ n-1 \end{Bmatrix} - \mathcal{M}_1 \begin{Bmatrix} 0 \\ n-1 \end{Bmatrix} \cdot \mathcal{M}_1 \begin{Bmatrix} n-1 \\ n-2 \end{Bmatrix}}{\mathcal{M}_1 \begin{Bmatrix} 0 & n-1 \\ n-2 & n-1 \end{Bmatrix}}. \tag{5.16}$$

Now using (5.7), (5.14) and (5.15) in the above expression we obtain
(5.17)

$$\mathcal{M}_1 = \frac{\mathcal{M}_2[z_1, \cdots, z_{m-2}, z_m] \cdot \mathcal{M}_1 \begin{Bmatrix} n-1 \\ n-1 \end{Bmatrix} - \mathcal{M}_2[z_1, \cdots, z_{m-2}, z_{m-1}] \cdot \mathcal{M}_1 \begin{Bmatrix} n-1 \\ n-2 \end{Bmatrix}}{\mathcal{M}_2 \begin{Bmatrix} n-1 \\ n-1 \end{Bmatrix}}$$

after straightforward simplifications. Now we notice that all objects on the right hand side of (5.17) either have $m-1$ or $m-2$ borders, and thus for each of them we can use the induction hypothesis, recalling (5.2) . To this end we already have equations (5.5) and (5.6) for $\mathcal{M}_1 \begin{Bmatrix} n-1 \\ n-1 \end{Bmatrix}$ and $\mathcal{M}_1 \begin{Bmatrix} n-1 \\ n-2 \end{Bmatrix}$, while for the other terms we can write

$$(5.18) \qquad \mathcal{M}_2 \begin{Bmatrix} n-1 \\ n-1 \end{Bmatrix} = H_{n-m+1}[\mu] \det \begin{pmatrix} P_{n-m+1}(z_1) & \cdots & P_{n-m+1}(z_{m-2}) \\ \vdots & \ddots & \vdots \\ P_{n-2}(z_1) & \cdots & P_{n-2}(z_{m-2}) \end{pmatrix},$$

$$(5.19) \qquad \mathcal{M}_2[z_1, \cdots, z_{m-2}, z_{m-1}]$$
$$= H_{n-m+1}[\mu] \det \begin{pmatrix} P_{n-m+1}(z_1) & \cdots & P_{n-m+1}(z_{m-2}) & P_{n-m+1}(z_{m-1}) \\ \vdots & \ddots & \vdots & \vdots \\ P_{n-1}(z_1) & \cdots & P_{n-1}(z_{m-2}) & P_{n-1}(z_{m-1}) \end{pmatrix},$$

and

$$(5.20) \qquad \mathcal{M}_2[z_1, \cdots, z_{m-2}, z_m]$$
$$= H_{n-m+1}[\mu] \det \begin{pmatrix} P_{n-m+1}(z_1) & \cdots & P_{n-m+1}(z_{m-2}) & P_{n-m+1}(z_m) \\ \vdots & \ddots & \vdots & \vdots \\ P_{n-1}(z_1) & \cdots & P_{n-1}(z_{m-2}) & P_{n-1}(z_m) \end{pmatrix}.$$

Plugging these, together with (5.5) and (5.6), into (5.17), we observe that the right hand side of (5.17) is nothing but

$$(5.21) \qquad H_{n-m}[\mu] \left(\frac{\mathcal{D}\begin{Bmatrix} 0 \\ m-2 \end{Bmatrix} \cdot \mathcal{D}\begin{Bmatrix} m-1 \\ m-1 \end{Bmatrix} - \mathcal{D}\begin{Bmatrix} 0 \\ m-1 \end{Bmatrix} \cdot \mathcal{D}\begin{Bmatrix} m-1 \\ m-2 \end{Bmatrix}}{\mathcal{D}\begin{Bmatrix} 0 & m-1 \\ m-2 & m-1 \end{Bmatrix}} \right),$$

with

$$(5.22) \qquad \mathcal{D} := \det \begin{pmatrix} P_{n-m}(z_1) & \cdots & P_{n-m}(z_{m-1}) & P_{n-m}(z_m) \\ \vdots & \ddots & \vdots & \vdots \\ P_{n-1}(z_1) & \cdots & P_{n-1}(z_{m-1}) & P_{n-1}(z_m) \end{pmatrix}.$$

Now, due to the following Dodgson Condensation identity
(5.23)

$$\mathcal{D} \cdot \mathcal{D}\begin{Bmatrix} 0 & m-1 \\ m-2 & m-1 \end{Bmatrix} = \mathcal{D}\begin{Bmatrix} 0 \\ m-2 \end{Bmatrix} \cdot \mathcal{D}\begin{Bmatrix} m-1 \\ m-1 \end{Bmatrix} - \mathcal{D}\begin{Bmatrix} 0 \\ m-1 \end{Bmatrix} \cdot \mathcal{D}\begin{Bmatrix} m-1 \\ m-2 \end{Bmatrix},$$

we finally arrive at the desired equality $\mathcal{M}_1 = H_{n-m}[\mu]\mathcal{D}$.

5.2. Proof of Propositions 1.13 and 1.16. We prove Proposition 1.13 in detail. The proof of Proposition 1.16 is similar.

The proof of the Proposition 1.13 relies on the LU decomposition of the Hankel matrix. Write the monic orthogonal polynomial (1.17) as

$$P_n(x) = \sum_{k=0}^{n} p_{n,k} x^k,$$

and introduce the triangular matrix

$$\boldsymbol{L}_n := \begin{pmatrix} \frac{p_{0,0}}{\sqrt{h_0}} & 0 & \cdots & 0 \\ \frac{p_{1,0}}{\sqrt{h_1}} & \frac{p_{1,1}}{\sqrt{h_1}} & \cdots & 0 \\ \vdots & \vdots & \ddots & \vdots \\ \frac{p_{n,0}}{\sqrt{h_n}} & \frac{p_{n,1}}{\sqrt{h_n}} & \cdots & \frac{p_{n,n}}{\sqrt{h_n}} \end{pmatrix}. \tag{5.24}$$

LEMMA 5.1 ([**54**]). *The LU decomposition of* $\boldsymbol{H}_{n+1}[\mu]$ *is given by*

$$\boldsymbol{H}_{n+1}[\mu] = [\boldsymbol{L}_n]^{-1} \left[\boldsymbol{L}_n^T\right]^{-1}. \tag{5.25}$$

PROOF OF LEMMA. We have

$$\begin{aligned} \delta_{jk} &= \int_{\mathbb{R}} \frac{P_j(x)}{\sqrt{h_j}} \frac{P_k(x)}{\sqrt{h_k}} \mathrm{d}\mu(x) = \sum_{m=0}^{j} \sum_{\ell=0}^{k} \frac{p_{j,m} p_{k,\ell}}{\sqrt{h_j h_k}} \int_{\mathbb{R}} x^{\ell+m} \mathrm{d}\mu(x) \\ &= \sum_{m=0}^{j} \sum_{\ell=0}^{k} \frac{p_{j,m} p_{k,\ell}}{\sqrt{h_j h_k}} \mu_{\ell+m} = \sum_{m=0}^{j} \sum_{\ell=0}^{k} (\boldsymbol{L}_n)_{j,m} (\boldsymbol{L}_n)_{k,\ell} (\boldsymbol{H}_{n+1}[\mu])_{\ell,m} \\ &= \sum_{m=0}^{j} \sum_{\ell=0}^{k} (\boldsymbol{L}_n)_{j,k} (\boldsymbol{H}_{n+1}[\mu])_{j,m} \left(\boldsymbol{L}_n^T\right)_{m,k} = \left(\boldsymbol{L}_n \boldsymbol{H}_{n+1}[\mu] \boldsymbol{L}_n^T\right)_{j,k}, \end{aligned} \tag{5.26}$$

which is equivalent to (5.25). □

Let

$$\mathbf{Z}_n(x_1, \cdots, x_m) = \begin{pmatrix} 1 & 1 & \cdots & 1 \\ x_1 & x_2 & \cdots & x_m \\ \vdots & \vdots & \ddots & \vdots \\ x_1^n & x_2^n & \cdots & x_m^n \end{pmatrix},$$

$$\mathbf{P}_n(x_1, x_2, \ldots, x_m) = \begin{pmatrix} \frac{P_0(x_1)}{\sqrt{h_0}} & \frac{P_0(x_2)}{\sqrt{h_0}} & \cdots & \frac{P_0(x_m)}{\sqrt{h_0}} \\ \frac{P_1(x_1)}{\sqrt{h_1}} & \frac{P_1(x_2)}{\sqrt{h_1}} & \cdots & \frac{P_1(x_m)}{\sqrt{h_1}} \\ \vdots & \vdots & \ddots & \vdots \\ \frac{P_n(x_1)}{\sqrt{h_n}} & \frac{P_n(x_2)}{\sqrt{h_n}} & \cdots & \frac{P_n(x_m)}{\sqrt{h_n}} \end{pmatrix},$$

and

$$\mathbf{F}_n(\mu;x_1,\dots,x_m;y_1,\dots,y_m;A) = \begin{pmatrix} \mu_0 & \mu_1 & \cdots & \mu_{n-m-1} & 1 & \cdots & 1 \\ \mu_1 & \mu_2 & \cdots & \mu_{n-m} & x_1 & \cdots & x_m \\ \mu_2 & \mu_3 & \cdots & \mu_{n-m+1} & x_1^2 & \cdots & x_m^2 \\ \vdots & \vdots & \ddots & \vdots & \vdots & \ddots & \vdots \\ \mu_{n-m-1} & \mu_{n-m} & \cdots & \mu_{2(n-m-1)} & x_1^{n-m-1} & \cdots & x_m^{n-m-1} \\ 1 & y_1 & \cdots & y_1^{n-m-1} & a_{11} & \cdots & a_{1m} \\ \vdots & \vdots & \ddots & \vdots & \vdots & \ddots & \vdots \\ 1 & y_m & \cdots & y_m^{n-m-1} & a_{m1} & \cdots & a_{mm} \end{pmatrix}.$$

We have

(5.27)

$$\begin{aligned} &\begin{pmatrix} \boldsymbol{L}_{n-m-1} & \mathbf{0}_{(n-m)\times m} \\ \mathbf{0}^T_{(n-m)\times m} & \mathbf{I}_m \end{pmatrix} \mathbf{F}_n(\mu;x_1,\dots,x_m;y_1,\dots,y_m;A) \begin{pmatrix} \boldsymbol{L}^T_{n-m-1} & \mathbf{0}_{(n-m)\times m} \\ \mathbf{0}^T_{(n-m)\times m} & \mathbf{I}_m \end{pmatrix} \\ &= \begin{pmatrix} \boldsymbol{L}_{n-m-1} & \mathbf{0}_{(n-m)\times m} \\ \mathbf{0}^T_{(n-m)\times m} & \mathbf{I}_m \end{pmatrix} \begin{pmatrix} \boldsymbol{H}_{n-m}[\mu] & \boldsymbol{Z}_{n-m-1}(x_1,\dots,x_m) \\ \boldsymbol{Z}^T_{n-m-1}(y_1,\dots,y_m) & A \end{pmatrix} \\ &\quad \times \begin{pmatrix} \boldsymbol{L}^T_{n-m-1} & \mathbf{0}_{(n-m)\times m} \\ \mathbf{0}^T_{(n-m)\times m} & \mathbf{I}_m \end{pmatrix} \\ &= \begin{pmatrix} \boldsymbol{L}_{n-m-1}\boldsymbol{H}_{n-m}[\mu]\boldsymbol{L}^T_{n-m-1} & \boldsymbol{L}_{n-m-1}\boldsymbol{Z}_{n-m-1}(x) \\ \boldsymbol{Z}^T_{n-m-1}(y)\boldsymbol{L}^T_{n-m-1} & A \end{pmatrix} \\ &= \begin{pmatrix} \boldsymbol{I}_{n-m} & \boldsymbol{P}_{n-m-1}(x) \\ \boldsymbol{P}^T_{n-m-1}(y) & A \end{pmatrix}. \end{aligned}$$

Taking the determinant of both sides of (5.27) yields (1.40), using the fact that

$$\det \mathbf{L}_{n-m-1} = \frac{1}{\sqrt{H_{n-m}[\mu]}}. \tag{5.28}$$

6. Proofs of Propositions 2.1 and 3.1

6.1. Proof of Proposition 2.1. By the reflection principle, the transition probability density for a standard Brownian motion to go from x to y during the time interval $[0,t]$ with an absorbing wall at zero is

$$p_{\text{abs}}(t;y|x) = \frac{1}{\sqrt{2\pi t}}\left[e^{-\frac{1}{2t}(x-y)^2} - e^{-\frac{1}{2t}(x+y)^2}\right]. \tag{6.1}$$

We will also consider a Brownian motion with an absorbing wall at zero and one at height $M>0$. By repeatedly applying the reflection principle, we find that the transition probability density for a Brownian motion to go from x to y during the time interval $[0,t]$ with absorbing walls at zero and M is

$$p^M_{\text{abs,abs}}(t;y|x) = \frac{1}{\sqrt{2\pi t}}\sum_{h=-\infty}^{\infty}\left[e^{-\frac{1}{2t}(x-y+2Mh)^2} - e^{-\frac{1}{2t}(x+y+2Mh)^2}\right]. \tag{6.2}$$

We can rewrite the two-wall formula using the Poisson resummation formula. This gives

$$
(6.3)\qquad
\begin{aligned}
p^M_{\mathrm{abs,abs}}(t;y|x) &= \frac{1}{2M}\sum_{h=-\infty}^{\infty} e^{-\frac{t\pi^2h^2}{2M^2}} e^{\frac{ih\pi x}{M}}\left(e^{-\frac{ih\pi y}{M}} - e^{\frac{ih\pi y}{M}}\right)\\
&= \frac{2}{M}\sum_{h=1}^{\infty} e^{-\frac{t\pi^2h^2}{2M^2}} \sin\left(\frac{h\pi x}{M}\right)\sin\left(\frac{h\pi y}{M}\right).
\end{aligned}
$$

Let

$$
(6.4)\qquad \vec{x} = (x_1, \dots x_n), \quad \vec{y} = (y_1, \dots y_n), \quad 0 < x_1 < \cdots < x_n, \quad 0 < y_1 < \cdots < y_n.
$$

Then by the Karlin–McGregor formula [64] , the transition probability for n non-intersecting BMs to go from $\vec{x}$ to $\vec{y}$ over the period of time from 0 to 1 with an absorbing wall at 0 is

$$
(6.5)\qquad q_{\mathrm{abs}}(\vec{y}|\vec{x}) = \det\left[p_{\mathrm{abs}}(1; y_k|x_j)\right]_{j,k=1}^{n},
$$

and the analogous transition probability with absorbing walls at 0 and M is

$$
(6.6)\qquad q^M_{\mathrm{abs,abs}}(\vec{y}|\vec{x}) = \det\left[p^M_{\mathrm{abs,abs}}(1; y_k|x_j)\right]_{j,k=1}^{n}.
$$

Then the cumulative distribution function for $\mathcal{M}_n$ is

$$
(6.7)\qquad \mathbb{P}_m^{\vec{\alpha},\vec{\beta}}(\mathcal{M}_n < M) = \lim_{\substack{y_1,y_2,\dots y_{n-m},x_1,x_2,\dots x_{n-m}\to 0\\ x_n\to\alpha_1,\dots,x_{n-m+1}\to\alpha_m\\ y_n\to\beta_1,\dots,y_{n-m+1}\to\beta_m}} \frac{q^M_{\mathrm{abs,abs}}(\vec{y}|\vec{x})}{q_{\mathrm{abs}}(\vec{y}|\vec{x})},
$$

where the limits $y_1, y_2, \dots y_{n-m}, x_1, x_2, \dots x_{n-m} \to 0$ are taken sequentially to respect the order $x_1 < x_2 < \cdots < x_{n-m}$ and $y_1 < y_2 < \cdots < y_{n-m}$.

We have

$$
(6.8)\qquad
\begin{aligned}
q_{\mathrm{abs}}(\vec{y}|\vec{x}) &= \left(\frac{1}{2\pi}\right)^{n/2} \det\left[e^{-\frac{1}{2}(y_k-x_j)^2} - e^{-\frac{1}{2}(y_k+x_j)^2}\right]_{j,k=1}^{n}\\
&= \left(\frac{2}{\pi}\right)^{n/2} \prod_{j=1}^{n} e^{-\frac{1}{2}(x_j^2+y_j^2)} \det\left[\sinh(x_j y_k)\right]_{j,k=1}^{n}.
\end{aligned}
$$

Expanding the first $n-m$ rows as a Taylor series about $x_j = 0$, and using row

operations we easily find that as $x_1, x_2, \ldots, x_{n-m} \to 0$ (respecting the order $x_1 < x_2 < \cdots < x_{n-m}$),
(6.9)

$$q_{\text{abs}}(\vec{y}|\vec{x}) = \left(\frac{2}{\pi}\right)^{n/2} \prod_{j=1}^{n} x_j y_j \prod_{j=1}^{n} e^{-\frac{1}{2}y_j^2} \prod_{j=n-m+1}^{n} e^{-\frac{1}{2}x_j^2} \prod_{k=1}^{n-m} \frac{x_k^{2(k-1)}}{(2k-1)!}$$

$$\times \det \begin{pmatrix} 1 & 1 & \cdots & 1 \\ y_1^2 & y_2^2 & \cdots & y_n^2 \\ \vdots & \vdots & \ddots & \vdots \\ y_1^{2(n-m-1)} & y_2^{2(n-m-1)} & \cdots & y_n^{2(n-m-1)} \\ \frac{\sinh(x_{n-m+1}y_1)}{x_{n-m+1}y_1} & \frac{\sinh(x_{n-m+1}y_2)}{x_{n-m+1}y_2} & \cdots & \frac{\sinh(x_{n-m+1}y_n)}{x_{n-m+1}y_n} \\ \vdots & \vdots & \ddots & \vdots \\ \frac{\sinh(x_n y_1)}{x_n y_1} & \frac{\sinh(x_n y_2)}{x_n y_2} & \cdots & \frac{\sinh(x_n y_n)}{x_n y_n} \end{pmatrix} (1+O(|\vec{x}_{n-m}|^2)),$$

where $\vec{x}_{n-m} = (x_1, x_2, \ldots, x_{n-m})$. We can now take $y_1, \ldots, y_{n-m} \to 0$, and $y_n = \beta_1$, $y_{n-1} = \beta_2, \cdots, y_{n-m+1} = \beta_m$ as well as $x_n = \alpha_1$, $x_{n-1} = \alpha_2, \cdots, x_{n-m+1} = \alpha_m$ to obtain

$$(6.10)\quad q_{\text{abs}}(\vec{y}|\vec{x}) = \left(\frac{2}{\pi}\right)^{n/2} \prod_{j=1}^{n-m} x_j y_j \prod_{j=1}^{m} \alpha_j \beta_j e^{-\frac{1}{2}(\alpha_j^2+\beta_j^2)} \prod_{k=1}^{n-m} \frac{x_k^{2(k-1)}}{(2k-1)!}$$

$$\times \det \begin{pmatrix} 1 & \cdots & 1 & 1 & \cdots & 1 \\ y_1^2 & \cdots & y_{n-m}^2 & \beta_m^2 & \cdots & \beta_1^2 \\ \vdots & \ddots & \vdots & \vdots & \ddots & \vdots \\ y_1^{2(n-m-1)} & \cdots & y_{n-m}^{2(n-m-1)} & \beta_m^{2(n-m-1)} & \cdots & \beta_1^{2(n-m-1)} \\ 0 & \cdots & 0 & \frac{\sinh(\alpha_m\beta_m)}{\alpha_m\beta_m} - \sum_{\ell=0}^{n-m-1} \frac{(\alpha_m\beta_m)^{2\ell}}{(2\ell+1)!} & \cdots & \frac{\sinh(\alpha_m\beta_1)}{\alpha_m\beta_1} - \sum_{\ell=0}^{n-m-1} \frac{(\alpha_m\beta_1)^{2\ell}}{(2\ell+1)!} \\ \vdots & \ddots & \vdots & \vdots & \ddots & \vdots \\ 0 & \cdots & 0 & \frac{\sinh(\alpha_1\beta_m)}{\alpha_1\beta_m} - \sum_{\ell=0}^{n-m-1} \frac{(\alpha_1\beta_m)^{2\ell}}{(2\ell+1)!} & \cdots & \frac{\sinh(\alpha_1\beta_1)}{\alpha_1\beta_1} - \sum_{\ell=0}^{n-m-1} \frac{(\alpha_1\beta_1)^{2\ell}}{(2\ell+1)!} \end{pmatrix}$$

$$\times \left(1 + O(|\vec{x}_{n-m}|^2) + O(|\vec{y}_{n-m}|^2)\right).$$

The above determinant is block upper-triangular and is the product of the two diagonal blocks, one of which is a Vandermonde determinant:
(6.11)

$$q_{\text{abs}}(\vec{y}|\vec{x}) = \left(\frac{2}{\pi}\right)^{n/2} \prod_{j=1}^{n-m} x_j y_j \prod_{j=1}^{m} \alpha_j \beta_j e^{-\frac{1}{2}(\alpha_j^2+\beta_j^2)} \prod_{k=1}^{n-m} \frac{x_k^{2(k-1)}}{(2k-1)!} \Delta(\vec{y}^2_{n-m})$$

$$\times \det \left[\frac{\sinh(\alpha_{m-j+1}\beta_{m-k+1})}{\alpha_{m-j+1}\beta_{m-k+1}} - \sum_{\ell=0}^{n-m-1} \frac{(\alpha_{m-j+1}\beta_{m-k+1})^{2\ell}}{(2\ell+1)!} \right]_{j,k=1}^{m}$$

$$\times \left(1 + O(|\vec{x}_{n-m}|^2) + O(|\vec{y}_{n-m}|^2)\right),$$

where

$$\Delta(\vec{y}^2_{n-m}) = \prod_{1 \le j < k \le n-m} (y_k^2 - y_j^2).$$

Now consider $q^M_{\text{abs,abs}}(\vec{y}|\vec{x})$. It is given as

$$q^M_{\text{abs,abs}}(\vec{y}|\vec{x}) = \left(\frac{2}{M}\right)^n \det\left[\sum_{h=1}^{\infty} e^{-\frac{\pi^2 h^2}{2M^2}} \sin\left(\frac{h\pi x_j}{M}\right)\sin\left(\frac{h\pi y_k}{M}\right)\right]_{j,k=1}^n. \tag{6.12}$$

Taking $x_1, x_2, \ldots, x_{n-m} \to 0$, using Taylor expansion and row operations, we find

(6.13)

$$q^M_{\text{abs,abs}}(\vec{y}|\vec{x}) = (-1)^{(n-m-1)(n-m)/2}\left(\frac{2}{M}\right)^n\left(\frac{\pi}{M}\right)^{(n-m)^2}\prod_{k=1}^{n-m}\frac{x_k^{2(k-1)}}{(2k-1)!}\prod_{k=1}^{n} x_k y_k$$

$$\times\det\begin{pmatrix} \nu_0(y_1) & \nu_0(y_2) & \cdots & \nu_0(y_n) \\ \nu_1(y_1) & \nu_1(y_2) & \cdots & \nu_1(y_n) \\ \vdots & \vdots & \ddots & \cdots \\ \nu_{n-m-1}(y_1) & \nu_{n-m-1}(y_2) & \cdots & \nu_{n-m-1}(y_n) \\ s(x_{n-m+1}, y_1) & s(x_{n-m+1}, y_2) & \cdots & s(x_{n-m+1}, y_n) \\ \vdots & \vdots & \ddots & \cdots \\ s(x_n, y_1) & s(x_n, y_2) & \cdots & s(x_n, y_n) \end{pmatrix}\left(1 + O(|\vec{x}_{n-m}|^2)\right),$$

where $\nu_k(\alpha)$ is the kth moment of the measure $\nu(\alpha)$ defined in (2.2), and the constants $s(\alpha, \beta)$ are as defined in (2.3). Now taking $y_1, y_2, \ldots, y_{n-m} \to 0$, and $y_n = \beta_1, y_{n-1} = \beta_2, \cdots, y_{n-m+1} = \beta_m$ as well as $x_n = \alpha_1, x_{n-1} = \alpha_2, \cdots, x_{n-m+1} = \alpha_m$ we find

(6.14)

$$q^M_{\text{abs,abs}}(\vec{y}|\vec{x}) = \left(\frac{2}{M}\right)^n\left(\frac{\pi}{M}\right)^{2(n-m)^2}\prod_{k=1}^{n-m}\frac{x_k^{2(k-1)}y_k^{2(k-1)}}{((2k-1)!)^2}\prod_{k=1}^{n-m} x_k y_k\prod_{k=1}^{m}\alpha_k\beta_k$$

$$\det\begin{pmatrix} \mu_0 & \mu_1 & \cdots & \mu_{n-m-1} & \nu_0(\beta_m) & \cdots & \nu_0(\beta_1) \\ \mu_1 & \mu_2 & \cdots & \mu_{n-m} & \nu_1(\beta_m) & \cdots & \nu_1(\beta_1) \\ \vdots & \vdots & \ddots & \vdots & \vdots & \ddots & \cdots \\ \mu_{n-m-1} & \mu_{n-m} & \cdots & \mu_{2(n-m-1)} & \nu_{n-m-1}(\beta_m) & \cdots & \nu_{n-m-1}(\beta_1) \\ \nu_0(\alpha_m) & \nu_1(\alpha_m) & \cdots & \nu_{n-m-1}(\alpha_m) & s(\alpha_m, \beta_m) & \cdots & s(\alpha_m, \beta_m) \\ \vdots & \vdots & \ddots & \vdots & \vdots & \ddots & \cdots \\ \nu_0(\alpha_1) & \nu_1(\alpha_1) & \cdots & \nu_{n-m-1}(\alpha_1) & s(\alpha_1, \beta_m) & \cdots & s(\alpha_1, \beta_1) \end{pmatrix}$$
$$\times\left(1 + O(|\vec{x}_{n-m}|^2) + O(|\vec{y}_{n-m}|^2)\right)$$

where μ_k is the kth moment of the measure μ defined in (2.1). The ratio of (6.10) and (6.14), taking $y_1, \ldots, y_{n-m}, x_1, x_2, \ldots x_{n-m} \to 0$ then gives (2.4).

The formula for $P_m^{\vec{0},\vec{\beta}}(\mathcal{M}_n < M)$ in (2.5) then obtained from (2.4) in a similar way by taking $\alpha_1, \ldots, \alpha_m \to 0$. We omit the details of this limit.

6.2. Proof of Proposition 3.1. We start with the distribution function for $\mathcal{W}_n$ when all starting and ending points are distinct. This formula was given in [**8**], and we present it here as a lemma.

LEMMA 6.1 (Equation (37) from [**8**]). *For n non-intersecting Browninan bridges $X_1(t) < X_2(t) < \cdots < X_n(t)$ on the time interval $[0, 1]$ with distinct starting points*

$X_j(0) = x_j$ with $x_1 < x_2 < \cdots < x_n$ and distinct ending points $X_j(1) = y_j$ with $y_1 < \cdots < y_n$, the width $\mathcal{W}_n := \sup_{0\le t\le 1} (X_n(t) - X_1(t))$ is given by the formula

$$\mathbb{P}(\mathcal{W}_n < M) = \frac{\int_0^1 \det[g(x_j - y_k, \tau)]_{j,k=1}^n \, d\tau}{\det[p(x_j - y_k)]_{j,k=1}^n}, \tag{6.15}$$

where

$$p(x) = \frac{1}{\sqrt{2\pi}} e^{-x^2/2}, \quad g(x,\tau) = \sum_{h\in\mathbb{Z}} p(x + hM) e^{2\pi i h\tau}.$$

The expression (6.15) has a well defined limit as $x_j \to 0$ and $y_j \to 0$ for $j = m_1+1, \ldots, n-m_2$. Consider first the denominator $\det[p(x_j - y_k)]_{j,k=1}^n$. Using the formula $p(x) = \frac{1}{2\pi}\int_{\mathbb{R}} e^{-y^2/2+\mathrm{i}xy}\, dy$, we can use the Maclaurin expansion of $p(x_j - y_k)$ and row operations to take the limit

$$\lim_{x_{m_1+1}, x_{m_1+2}, \ldots, x_{n-m_2}\to 0} \frac{\det[p(x_j - y_k)]_{j,k=1}^n}{\prod_{j=1}^{n-m_1-m_2} \frac{(x_{m_1+j} y_{m_1+j})^{j-1}}{((j-1)!)^2}} = \tag{6.16}$$

$$\frac{1}{\prod_{j=1}^{n-m_1-m_2} \frac{(y_j)^{j-1}}{(j-1)!}} \det \begin{bmatrix} \big(p(x_j - y_k)\big)_{\substack{j=1\ldots,m_1\\ k=1\ldots,n}} \\ \left(\frac{1}{2\pi}\int_{\mathbb{R}} e^{-y^2/2-\mathrm{i}y y_k}(\mathrm{i}y)^{j-m_1-1}\mathrm{d}y\right)_{\substack{j=m_1+1\ldots,n-m_1-m_2\\ k=1\ldots,n}} \\ \big(p(x_j - y_k)\big)_{\substack{j=n-m_1-m_2+1\ldots,n\\ k=1\ldots,n}} \end{bmatrix},$$

where the limit is taken sequentially, first $x_{m_1+1} \to 0$, then $x_{m_1+2} \to 0$, etc. Now taking the limit as $y_{m_1+1}, \ldots, y_{n-m_2} \to 0$ (again sequentially), using Maclaurin expansion and column operations we find

(6.17)

$$\lim_{\substack{x_{m_1+1}, x_{m_1+2}, \ldots, x_{n-m_2}\to 0\\ y_{m_1+1}, y_{m_1+2}, \ldots, y_{n-m_2}\to 0}} \frac{\det[p(x_j - y_k)]_{j,k=1}^n}{\prod_{j=1}^{n-m_1-m_2} \frac{(x_{m_1+j} y_{m_1+j} j)^{j-1}}{((j-1)!)^2}} =$$

$$\det\begin{bmatrix} \big(p(x_j-y_k)\big)_{j,k=1}^{m_1} & \left(\frac{1}{2\pi}\int_{\mathbb{R}} e^{-y^2/2+\mathrm{i}x_k}(-\mathrm{i}y)^{k-m_1-1}\mathrm{d}y\right)_{\substack{j=1,\ldots,m_1\\ k=m_1+1\ldots,n-m_2}} & \big(p(x_j-y_k)\big)_{\substack{j=1,\ldots,m_1\\ k=n-m_2+1,\ldots,n}} \\ \left(\frac{1}{2\pi}\int_{\mathbb{R}} e^{-y^2/2-\mathrm{i}y_k}(\mathrm{i}y)^{j-m_1-1}\mathrm{d}y\right)_{\substack{j=m_1+1,\ldots,n-m_2\\ k=1,\ldots,m_1}} & \left(\frac{1}{2\pi}\int_{\mathbb{R}} e^{-y^2/2}(\mathrm{i}y)^{j-m_1-1}(-\mathrm{i}y)^{k-m_1-1}\right)_{j,k=m_1+1}^{n-m} & \left(\frac{1}{2\pi}\int_{\mathbb{R}} e^{-y^2/2-\mathrm{i}y_k}(\mathrm{i}y)^{j-m_1-1}\mathrm{d}y\right)_{\substack{j=m_1+1,\ldots,n-m_2\\ k=n-m_2+1,\ldots,n}} \\ \big(p(x_j-y_k)\big)_{\substack{j=n-m_2+1,\ldots,n\\ k=1,\ldots,m_1}} & \left(\frac{1}{2\pi}\int_{\mathbb{R}} e^{-y^2/2+\mathrm{i}x_k}(-\mathrm{i}y)^{k-m_1-1}\mathrm{d}y\right)_{\substack{j=n-m_2+1,\ldots,n\\ k=m_1+1,\ldots,n-m_2}} & \big(p(x_j-y_k)\big)_{\substack{j=n-m_2+1,\ldots,n\\ k=n-m_2+1,\ldots,n}} \end{bmatrix}$$

$$= \det\begin{bmatrix} \big(p(x_j-y_k)\big)_{j,k=1}^{m_1} & \left(\frac{1}{2\pi}\int_{\mathbb{R}} e^{-y^2/2+\mathrm{i}x_k}y^{k-m_1-1}\mathrm{d}y\right)_{\substack{j=1,\ldots,m_1\\ k=m_1+1\ldots,n-m_2}} & \big(p(x_j-y_k)\big)_{\substack{j=1,\ldots,m_1\\ k=n-m+1,\ldots,n}} \\ \left(\frac{1}{2\pi}\int_{\mathbb{R}} e^{-y^2/2-\mathrm{i}y_k}y^{j-m_1-1}\mathrm{d}y\right)_{\substack{j=m_1+1,\ldots,n-m\\ k=1,\ldots,m_1}} & \left(\frac{1}{2\pi}\int_{\mathbb{R}} e^{-y^2/2}y^{j+k-2(m_1+1)}\right)_{j,k=m_1+1}^{n-m} & \left(\frac{1}{2\pi}\int_{\mathbb{R}} e^{-y^2/2-\mathrm{i}y_k}y^{j-m_1-1}\mathrm{d}y\right)_{\substack{j=m_1+1,\ldots,n-m\\ k=n-m_2+1,\ldots,n}} \\ \big(p(x_j-y_k)\big)_{\substack{j=n-m_2+1,\ldots,n\\ k=1,\ldots,m_1}} & \left(\frac{1}{2\pi}\int_{\mathbb{R}} e^{-y^2/2+\mathrm{i}x_k}y^{k-m_1-1}\mathrm{d}y\right)_{\substack{j=n-m_2+1,\ldots,n\\ k=m_1+1,\ldots,n-m_2}} & \big(p(x_j-y_k)\big)_{\substack{j=n-m_2+1,\ldots,n\\ k=n-m_2+1,\ldots,n}} \end{bmatrix}.$$

The above determinant is a framed Hankel determinant. Factoring out $(2\pi)^{-n}$ and writing $x_j = \alpha_j$ and $y_j = \beta_j$ for $j = 1, \ldots, m_1$, as well as $x_{n-m_2+j} = \alpha_{m_1+j}$ and $y_{n-m_2+j} = \beta_{m_1+j}$, it becomes

$$\left(\frac{1}{2\pi}\right)^n H_n^F[\mu; \mu_{\beta_1}\mu_{\beta_2}, \ldots, \mu_{\beta_m}; \mu_{(-\alpha_1)}, \ldots, \mu_{(-\alpha_m)}; A]$$

after simple row and column operations, where μ and μ_β are defined in (3.1) and (3.3), the matrix A is defined in (3.5).

Now consider $\det[g(\alpha_j - \beta_k, \theta)]_{j,k=1}^n$. It is helpful to use the Poisson resummation formula to write $g(x,\theta)$ as

$$g(x,\theta) = \frac{1}{M}\sum_{h\in\mathbb{Z}} e^{-\frac{1}{2}(\frac{2\pi}{M}(h-\tau))^2 + \frac{2\pi i x}{M}(h-\tau)}.$$

Furthermore, writing $y = \frac{2\pi}{M}(h-\tau)$, the above formula becomes (recall the discrete set $\mathcal{D}_{M,\tau}$ defined in (3.2))

$$g(x,\theta) = \frac{1}{M}\sum_{y\in\mathcal{D}_{M,\tau}} e^{-y^2/2 + ixy},$$

which has a very similar form to $p(x) = \frac{1}{2\pi}\int_{\mathbb{R}} e^{-y^2/2+ixy}\,dy$. Similar to (6.17) then, we find

$$\lim_{\substack{x_{m_1+1},x_{m_1+2},\dots,x_{n-m_2}\to 0\\ y_{m_1+1},y_{m_1+2},\dots,y_{n-m_2}\to 0}} \frac{\det[p(x_j-y_k)]_{j,k=1}^n}{\prod_{j=1}^{n-m_1-m_2}\frac{(x_{m_1+j}y_{m_1+j}j)^{j-1}}{((j-1)!)^2}} = \tag{6.18}$$
$$= \left(\frac{1}{M}\right)^n H_n^F[\mu;\mu_{\beta_1}^{M,\tau}\mu_{\beta_2}^{M,\tau},\dots,\mu_{\beta_m}^{M,\tau};\mu_{(-\alpha_1)}^{M,\tau},\dots,\mu_{(-\alpha_m)}^{M,\tau};A^{M,\tau}],$$

after writing $x_j = \alpha_j$ and $y_j = \beta_j$ for $j = 1,\dots,m_1$, as well as $x_{n-m_2+j} = \alpha_{m_1+j}$ and $y_{n-m_2+j} = \beta_{m_1+j}$, where $\mu^{M,\tau}$ and $\mu_\beta^{M,\tau}$ are defined in (3.1) and (3.3), the matrix $A^{M,\tau}$ is defined in (3.5).

Combining (6.15), (6.17), and (6.18), we obtain (3.4). The proof of (3.6) is similar and we omit the details. This completes the proof of Proposition 3.1.

7. An illustration of asymptotics: continuous time non-intersecting paths

Each one of the applications discussed in sections 2, 3, and 4 require its own careful treatment in a separate work. However, in this section we showcase the asymptotic analysis associated with one of the aforementioned applications, that is finding the transition probability of the non-intersecting paths modeled by (3.15), with prescribed starting and ending points. To this end, for certain choices of starting and ending points, we use some ideas from **[10]** and **[52]** to illustrate some calculations.

The asymptotic behavior of Toeplitz determinants with respect to smooth functions can be described by the Szegő–Widom theorem **[28, 74, 78]**, which is formulated as

$$D_n[\phi] \sim G[\phi]^n E[\phi], \qquad n\to\infty, \tag{7.1}$$

where the terms $G[\phi]$ and $E[\phi]$ are defined by:

$$G[\phi] = \exp\left([\log\phi]_0\right) \quad \text{and} \quad E[\phi] = \exp\left(\sum_{k\ge 1} k[\log\phi]_k[\log\phi]_{-k}\right). \tag{7.2}$$

This theorem holds true when the function ϕ is suitably smooth, does not vanish on the unit circle, and possesses a winding number of zero. We refer to **[46]** for a comprehensive survey of the Szegö theorem, including an intriguing account of its historical developments.

Let us consider the Toeplitz determinant generated by the symbol

$$P_X(z;T) = e^{\frac{T}{2}(z+1/z-2)}, \tag{7.3}$$

first introduced in (3.15). This is a Szegő-type symbol, and thus the Szegő limit theorem provides us with the large-size asymptotics. Indeed,

$$\begin{aligned} D_n[e^{\frac{T}{2}(z+1/z-2)}] &= e^{-nT} D_n[e^{\frac{T}{2}(z+1/z)}] \\ &= e^{-nT} G[\phi]^n E[\phi] \left(1 + O(e^{-cn})\right), \qquad \text{for any} \quad c > 0, \end{aligned} \tag{7.4}$$

where

$$G[\phi] = [\log \phi]_0 = \int_{\mathbb{T}} \log(\phi(\tau;t)) \frac{\mathrm{d}\tau}{2\pi \mathrm{i} \tau}, \tag{7.5}$$

$$E[\phi] = \sum_{k=0}^{\infty} k [\log \phi]_k [\log \phi]_{-k}, \tag{7.6}$$

and

$$\phi \equiv \phi(z;T) := e^{\frac{T}{2}(z+1/z)}. \tag{7.7}$$

REMARK 7.1. The error term in (7.4) is $O(e^{-cn})$ for *any* $c > 0$, because e^{-c} is the radius of the contour Γ_0 shown in Figure 7 and due to the analytic properties of ϕ this radius can be made as small as desired.

The *Szegő function* associated with ϕ (see (A.8) and (A.9) in the Appendix) can be explicitly calculated as

$$\alpha(z) = \begin{cases} e^{\frac{Tz}{2}} & |z| < 1, \\ e^{-\frac{T}{2z}} & |z| > 1. \end{cases} \tag{7.8}$$

Noticing that $G[\phi] = \alpha(0)$, we see that

$$G[\phi] = 1 \tag{7.9}$$

in (7.4). Due to the simplicity of the symbol, it is also straightforward to arrive at

$$E[\phi] = e^{\frac{T^2}{4}}, \tag{7.10}$$

via (7.6). So rewriting (7.4) we get

$$D_n[e^{\frac{T}{2}(z+1/z-2)}] = e^{\frac{T^2}{4}} e^{-nT} \left(1 + O(e^{-cn})\right), \qquad \text{for any} \quad c > 0. \tag{7.11}$$

7.1. Top two paths with non-consecutive ending points. Proof of Theorem 1.19. Consider n continuous non-intersecting paths with starting points $x_1, \cdots, x_n$ and ending points $y_1, \cdots, y_n$, where $x_j = j$, $j = 1, \cdots, n$, $y_k = k$, for $k = 1, \cdots, n-2$, and $n - 1 < y_{n-1} < y_n$, are arbitrary. For this calculation we assume

- T is finite, and
- the gaps $y_n - n$ and $y_{n-1} - n$ are finite as well.

As described in §3.2, the transition probability of the non-intersecting paths modeled by (3.15), with the above starting and ending points, is given by the two-bordered Toeplitz determinant:

$$e^{-nT} D_n^B[\phi; \boldsymbol{\psi}_2], \tag{7.12}$$

where

$$\psi_1(z;n) = \phi(z)z^{-(y_{n-1}-n)} \quad \text{and} \quad \psi_2(z;n) = \phi(z)z^{-(y_n-n)}. \tag{7.13}$$

Let us denote

$$D_n^B[\phi;\boldsymbol{\psi}_2] \equiv \mathcal{D},$$

and consider the Dodgson Condensation identity

$$\mathcal{D}\cdot\mathcal{D}\begin{Bmatrix} 0 & n-1 \\ n-2 & n-1 \end{Bmatrix} = \mathcal{D}\begin{Bmatrix} 0 \\ n-2 \end{Bmatrix}\cdot\mathcal{D}\begin{Bmatrix} n-1 \\ n-1 \end{Bmatrix} - \mathcal{D}\begin{Bmatrix} 0 \\ n-1 \end{Bmatrix}\cdot\mathcal{D}\begin{Bmatrix} n-1 \\ n-2 \end{Bmatrix}, \tag{7.14}$$

which can be written as

$$\begin{aligned} D_n^B[\phi;\boldsymbol{\psi}_2(z;n)] &= D_{n-1}^B[\phi;z^{-1}\psi_1(z;n)]\frac{D_{n-1}^B[z\phi;\psi_2(z;n)]}{D_{n-2}[z\phi]} \\ &\quad - D_{n-1}^B[\phi;z^{-1}\psi_2(z;n)]\frac{D_{n-1}^B[z\phi;\psi_1(z;n)]}{D_{n-2}[z\phi]}, \end{aligned} \tag{7.15}$$

or

$$\begin{aligned} D_{n+2}^B[\phi;\boldsymbol{\psi}_2(z;n+2)] &= D_{n+1}^B[\phi;z^{-1}\psi_1(z;n+2)]\frac{D_{n+1}^B[z\phi;\psi_2(z;n+2)]}{D_n[z\phi]} \\ &\quad - D_{n+1}^B[\phi;z^{-1}\psi_2(z;n+2)]\frac{D_{n+1}^B[z\phi;\psi_1(z;n+2)]}{D_n[z\phi]}. \end{aligned} \tag{7.16}$$

As illustrated in **[52]**, the two bordered Toeplitz determinants

$$D_{n+1}^B[\phi;z^{-1}\psi_1] \quad \text{and} \quad D_{n+1}^B[\phi;z^{-1}\psi_2]$$

can be found from

$$D_{n+1}^B[\phi;z^{-\ell}\phi] = \frac{D_n[\phi]}{\ell!}\frac{\mathrm{d}^\ell}{\mathrm{d}z^\ell}X_{12}(z;n)\bigg|_{z=0}, \tag{7.17}$$

where $X_{12}(z;n)$ is the 12-entry in the solution of the Riemann–Hilbert problem **RH-X1**–**RH-X3**. Thus it can be shown that for a Szegő-type symbol ϕ we have

$$D_{n+1}^B[\phi;z^{-\ell}\phi] = G[\phi]^n E[\phi]\left(\frac{\alpha^{(\ell)}(0)}{\ell!} + O(e^{-\mathfrak{c}n})\right), \quad \text{as} \quad n\to\infty, \tag{7.18}$$

where α is given by (A.9), and the constants $G[\phi]$ and $E[\phi]$ are given by (7.2) and $\mathfrak{c}$ is some positive constant (see Corollary 2.7.1 of **[52]**) depending on the analytic properties of ϕ.

In view of (7.8), (7.9), (7.10), and (7.18) we can write

(7.19)

$$D_{n+1}^B[\phi;z^{-\ell}\phi] = G[\phi]^n E[\phi]\left(\frac{\alpha^{(\ell)}(0)}{\ell!} + O(e^{-\mathfrak{c}n})\right) = e^{\frac{T^2}{4}}\left(\frac{1}{\ell!}\left(\frac{T}{2}\right)^\ell + O(e^{-\mathfrak{c}n})\right).$$

Therefore

$$\begin{aligned} D_{n+1}^B[\phi;z^{-1}\psi_1(z;n+2)] &= D_{n+1}^B[\phi;\phi(z)z^{-(y_{n+1}-n-1)}] \\ &= e^{\frac{T^2}{4}}\left(\frac{\left(\frac{T}{2}\right)^{y_{n+1}-n-1}}{(y_{n+1}-n-1)!} + O(e^{-\mathfrak{c}n})\right), \end{aligned} \tag{7.20}$$

and similarly

$$\begin{aligned} D^B_{n+1}[\phi; z^{-1}\psi_2(z;n+2)] &= D^B_{n+1}[\phi; \phi(z) z^{-(y_{n+2}-n-1)}] \\ &= e^{\frac{T^2}{4}} \left(\frac{\left(\frac{T}{2}\right)^{y_{n+2}-n-1}}{(y_{n+2}-n-1)!} + O(e^{-cn}) \right). \end{aligned} \tag{7.21}$$

We have just arrived at the following result (compare with Theorem 1.19) when only the top end-point is not adjacent to the rest.

THEOREM 7.2. *Consider $n+1$ independent continuous time random walks $X_1(t), \dots, X_{n+1}(t)$ which begin at locations $x_1 < x_2 < \cdots < x_{n+1}$ at time $t=0$. For a given $y_1 < y_2 < \cdots < y_{n+1}$, let $\mathbb{P}_{n+1}(x_1, \dots, x_{n+1}; y_1, \dots, y_{n+1}; T)$ be the probability that $X_1(t), \dots, X_{n+1}(t)$ satisfy $X_j(T) = y_j$ for $j = 1, \dots, n+1$, and $X_1(t), \dots, X_{n+1}(t)$ do not intersect for $0 \le t \le T$. Suppose that $x_j = j$ for $1 \le j \le n+1$, $y_k = k$ for $1 \le k \le n$, while y_{n+1} is arbitrary. If T and $\beta_1 := y_{n+1} - n$ are bounded as $n \to \infty$, $\mathbb{P}_{n+1}(x_1, \dots, x_{n+1}; y_1, \dots, y_{n+1}; T)$ satisfies as $n \to \infty$,*

$$\mathbb{P}_{n+1}(x_1, \dots, x_{n+1}; y_1, \dots, y_{n+1}; T) = \frac{e^{-(n+1)T} \left(\frac{T}{2}\right)^{\beta_1 - 1} e^{\frac{T^2}{4}}}{(\beta_1 - 1)!} \left(1 + O(e^{-cn})\right), \tag{7.22}$$

where $c > 0$ is any positive number.

Some of the terms on the right hand side of (7.16) involve the symbol $z\phi$, which is not a Szegő-type (due to non-zero winding number) nor a non-degenerate Fisher–Hartwig symbol (see e.g. [**45**]). This implies that one needs to analyze a separate Riemann–Hilbert problem. We write $Z(z;n)$ to refer to the solution of the X-RHP when ϕ is replaced by $z\phi$. More precisely, $Z(z;n)$ satisfies

- **RH-Z1** $Z(\cdot;n) : \mathbb{C} \setminus \mathbb{T} \to \mathbb{C}^{2\times 2}$ is analytic,
- **RH-Z2** The limits of $Z(\zeta;n)$ as ζ tends to $z \in \mathbb{T}$ from the inside and outside of the unit circle exist, which are denoted by $Z_\pm(z;n)$ respectively. Moreover, the functions $z \mapsto Z_\pm(z)$ are continuous on $\mathbb{T}$ and are related by

$$Z_+(z;n) = Z_-(z;n) \begin{pmatrix} 1 & z^{-n+1}\phi(z) \\ 0 & 1 \end{pmatrix}, \qquad z \in \mathbb{T}, \tag{7.23}$$

- **RH-Z3** As $z \to \infty$

$$Z(z;n) = \left(I + O(z^{-1})\right) z^{n\sigma_3}. \tag{7.24}$$

The usual Deift–Zhou nonlinear steepest descent method would fail for the Z-RHP (at the level of constructing the so-called *global parametrix*, see the Appendix). However, for all n, it is in fact possible to find algebraic relationships between the solutions $X(z;n)$ and $Z(z;n)$. Indeed, in [**52**] it was shown that one has the following two relationships between these solutions.

THEOREM 7.3 ([**52**]). *The solution $Z(z;n)$ to the Riemann–Hilbert problem* **RH-Z1** *through* **RH-Z3** *can be expressed in terms of the data extracted from the* *solution $X(z;n)$ of the Riemann–Hilbert problem* **RH-X1** *through* **RH-X3** *as*

(7.25)
$$Z(z;n)=\left[\begin{pmatrix}\dfrac{\overset{\infty}{X}_{1,12}(n)X_{21}(0;n)}{X_{11}(0;n)} & -\overset{\infty}{X}_{1,12}(n)\\ -\dfrac{X_{21}(0;n)}{X_{11}(0;n)} & 1\end{pmatrix}z^{-1}+\begin{pmatrix}1&0\\0&0\end{pmatrix}\right]X(z;n)\begin{pmatrix}1&0\\0&z\end{pmatrix},$$

where $\overset{\infty}{X}_{1,12}(n)$ is the 12-entry of the matrix $\overset{\infty}{X}_1(n)$ in the asymptotic expansion (1.30).

THEOREM 7.4 ([**52**]). *The solution $Z(z;n)$ to the Riemann–Hilbert problem* **RH-Z1** *through* **RH-Z3** *can be expressed in terms of the data extracted from the solution $X(z;n)$ of the Riemann–Hilbert problem* **RH-X1** *through* **RH-X3** *as*

(7.26)
$$Z(z;n)=\begin{pmatrix} z+\overset{\infty}{X}_{1,22}(n-1)-\dfrac{\overset{\infty}{X}_{2,12}(n-1)}{\overset{\infty}{X}_{1,12}(n-1)} & -\overset{\infty}{X}_{1,12}(n-1)\\ \dfrac{1}{\overset{\infty}{X}_{1,12}(n-1)} & 0\end{pmatrix}X(z;n-1),$$

where $\overset{\infty}{X}_{1,jk}(n)$ and $\overset{\infty}{X}_{2,jk}(n)$ are the jk-entries of the matrices $\overset{\infty}{X}_1(n)$ and $\overset{\infty}{X}_2(n)$ in the asymptotic expansion (1.30).

REMARK 7.5. The compatibility of (7.25) and (7.26) yields the well known three-term recurrence relations for the system of bi-orthogonal polynomials on the unit circle, see Lemma 2.16 of [**52**].

Now let us focus on the other two terms, i.e. $\frac{D^B_{n+1}[z\phi;\psi_2]}{D_n[z\phi]}$ and $\frac{D^B_{n+1}[z\phi;\psi_1]}{D_n[z\phi]}$, on the right hand side of (7.16). From Lemma 2.10 of [**52**] , in particular, we have

$$\frac{D^B_{n+1}[z\phi;z^{1-\ell}\phi]}{D_n[z\phi]}=\frac{1}{\ell!}\frac{\mathrm{d}^\ell}{\mathrm{d}z^\ell}Z_{12}(z;n)\bigg|_{z=0}. \tag{7.27}$$

Using (7.25) we have

$$Z_{12}(z;n)=(\mathcal{B}(n)+z)\,X_{12}(z;n)-\overset{\infty}{X}_{1,12}(n)X_{22}(z;n), \tag{7.28}$$

where

$$\mathcal{B}(n):=\frac{\overset{\infty}{X}_{1,12}(n)X_{21}(0;n)}{X_{11}(0;n)}. \tag{7.29}$$

Using (A.19), as $n\to\infty$, we have

$$\mathcal{B}(n)=-\frac{C_n[\phi]}{C_{n-1}[\phi]}\times\left(1+O(\rho^{-2n})\right), \tag{7.30}$$

where

$$C_n[\phi]:=\frac{1}{2\pi\mathrm{i}}\int_{\Gamma_0}\tau^n\phi^{-1}(\tau)\alpha^2(\tau)\mathrm{d}\tau\equiv -R_{1,12}(0;n+1)=O(\rho^{-n}), \tag{7.31}$$

α is given by (A.9), ρ^{-1} is the radius of the circle Γ_0 shown in Figure 7, and $R_{1,12}$ is given by (A.18). Let us compute the asymptotics of $\mathcal{B}(n)$ via (7.30) and (7.31). For ϕ given by (7.3), and using (7.8) we have

$$
\begin{aligned}
C_n[\phi] &= \frac{1}{2\pi \mathrm{i}}\int_{\Gamma_0} \tau^n e^{-\frac{T}{2}(\tau+1/\tau)} e^{T\tau} \mathrm{d}\tau = \frac{1}{2\pi \mathrm{i}}\int_{\Gamma_0} \tau^n e^{\frac{T}{2}(\tau-1/\tau)} \mathrm{d}\tau \\
&= \frac{1}{2\pi \mathrm{i}} \sum_{k=0}^{\infty} \left(\frac{T}{2}\right)^k \frac{1}{k!} \int_{\Gamma_0} \tau^n (\tau-\tau^{-1})^k \mathrm{d}\tau \\
&= \frac{1}{2\pi \mathrm{i}} \sum_{k=0}^{\infty} \left(\frac{T}{2}\right)^k \frac{1}{k!} \int_{\Gamma_0} \tau^n \sum_{\ell=0}^{k} \binom{k}{\ell} \tau^{k-\ell}(-\tau^{-1})^{\ell} \mathrm{d}\tau \\
&= \sum_{k=0}^{\infty} \left(\frac{T}{2}\right)^k \frac{1}{k!} \sum_{\ell=0}^{k} (-1)^{\ell} \binom{k}{\ell} \int_{\Gamma_0} \tau^{n+k-2\ell+1} \frac{\mathrm{d}\tau}{2\pi \mathrm{i}\tau}.
\end{aligned} \tag{7.32}
$$

Only the pairs (k,ℓ) for which

$$
n+k-2\ell+1=0, \tag{7.33}
$$

contribute. Notice that if $k \leq n$, no such pair exists as the above equality implies that $\ell > k$. Moreover, if $k > n$ and $k = n+2j$ for $j \in \mathbb{N}$, the above equality does not yield an integer value for ℓ, so such values of k also do not contribute. Hence

$$
\begin{aligned}
C_n[\phi] &= \left(\frac{T}{2}\right)^{n+1} \frac{1}{(n+1)!} \sum_{\ell=0}^{n+1} (-1)^{\ell} \binom{n+1}{\ell} \int_{\Gamma_0} \tau^{2n-2\ell+2} \frac{\mathrm{d}\tau}{2\pi \mathrm{i}\tau} \\
&+ \sum_{k=n+3}^{\infty} \left(\frac{T}{2}\right)^k \frac{1}{k!} \sum_{\ell=0}^{k} (-1)^{\ell} \binom{k}{\ell} \int_{\Gamma_0} \tau^{n+k-2\ell+1} \frac{\mathrm{d}\tau}{2\pi \mathrm{i}\tau}.
\end{aligned} \tag{7.34}
$$

Let us focus on the first term on the right hand side of the above equality. Obviously, only $\ell = n+1$ contributes and we thus have

(7.35)

$$
\begin{aligned}
C_n[\phi] &= \left(\frac{T}{2}\right)^{n+1} \frac{(-1)^{n+1}}{(n+1)!} + \sum_{k=n+3}^{\infty} \left(\frac{T}{2}\right)^k \frac{1}{k!} \sum_{\ell=0}^{k} (-1)^{\ell} \binom{k}{\ell} \int_{\Gamma_0} \tau^{n+k-2\ell+1} \frac{\mathrm{d}\tau}{2\pi \mathrm{i}\tau} \\
&= \left(\frac{T}{2}\right)^{n+1} \frac{(-1)^{n+1}}{(n+1)!} \left(1+O\left(\frac{1}{n}\right)\right),
\end{aligned}
$$

and therefore

$$
C_n[\phi] = -\frac{T}{2\sqrt{2\pi}} \frac{\left(-\frac{Te}{2n}\right)^n}{(n+2)^{3/2}} \left(1+O\left(\frac{1}{n}\right)\right), \tag{7.36}
$$

where we have used the Stirling's formula. Therefore from (7.35)

$$
\mathcal{B}(n) = -\frac{C_n[\phi]}{C_{n-1}[\phi]} \times \left(1+O(\rho^{-2n})\right) = \frac{T}{2(n+1)} \left(1+O\left(\frac{1}{n}\right)\right). \tag{7.37}
$$

Now, let us focus on $\overset{\infty}{X}_{1,12}(n)$ in (7.28). From **RH-X3** we have

$$
\overset{\infty}{X}_1(n) = \lim_{z\to\infty} z\left(X(z;n) z^{-n\sigma_3} - I\right). \tag{7.38}
$$

From this, and recalling (A.19) for $z \in \Omega_\infty$, (A.20) and the fact that $\alpha(z) \to 1$ as $z \to \infty$, we find

(7.39)
$$\overset{\infty}{X}_{1,12}(n) = \lim_{z\to\infty} zR_{1,12}(z) + O(\rho^{-3n}) = \frac{1}{2\pi i}\int_{\Gamma_0} \tau^n \phi^{-1}(\tau)\alpha^2(\tau)d\tau \times (1+O(\rho^{-2n}))$$
$$= C_n[\phi] \times (1+O(\rho^{-2n})) = O\left(\frac{\left(-\frac{Te}{2n}\right)^n}{(n+2)^{3/2}}\right),$$

by (7.36), where leading up to the last asymptotic equality we also have used (A.16), (A.18) and the fact that $R_{2\ell}(z;n)$ is diagonal and $R_{2\ell+1}(z;n)$ is off-diagonal, $\ell \in \mathbb{N} \cup \{0\}$. Now, in view of (7.28) let us focus on $X_{12}(z;n)$ and $X_{22}(z;n)$ in a neighborhood of $z=0$, using (A.19), as $n \to \infty$. We have

$$X_{12}(z;n) = \alpha(z)\left(1+O\left(\frac{\rho^{-2n}}{1+|z|}\right)\right), \tag{7.40}$$

$$X_{22}(z;n) = R_{1,21}(z;n)\alpha(z)\left(1+O\left(\frac{\rho^{-2n}}{1+|z|}\right)\right) = O\left(\frac{\rho^{-n}}{1+|z|}\right), \tag{7.41}$$

where we have used (A.16), α and R_1 are respectively given by (A.9) and (A.18), and ρ^{-1} is the radius of the circle Γ_0 shown in Figure 7. Combining these we conclude that
(7.42)
$$Z_{12}(z;n) = (\mathcal{B}(n)+z)\,\alpha(z)\left(1+O\left(\frac{\rho^{-2n}}{1+|z|}\right)\right) + O\left(\frac{\left(-\frac{Te}{2\rho n}\right)^n}{(n+2)^{3/2}(1+|z|)}\right),$$

where, in Ω_0 and as $n \to \infty$, the first term is $O(1)$ and the second term decays faster than $O(e^{-cn})$, for any $c > 0$. Therefore, the asymptotically relevant term for computing the RHS of (7.27) is the first term on the RHS of (7.42). Indeed we find

$$\left.\frac{d^\ell}{dz^\ell} Z_{12}(z;n)\right|_{z=0} = \mathcal{B}(n)\alpha^{(\ell)}(0) + \ell\alpha^{(\ell-1)}(0) + O(\rho^{-2n}), \tag{7.43}$$

therefore from (7.27) and (7.8) we have

(7.44)
$$\frac{D^B_{n+1}[z\phi; z^{1-\ell}\phi]}{D_n[z\phi]} = \frac{1}{\ell!}\mathcal{B}(n)\alpha^{(\ell)}(0) + \frac{1}{(\ell-1)!}\alpha^{(\ell-1)}(0) + O(\rho^{-2n})$$
$$= \frac{1}{(\ell-1)!}\left(\frac{T}{2}\right)^{\ell-1} + \frac{1}{(\ell)!}\left(\frac{T}{2}\right)^{\ell+1}\frac{1}{(n+1)}\left(1+O\left(\frac{1}{n}\right)\right).$$

Therefore

(7.45)

$$\frac{D^B_{n+1}[z\phi;\psi_1(z;n+2)]}{D_n[z\phi]} = \frac{1}{(y_{n+1}-n-2)!}\left(\frac{T}{2}\right)^{y_{n+1}-n-2} + \frac{1}{(y_{n+1}-n-1)!}\left(\frac{T}{2}\right)^{y_{n+1}-n}\frac{1}{(n+1)}\left(1+O\left(\frac{1}{n}\right)\right),$$

(7.46)

$$\frac{D^B_{n+1}[z\phi;\psi_2(z;n+2)]}{D_n[z\phi]} = \frac{1}{(y_{n+2}-n-2)!}\left(\frac{T}{2}\right)^{y_{n+2}-n-2} + \frac{1}{(y_{n+2}-n-1)!}\left(\frac{T}{2}\right)^{y_{n+2}-n}\frac{1}{(n+1)}\left(1+O\left(\frac{1}{n}\right)\right).$$

Plugging these, together with (7.20) and (7.21), into (7.16) and subsequently in (7.12) yields the result of Theorem 1.19.

7.2. Top path with non-consecutive starting and ending points. Proof of Theorem 1.20. Consider n continuous non-intersecting paths with starting points $x_1, \cdots, x_n$ and ending points $y_1, \cdots, y_n$, where $x_j = j$, $j = 1, \cdots, n-1$, $y_k = k$, for $k = 1, \cdots, n-1$, while both x_n and y_n are arbitrary. In this section, like the previous one, we assume

- T is finite, and
- the gaps $y_n - n$ and $x_n - n$ are finite as well.

From (3.16) we see that the transition probability of the non-intersecting paths modeled by (3.15), with the above starting and ending points, is given by the semi-framed Toeplitz determinant

$$\mathcal{L}_n[\phi;\psi,\eta;a] := \det\begin{pmatrix} \phi_0 & \phi_{-1} & \cdots & \phi_{-n+2} & \psi_{n-2} \\ \phi_1 & \phi_0 & \cdots & \phi_{-n+3} & \psi_{n-3} \\ \vdots & \vdots & \ddots & \vdots & \vdots \\ \phi_{n-2} & \phi_{n-3} & \cdots & \phi_0 & \psi_0 \\ \eta_0 & \eta_1 & \cdots & \eta_{n-2} & a \end{pmatrix}, \tag{7.47}$$

where ϕ is given by (7.7),

$$\psi(\zeta;n) = \phi(\zeta)\zeta^{-(y_n-n+1)}, \quad \eta(\zeta;n) = \phi(\zeta)\zeta^{x_n-1}, \tag{7.48}$$

and $a = \phi_{y_n-x_n}$. Above we have used the fact that the symbol ϕ is *even*, that is $\phi(z^{-1}) = \phi(z)$, which implies that $\phi_j = \phi_{-j}$. Let us now recall the following results from Theorem 1.9 of [**52**] and its corollary (see also Corollary 1.17 of the current work).

THEOREM 7.6 ([**52**]). *The semi-framed Toeplitz determinant $\mathcal{L}_n[\phi;\psi,\eta;a]$ can be represented in terms of the reproducing kernel of the system of bi-orthogonal polynomials on the unit circle associated with ϕ given by (1.46) and (1.47) as*

$$\frac{\mathcal{L}_{n+2}[\phi;\psi,\eta;a]}{D_{n+1}[\phi]} = a - \int_{\mathbb{T}}\left[\int_{\mathbb{T}} \mathcal{K}_n(z_1,z_2^{-1})\eta(z_2)\frac{dz_2}{2\pi i z_2}\right] z_1^{-n}\psi(z_1)\frac{dz_1}{2\pi i z_1}, \tag{7.49}$$

where $D_n[\phi]$ is given by (1.1). Using the Christoffel-Darboux identity we have

(7.50)
$$\frac{\mathcal{L}_{n+2}\left[\phi;\psi,\eta;a\right]}{D_{n+1}[\phi]} = a - \int_{\mathbb{T}}\int_{\mathbb{T}} \frac{\tilde{\eta}(z_2)\tilde{\psi}(z_1)}{z_1-z_2} \det\begin{pmatrix} X_{11}(z_2;n+1) & X_{21}(z_2;n+2) \\ X_{11}(z_1;n+1) & X_{21}(z_1;n+2)\end{pmatrix} \frac{dz_2}{2\pi i z_2}\frac{dz_1}{2\pi i z_1},$$

where $D_n[\phi]$ is given by (1.1), and X_{11} and X_{21} are respectively the 11 and 21 entries of the solution to **RH-X1** *to* **RH-X3**.

We need to analyze the semi-framed Toeplitz determinant

$$e^{-(n+1)T}\mathcal{L}_{n+1}[\phi;\psi(z;n+1),\eta(z;n+1);\phi_{y_{n+1}-x_{n+1}}],$$

where ψ and η are given by (7.48) and ϕ is given by (7.7). Theorem 7.6 gives a representation of semi-framed Toeplitz determinants in terms of the reproducing kernel of the biorthogonal polynomials on the unit circle. Indeed,

$$\mathcal{L}_{n+1}[\phi;\psi(z;n+1),\eta(z;n+1);\phi_{y_{n+1}-x_{n+1}}] = D_n[\phi]\Big\{\phi_{y_{n+1}-x_{n+1}} - \int_{\mathbb{T}}\int_{\mathbb{T}} \frac{\tilde{\eta}(z_2;n+1)\tilde{\psi}(z_1;n+1)}{z_1-z_2} \det\begin{pmatrix} X_{11}(z_2;n) & X_{21}(z_2;n+1) \\ X_{11}(z_1;n) & X_{21}(z_1;n+1)\end{pmatrix} \frac{\mathrm{d}z_2}{2\pi \mathrm{i} z_2}\frac{\mathrm{d}z_1}{2\pi \mathrm{i} z_1}\Big\}, \tag{7.51}$$

by (7.50). Recall that the integral representation for the modified Bessel function of integer order j is

$$I_j(T) = \frac{1}{\pi}\int_0^\pi e^{T\cos\theta}\cos(j\theta)\mathrm{d}\theta = \int_{\mathbb{T}} z^{-j}e^{\frac{T}{2}(z+z^{-1})}\frac{\mathrm{d}z}{2\pi \mathrm{i}z}, \tag{7.52}$$

which is precisely the j-th Fourier coefficient of ϕ given by (7.7), so we have

$$\phi_{y_{n+1}-x_{n+1}} = I_{y_{n+1}-x_{n+1}}(T). \tag{7.53}$$

Now we focus on computing the double integral in (7.51). From (A.19) in Ω_1 (see Figure 7) we have

$$X_{11}(z;n) = z^n\alpha(z)\phi^{-1}(z)\left(1+O(\rho^{-n})\right), \qquad n\to\infty, \tag{7.54}$$

$$X_{21}(z;n) = -\alpha^{-1}(z)\left(1+O(\rho^{-n})\right), \qquad n\to\infty. \tag{7.55}$$

Therefore, as $n\to\infty$, (7.51) can be written as

(7.56)
$$\frac{\mathcal{L}_{n+1}[\phi;\psi(z;n+1),\eta(z;n+1);\phi_{y_{n+1}-x_{n+1}}]}{D_n[\phi]} - \phi_{y_{n+1}-x_{n+1}}$$
$$= \int_{\mathbb{T}}\int_{\mathbb{T}} \frac{\tilde{\eta}(z_2;n+1)\tilde{\psi}(z_1;n+1)}{z_1-z_2}\left(-z_2^n\alpha_+(z_2)\phi^{-1}(z_2)\alpha_+^{-1}(z_1)\right)\frac{\mathrm{d}z_2}{2\pi \mathrm{i}z_2}\frac{\mathrm{d}z_1}{2\pi \mathrm{i}z_1}$$
$$+ \int_{\mathbb{T}}\int_{\mathbb{T}} \frac{\tilde{\eta}(z_2;n+1)\tilde{\psi}(z_1;n+1)}{z_1-z_2}\left(z_1^n\alpha_+(z_1)\phi^{-1}(z_1)\alpha_+^{-1}(z_2)\right)\frac{\mathrm{d}z_2}{2\pi \mathrm{i}z_2}\frac{\mathrm{d}z_1}{2\pi \mathrm{i}z_1} + O(\rho^{-n}).$$

Plugging in (7.48) and using the scalar jump condition (A.8) we can rewrite this after straightforward simplifications as

$$
\begin{aligned}
&\frac{\mathcal{L}_{n+1}[\phi;\psi(z;n+1),\eta(z;n+1);\phi_{y_{n+1}-x_{n+1}}]}{D_n[\phi]}-\phi_{y_{n+1}-x_{n+1}}\\
&=\int_{\mathbb{T}}\int_{\mathbb{T}}\frac{1}{z_1-z_2}\alpha_+(z_1)\alpha_-^{-1}(z_2)z_2^{1-x_{n+1}}z_1^{y_{n+1}}\frac{\mathrm{d}z_1}{2\pi\mathrm{i}z_1}\frac{\mathrm{d}z_2}{2\pi\mathrm{i}z_2}\\
&-\int_{\mathbb{T}}\int_{\mathbb{T}}\frac{1}{z_1-z_2}\left(\frac{z_2}{z_1}\right)^n\alpha_+(z_2)\alpha_-^{-1}(z_1)z_2^{1-x_{n+1}}z_1^{y_{n+1}}\frac{\mathrm{d}z_1}{2\pi\mathrm{i}z_1}\frac{\mathrm{d}z_2}{2\pi\mathrm{i}z_2}+O(\rho^{-n}).
\end{aligned}
\tag{7.57}
$$

For a fixed $\mathfrak{r}>1$, define $\mathbb{T}^{\mathfrak{r}}_{\mp}:=\left\{z\;:\;|z|=\mathfrak{r}^{\pm1}\right\}$. We now deform the contours of integration to rewrite the previous equation as

$$
\begin{aligned}
&\frac{\mathcal{L}_{n+1}[\phi;\psi(z;n+1),\eta(z;n+1);\phi_{y_{n+1}-x_{n+1}}]}{D_n[\phi]}-\phi_{y_{n+1}-x_{n+1}}\\
&=\int_{\mathbb{T}^{\mathfrak{r}}_-}\int_{\mathbb{T}^{\mathfrak{r}}_+}\frac{1}{z_1-z_2}\alpha(z_1)\alpha^{-1}(z_2)z_2^{1-x_{n+1}}z_1^{y_{n+1}}\frac{\mathrm{d}z_1}{2\pi\mathrm{i}z_1}\frac{\mathrm{d}z_2}{2\pi\mathrm{i}z_2}\\
&-\int_{\mathbb{T}^{\mathfrak{r}}_+}\int_{\mathbb{T}^{\mathfrak{r}}_-}\frac{1}{z_1-z_2}\left(\frac{z_2}{z_1}\right)^n\alpha(z_2)\alpha^{-1}(z_1)z_2^{1-x_{n+1}}z_1^{y_{n+1}}\frac{\mathrm{d}z_1}{2\pi\mathrm{i}z_1}\frac{\mathrm{d}z_2}{2\pi\mathrm{i}z_2}+O(\rho^{-n})\\
&=-\int_{\mathbb{T}^{\mathfrak{r}}_-}\int_{\mathbb{T}^{\mathfrak{r}}_+}\alpha(z_1)\alpha^{-1}(z_2)z_2^{-x_{n+1}}z_1^{y_{n+1}}\left(\sum_{k=0}^{\infty}\frac{z_1^k}{z_2^k}\right)\frac{\mathrm{d}z_1}{2\pi\mathrm{i}z_1}\frac{\mathrm{d}z_2}{2\pi\mathrm{i}z_2}\\
&-\int_{\mathbb{T}^{\mathfrak{r}}_+}\int_{\mathbb{T}^{\mathfrak{r}}_-}\left(\frac{z_2}{z_1}\right)^n\alpha(z_2)\alpha^{-1}(z_1)z_2^{1-x_{n+1}}z_1^{y_{n+1}-1}\left(\sum_{k=0}^{\infty}\frac{z_2^k}{z_1^k}\right)\frac{\mathrm{d}z_1}{2\pi\mathrm{i}z_1}\frac{\mathrm{d}z_2}{2\pi\mathrm{i}z_2}+O(\rho^{-n})\\
&=-\sum_{k=0}^{\infty}\left[\int_{\mathbb{T}^{\mathfrak{r}}_+}\alpha(z)z^{n+1+k-x_{n+1}}\frac{\mathrm{d}z}{2\pi\mathrm{i}z}\right]\left[\int_{\mathbb{T}^{\mathfrak{r}}_-}\alpha^{-1}(z)z^{y_{n+1}-n-k-1}\frac{\mathrm{d}z}{2\pi\mathrm{i}z}\right]\\
&-\sum_{k=0}^{\infty}\left[\int_{\mathbb{T}^{\mathfrak{r}}_+}\alpha(z)z^{y_{n+1}+k}\frac{\mathrm{d}z}{2\pi\mathrm{i}z}\right]\left[\int_{\mathbb{T}^{\mathfrak{r}}_-}\alpha^{-1}(z)z^{-x_{n+1}-k}\frac{\mathrm{d}z}{2\pi\mathrm{i}z}\right]+O(\rho^{-n}).
\end{aligned}
\tag{7.58}
$$

Using (7.8) we obtain

$$
\begin{aligned}
&\frac{\mathcal{L}_{n+1}[\phi;\psi(z;n+1),\eta(z;n+1);\phi_{y_{n+1}-x_{n+1}}]}{D_n[\phi]}-\phi_{y_{n+1}-x_{n+1}}\\
&=-\sum_{k=0}^{\infty}\left[\int_{\mathbb{T}^{\mathfrak{r}}_+}e^{\frac{Tz}{2}}z^{n+1+k-x_{n+1}}\frac{\mathrm{d}z}{2\pi\mathrm{i}z}\right]\left[\int_{\mathbb{T}^{\mathfrak{r}}_-}e^{\frac{T}{2z}}z^{y_{n+1}-n-k-1}\frac{\mathrm{d}z}{2\pi\mathrm{i}z}\right]\\
&-\sum_{k=0}^{\infty}\left[\int_{\mathbb{T}^{\mathfrak{r}}_+}e^{\frac{Tz}{2}}z^{y_{n+1}+k}\frac{\mathrm{d}z}{2\pi\mathrm{i}z}\right]\left[\int_{\mathbb{T}^{\mathfrak{r}}_-}e^{\frac{T}{2z}}z^{-x_{n+1}-k}\frac{\mathrm{d}z}{2\pi\mathrm{i}z}\right]+O(\rho^{-n}).
\end{aligned}
\tag{7.59}
$$

By the Cauchy Theorem we obtain

$$
\begin{aligned}
&\frac{\mathcal{L}_{n+1}[\phi;\psi(z;n+1),\eta(z;n+1);\phi_{y_{n+1}-x_{n+1}}]}{D_n[\phi]}-\phi_{y_{n+1}-x_{n+1}}\\
&=-\sum_{k=0}^{\infty}\left[\int_{\mathbb{T}^{\mathfrak{r}}_+}e^{\frac{Tz}{2}}z^{n+1+k-x_{n+1}}\frac{\mathrm{d}z}{2\pi\mathrm{i}z}\right]\left[\int_{\mathbb{T}^{\mathfrak{r}}_-}e^{\frac{T}{2z}}z^{y_{n+1}-n-k-1}\frac{\mathrm{d}z}{2\pi\mathrm{i}z}\right]+O(\rho^{-n}).
\end{aligned}
\tag{7.60}
$$

Residue calculation combined with (7.53) and the Szegő Theorem for ϕ (recall (7.9) and (7.10)) yields (1.51).

Appendix A. Solution of the Riemann–Hilbert problem for BOPUC with Szegő-type symbols

Let us recall the Riemann–Hilbert problem for BOPUC is due to J.Baik, P.Deift and K.Johansson [**6**].

- **RH-X1** $X(\cdot;n):\mathbb{C}\setminus\mathbb{T}\to\mathbb{C}^{2\times 2}$ is analytic,
- **RH-X2** The limits of $X(\zeta;n)$ as ζ tends to $z\in\mathbb{T}$ from the inside and outside of the unit circle exist, and are denoted $X_{\pm}(z;n)$ respectively and are related by

$$X_+(z;n)=X_-(z;n)\begin{pmatrix}1 & z^{-n}\phi(z)\\ 0 & 1\end{pmatrix},\qquad z\in\mathbb{T},\tag{A.1}$$

- **RH-X3** As $z\to\infty$

$$X(z;n)=\big(I+O(z^{-1})\big)z^{n\sigma_3}.\tag{A.2}$$

For the purposes of section 7 in this Appendix we show the standard steepest descent analysis to asymptotically solve this problem for a Szegő-type symbol which was first shown in [**44**]. We first normalize the behavior at ∞ by defining

$$T(z;n):=\begin{cases}X(z;n)z^{-n\sigma_3}, & |z|>1,\\ X(z;n), & |z|<1.\end{cases}\tag{A.3}$$

The function T defined above satisfies the following RH problem

- **RH-T1** $T(\cdot;n):\mathbb{C}\setminus\mathbb{T}\to\mathbb{C}^{2\times 2}$ is analytic,
- **RH-T2** $T_+(z;n)=T_-(z;n)\begin{pmatrix}z^n & \phi(z)\\ 0 & z^{-n}\end{pmatrix},\qquad z\in\mathbb{T},$
- **RH-T3** $T(z;n)=I+O(1/z),\qquad z\to\infty.$

So T has a highly-oscillatory jump matrix as $n\to\infty$. The next transformation yields a Riemann Hilbert problem, normalized at infinity, having an exponentially decaying jump matrix on the *lenses*. Note that we have the following factorization of the jump matrix of the T-RHP:

$$\begin{aligned}\begin{pmatrix}z^n & \phi(z)\\ 0 & z^{-n}\end{pmatrix}&=\begin{pmatrix}1 & 0\\ z^{-n}\phi(z)^{-1} & 1\end{pmatrix}\begin{pmatrix}0 & \phi(z)\\ -\phi(z)^{-1} & 0\end{pmatrix}\begin{pmatrix}1 & 0\\ z^{n}\phi(z)^{-1} & 1\end{pmatrix}\\ &\equiv J_0(z;n)J^{(\infty)}(z)J_1(z;n).\end{aligned}\tag{A.4}$$

Now, we define the following function:

$$S(z;n):=\begin{cases}T(z;n)J_1^{-1}(z;n), & z\in\Omega_1,\\ T(z;n)J_0(z;n), & z\in\Omega_2,\\ T(z;n), & z\in\Omega_0\cup\Omega_\infty.\end{cases}\tag{A.5}$$

Also introduce the following function on $\Gamma_S:=\Gamma_0\cup\Gamma_1\cup\mathbb{T}$:

$$J_S(z;n)=\begin{cases}J_1(z;n), & z\in\Gamma_0,\\ J^{(\infty)}(z), & z\in\mathbb{T},\\ J_0(z;n), & z\in\Gamma_1.\end{cases}\tag{A.6}$$

Therefore we have the following Riemann–Hilbert problem for $S(z;n)$

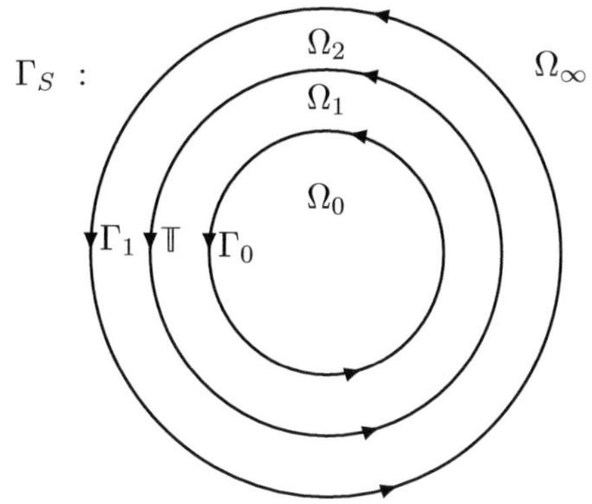

FIGURE 7. Opening of lenses: the jump contour for the S-RHP.

- **RH-S1** $S(\cdot;n):\mathbb{C}\setminus\Gamma_S\to\mathbb{C}^{2\times 2}$ is analytic.
- **RH-S2** $S_+(z;n)=S_-(z;n)J_S(z;n), \qquad z\in\Gamma_S.$
- **RH-S3** $S(z;n)=I+O(1/z), \qquad \text{as } z\to\infty.$

Note that the matrices $J_0(z;n)$ and $J_1(z;n)$ tend to the identity matrix uniformly on their respective contours, exponentially fast as $n\to\infty$. We are looking for a piecewise analytic function $P^{(\infty)}(z):\mathbb{C}\setminus\mathbb{T}:\to\mathbb{C}^{2\times 2}$ such that

- **RH-Global1** $P^{(\infty)}$ is holomorphic in $\mathbb{C}\setminus\mathbb{T}$.
- **RH-Global2** for $z\in\mathbb{T}$ we have

$$P_+^{(\infty)}(z)=P_-^{(\infty)}(z)\begin{pmatrix}0 & \phi(z)\\ -\phi^{-1}(z) & 0\end{pmatrix}. \tag{A.7}$$

- **RH-Global3** $P^{(\infty)}(z)=I+O(1/z), \qquad \text{as } z\to\infty.$

We can find a piecewise analytic function α which solves the following scalar multiplicative Riemann–Hilbert problem

$$\alpha_+(z)=\alpha_-(z)\phi(z) \qquad z\in\mathbb{T}. \tag{A.8}$$

By Plemelj–Sokhotski formula we have

$$\alpha(z)=\exp\left[\frac{1}{2\pi \mathrm{i}}\int_{\mathbb{T}}\frac{\ln(\phi(\tau))}{\tau-z}d\tau\right], \tag{A.9}$$

Now, using (A.8) we have the following factorization

$$\begin{pmatrix}0 & \phi(z)\\ -\phi^{-1}(z) & 0\end{pmatrix}=\begin{pmatrix}\alpha_-^{-1}(z) & 0\\ 0 & \alpha_-(z)\end{pmatrix}\begin{pmatrix}0 & 1\\ -1 & 0\end{pmatrix}\begin{pmatrix}\alpha_+^{-1}(z) & 0\\ 0 & \alpha_+(z)\end{pmatrix}. \tag{A.10}$$

So, the function

$$P^{(\infty)}(z):=\begin{cases}\begin{pmatrix}0 & \alpha(z)\\ -\alpha^{-1}(z) & 0\end{pmatrix}, & |z|<1,\\[2mm] \begin{pmatrix}\alpha(z) & 0\\ 0 & \alpha^{-1}(z)\end{pmatrix}, & |z|>1,\end{cases} \tag{A.11}$$

satisfies (A.7). Also, by the properties of the Cauchy integral, $P^{(\infty)}(z)$ is holomorphic in $\mathbb{C}\setminus\mathbb{T}$. Moreover, $\alpha(z)=1+O(z^{-1})$, as $z\to\infty$ and hence

$$P^{(\infty)}(z)=I+O(1/z), \qquad z\to\infty. \tag{A.12}$$

Therefore $P^{(\infty)}$ given by (A.11) is the unique solution of the global parametrix Riemann–Hilbert problem. Let us now consider the ratio

$$R(z;n) := S(z;n)\left[P^{(\infty)}(z)\right]^{-1}. \tag{A.13}$$

We have the following Riemann–Hilbert problem for $R(z;n)$

- **RH-R1** R is holomorphic in $\mathbb{C}\setminus(\Gamma_0\cup\Gamma_1)$.
- **RH-R2** $R_+(z;n) = R_-(z;n)J_R(z;n), \qquad z\in\Gamma_0\cup\Gamma_1 =: \Sigma_R,$
- **RH-R3** $R(z;n) = I + O(1/z) \qquad \text{as } z\to\infty.$

This Riemann Hilbert problem is solvable for large n (see e.g. **[42]**) and $R(z;n)$ can be written as

$$R(z;n) = I + R_1(z;n) + R_2(z;n) + R_3(z;n) + \cdots, \qquad n\geq n_0 \tag{A.14}$$

where R_k can be found recursively. Indeed

(A.15)

$$R_k(z;n) = \frac{1}{2\pi \mathrm{i}}\int_{\Sigma_R}\frac{[R_{k-1}(\mu;n)]_-\left(J_R(\mu;n)-I\right)}{\mu-z}\mathrm{d}\mu, \qquad z\in\mathbb{C}\setminus\Sigma_R, \qquad k\geq 1.$$

It is easy to check that $R_{2\ell}(z;n)$ is diagonal and $R_{2\ell+1}(z;n)$ is off-diagonal; $\ell\in\mathbb{N}\cup\{0\}$, and that

$$R_{k,ij}(z;n) = O\left(\frac{\rho^{-kn}}{1+|z|}\right), \qquad n\to\infty, \qquad k\geq 1, \tag{A.16}$$

uniformly in $z\in\mathbb{C}\setminus\Sigma_R$, where ρ (resp. ρ^{-1}) is the radius of Γ_1(resp. Γ_0). Let us compute $R_1(z;n)$; we have

$$\begin{aligned} J_R(z)-I &= \begin{cases} P^{(\infty)}(z)\begin{pmatrix} 0 & 0 \\ z^n\phi^{-1}(z) & 0\end{pmatrix}\left[P^{(\infty)}(z)\right]^{-1}, & z\in\Gamma_0, \\ P^{(\infty)}(z)\begin{pmatrix} 0 & 0 \\ z^{-n}\phi^{-1}(z) & 0\end{pmatrix}\left[P^{(\infty)}(z)\right]^{-1}, & z\in\Gamma_1, \end{cases} \\ &= \begin{cases} \begin{pmatrix} 0 & -z^n\phi^{-1}(z)\alpha^2(z) \\ 0 & 0\end{pmatrix}, & z\in\Gamma_0, \\ \begin{pmatrix} 0 & 0 \\ z^{-n}\phi^{-1}(z)\alpha^{-2}(z) & 0\end{pmatrix}, & z\in\Gamma_1. \end{cases} \end{aligned} \tag{A.17}$$

Therefore

(A.18)

$$R_1(z;n) = \begin{pmatrix} 0 & -\dfrac{1}{2\pi \mathrm{i}}\displaystyle\int_{\Gamma_0}\frac{\tau^n\phi^{-1}(\tau)\alpha^2(\tau)}{\tau-z}d\tau \\ \dfrac{1}{2\pi \mathrm{i}}\displaystyle\int_{\Gamma_1}\frac{\tau^{-n}\phi^{-1}(\tau)\alpha^{-2}(\tau)}{\tau-z}d\tau & 0\end{pmatrix}.$$

If we trace back the Riemann–Hilbert problems $R \mapsto S \mapsto T \mapsto X$ we will obtain

$$X(z;n) = R(z;n) \begin{cases} \begin{pmatrix} \alpha(z) & 0 \\ 0 & \alpha^{-1}(z) \end{pmatrix} z^{n\sigma_3}, & z \in \Omega_\infty, \\ \begin{pmatrix} \alpha(z) & 0 \\ -z^{-n}\alpha^{-1}(z)\phi^{-1}(z) & \alpha^{-1}(z) \end{pmatrix} z^{n\sigma_3}, & z \in \Omega_2, \\ \begin{pmatrix} z^n\alpha(z)\phi^{-1}(z) & \alpha(z) \\ -\alpha^{-1}(z) & 0 \end{pmatrix}, & z \in \Omega_1, \\ \begin{pmatrix} 0 & \alpha(z) \\ -\alpha^{-1}(z) & 0 \end{pmatrix}, & z \in \Omega_0, \end{cases} \tag{A.19}$$

where for $z \in \mathbb{C} \setminus \Sigma_R$, as $n \to \infty$, we have

$$R(z;n) = \begin{pmatrix} 1 + O\left(\frac{\rho^{-2n}}{1+|z|}\right) & R_{1,12}(z;n) + O\left(\frac{\rho^{-3n}}{1+|z|}\right) \\ R_{1,21}(z;n) + O\left(\frac{\rho^{-3n}}{1+|z|}\right) & 1 + O\left(\frac{\rho^{-2n}}{1+|z|}\right) \end{pmatrix}. \tag{A.20}$$

Acknowledgments

This work was initiated during the special session on *Orthogonal Polynomials and their Applications* at the Joint Mathematics Meetings (JMM) in Boston in January, 2023. The authors are grateful to JMM and to the organizers of that session. We also thank Robert Buckingham for help with the simulations.

References

[1] Francine F. Abeles, *Dodgson condensation: the historical and mathematical development of an experimental method*, Linear Algebra Appl. **429** (2008), no. 2-3, 429–438, DOI 10.1016/j.laa.2007.11.022. MR2419937 ↑126

[2] Mark Adler, Jonathan Delépine, and Pierre van Moerbeke, *Dyson's nonintersecting Brownian motions with a few outliers*, Comm. Pure Appl. Math. **62** (2009), no. 3, 334–395, DOI 10.1002/cpa.20264. MR2487852 ↑109

[3] Mark Adler, Patrik L. Ferrari, and Pierre van Moerbeke, *Airy processes with wanderers and new universality classes*, Ann. Probab. **38** (2010), no. 2, 714–769, DOI 10.1214/09-AOP493. MR2642890 ↑109

[4] Helen Au-Yang and Jacques H. H. Perk, *Critical correlations in a Z-invariant inhomogeneous Ising model*, Phys. A **144** (1987), no. 1, 44–104, DOI 10.1016/0378-4371(87)90145-2. MR899036 ↑92, 96

[5] J. Baik, T. Kriecherbauer, K. T.-R. McLaughlin, and P. D. Miller, *Discrete orthogonal polynomials*, Annals of Mathematics Studies, vol. 164, Princeton University Press, Princeton, NJ, 2007. Asymptotics and applications. MR2283089 ↑98, 111, 115

[6] Jinho Baik, Percy Deift, and Kurt Johansson, *On the distribution of the length of the longest increasing subsequence of random permutations*, J. Amer. Math. Soc. **12** (1999), no. 4, 1119–1178, DOI 10.1090/S0894-0347-99-00307-0. MR1682248 ↑92, 99, 120, 146

[7] Jinho Baik and Robert Jenkins, *Limiting distribution of maximal crossing and nesting of Poissonized random matchings*, Ann. Probab. **41** (2013), no. 6, 4359–4406, DOI 10.1214/12-AOP781. MR3161478 ↑120

[8] Jinho Baik and Zhipeng Liu, *Discrete Toeplitz/Hankel determinants and the width of nonintersecting processes*, Int. Math. Res. Not. IMRN **20** (2014), 5737–5768, DOI 10.1093/imrn/rnt143. MR3271188 ↑104, 105, 111, 112, 113, 115, 116, 117, 134

[9] Estelle Basor and Torsten Ehrhardt, *Asymptotic formulas for determinants of a special class of Toeplitz + Hankel matrices*, Large truncated Toeplitz matrices, Toeplitz operators, and related topics, Oper. Theory Adv. Appl., vol. 259, Birkhäuser/Springer, Cham, 2017, pp. 125–154. MR3644515 ↑92

[10] Estelle Basor, Torsten Ehrhardt, Roozbeh Gharakhloo, Alexander Its, and Yuqi Li, *Asymptotics of bordered Toeplitz determinants and next-to-diagonal Ising correlations*, J. Stat. Phys. **187** (2022), no. 1, Paper No. 8, 49, DOI 10.1007/s10955-022-02894-7. MR4386372 ↑92, 104, 136

[11] Estelle L. Basor and Torsten Ehrhardt, *On a class of Toeplitz + Hankel operators*, New York J. Math. **5** (1999), 1–16. MR1679634 ↑92

[12] Estelle L. Basor and Torsten Ehrhardt, *Asymptotic formulas for determinants of a sum of finite Toeplitz and Hankel matrices*, Math. Nachr. **228** (2001), 5–45, DOI 10.1002/1522-2616(200108)228:1<5::AID-MANA5>3.3.CO;2-5. MR1845906 ↑92

[13] Estelle L. Basor and Torsten Ehrhardt, *Asymptotic formulas for the determinants of symmetric Toeplitz plus Hankel matrices*, Toeplitz matrices and singular integral equations (Pobershau, 2001), Oper. Theory Adv. Appl., vol. 135, Birkhäuser, Basel, 2002, pp. 61–90. MR1935758 ↑92

[14] Estelle L. Basor and Torsten Ehrhardt, *Determinant computations for some classes of Toeplitz-Hankel matrices*, Oper. Matrices **3** (2009), no. 2, 167–186, DOI 10.7153/oam-03-09. MR2522773 ↑92

[15] Elliot Blackstone, Christophe Charlier, and Jonatan Lenells, *Toeplitz determinants with a one-cut regular potential and Fisher-Hartwig singularities I. Equilibrium measure supported on the unit circle*, Proc. Roy. Soc. Edinburgh Sect. A **154** (2024), no. 5, 1431–1472, DOI 10.1017/prm.2023.73. MR4806283 ↑99

[16] Pavel Bleher and Thomas Bothner, *Exact solution of the six-vertex model with domain wall boundary conditions: critical line between disordered and antiferroelectric phases*, Random Matrices Theory Appl. **1** (2012), no. 4, 1250012, 43, DOI 10.1142/S2010326312500128. MR3039373 ↑123

[17] Pavel Bleher, Roozbeh Gharakhloo, and Kenneth T-R McLaughlin, *Phase diagram and topological expansion in the complex quartic random matrix model*, Comm. Pure Appl. Math. **77** (2024), no. 2, 1405–1485, DOI 10.1002/cpa.22164. MR4673885 ↑92

[18] Pavel Bleher and Karl Liechty, *Exact solution of the six-vertex model with domain wall boundary conditions. Ferroelectric phase*, Comm. Math. Phys. **286** (2009), no. 2, 777–801, DOI 10.1007/s00220-008-0709-9. MR2472044 ↑123

[19] Pavel Bleher and Karl Liechty, *Exact solution of the six-vertex model with domain wall boundary conditions: antiferroelectric phase*, Comm. Pure Appl. Math. **63** (2010), no. 6, 779–829, DOI 10.1002/cpa.20311. MR2649363 ↑123, 125

[20] Pavel Bleher and Karl Liechty, *Uniform asymptotics for discrete orthogonal polynomials with respect to varying exponential weights on a regular infinite lattice*, Int. Math. Res. Not. IMRN **2** (2011), 342–386, DOI 10.1093/imrn/rnq081. MR2764867 ↑98, 115

[21] Pavel Bleher and Karl Liechty, *Random matrices and the six-vertex model*, CRM Monograph Series, vol. 32, American Mathematical Society, Providence, RI, 2014, DOI 10.1090/crmm/032. MR3155177 ↑92, 98, 105, 111, 115, 121, 123, 125

[22] Pavel M. Bleher and Alfredo Deaño, *Topological expansion in the cubic random matrix model*, Int. Math. Res. Not. IMRN **12** (2013), 2699–2755, DOI 10.1093/imrn/rns126. MR3071662 ↑92

[23] Pavel M. Bleher and Vladimir V. Fokin, *Exact solution of the six-vertex model with domain wall boundary conditions. Disordered phase*, Comm. Math. Phys. **268** (2006), no. 1, 223–284, DOI 10.1007/s00220-006-0097-y. MR2249800 ↑123, 125

[24] N. M. Bogoliubov, A. G. Pronko, and M. B. Zvonarev, *Boundary correlation functions of the six-vertex model*, J. Phys. A **35** (2002), no. 27, 5525–5541, DOI 10.1088/0305-4470/35/27/301. MR1917248 ↑105, 124

[25] Alexei Borodin, Ivan Corwin, and Vadim Gorin, *Stochastic six-vertex model*, Duke Math. J. **165** (2016), no. 3, 563–624, DOI 10.1215/00127094-3166843. MR3466163 ↑125

[26] Alexei Borodin and Sandrine Péché, *Airy kernel with two sets of parameters in directed percolation and random matrix theory*, J. Stat. Phys. **132** (2008), no. 2, 275–290, DOI 10.1007/s10955-008-9553-8. MR2415103 ↑109

[27] Gaëtan Borot, Alice Guionnet, and Karol K. Kozlowski, *Large-N asymptotic expansion for mean field models with Coulomb gas interaction*, Int. Math. Res. Not. IMRN **20** (2015), 10451–10524, DOI 10.1093/imrn/rnu260. MR3455872 ↑92

[28] Albrecht Böttcher and Bernd Silbermann, *Analysis of Toeplitz operators*, 2nd ed., Springer Monographs in Mathematics, Springer-Verlag, Berlin, 2006. Prepared jointly with Alexei Karlovich. MR2223704 ↑136

[29] David M. Bressoud, *Proofs and confirmations*, MAA Spectrum, Mathematical Association of America, Washington, DC; Cambridge University Press, Cambridge, 1999. The story of the alternating sign matrix conjecture. MR1718370 ↑126

[30] L. Brightmore, G. P. Gehér, A. R. Its, V. E. Korepin, F. Mezzadri, M. Y. Mo, and J. A. Virtanen, *Entanglement entropy of two disjoint intervals separated by one spin in a chain of free fermion*, J. Phys. A **53** (2020), no. 34, 345303, 31, DOI 10.1088/1751-8121/ab9cf2. MR4137534 ↑97

[31] Christophe Charlier, *Asymptotics of Hankel determinants with a one-cut regular potential and Fisher-Hartwig singularities*, Int. Math. Res. Not. IMRN **24** (2019), 7515–7576, DOI 10.1093/imrn/rny009. MR4043828 ↑92, 98

[32] Christophe Charlier, *Asymptotics of Muttalib-Borodin determinants with Fisher-Hartwig singularities*, Selecta Math. (N.S.) **28** (2022), no. 3, Paper No. 50, 60, DOI 10.1007/s00029-022-00762-6. MR4393137 ↑92

[33] Christophe Charlier and Tom Claeys, *Thinning and conditioning of the circular unitary ensemble*, Random Matrices Theory Appl. **6** (2017), no. 2, 1750007, 51, DOI 10.1142/S2010326317500071. MR3656965 ↑92

[34] Christophe Charlier and Roozbeh Gharakhloo, *Asymptotics of Hankel determinants with a Laguerre-type or Jacobi-type potential and Fisher-Hartwig singularities*, Adv. Math. **383** (2021), Paper No. 107672, 69, DOI 10.1016/j.aim.2021.107672. MR4232544 ↑92, 98

[35] Tom Claeys and Stefano Romano, *Biorthogonal ensembles with two-particle interactions*, Nonlinearity **27** (2014), no. 10, 2419–2444, DOI 10.1088/0951-7715/27/10/2419. MR3265719 ↑92

[36] F. Colomo and A. G. Pronko, *Emptiness formation probability in the domain-wall six-vertex model*, Nuclear Phys. B **798** (2008), no. 3, 340–362, DOI 10.1016/j.nuclphysb.2007.12.016. MR2411855 ↑100, 105, 124

[37] F. Colomo and A. G. Pronko, *The arctic curve of the domain-wall six-vertex model*, J. Stat. Phys. **138** (2010), no. 4-5, 662–700, DOI 10.1007/s10955-009-9902-2. MR2594917 ↑105, 124

[38] F. Colomo, A. G. Pronko, and P. Zinn-Justin, *The arctic curve of the domain wall six-vertex model in its antiferroelectric regime*, J. Stat. Mech. Theory Exp. **3** (2010), L03002, 11, DOI 10.1088/1742-5468/2010/03/l03002. MR2629602 ↑105, 124

[39] Ivan Corwin and Alan Hammond, *Brownian Gibbs property for Airy line ensembles*, Invent. Math. **195** (2014), no. 2, 441–508, DOI 10.1007/s00222-013-0462-3. MR3152753 ↑109

[40] Ivan Corwin, Jeremy Quastel, and Daniel Remenik, *Continuum statistics of the* Airy_2 *process*, Comm. Math. Phys. **317** (2013), no. 2, 347–362, DOI 10.1007/s00220-012-1582-0. MR3010187 ↑109

[41] P. Deift, A. Its, and I. Krasovsky, *Eigenvalues of Toeplitz matrices in the bulk of the spectrum*, Bull. Inst. Math. Acad. Sin. (N.S.) **7** (2012), no. 4, 437–461. MR3077466 ↑92

[42] P. Deift, T. Kriecherbauer, K. T.-R. McLaughlin, S. Venakides, and X. Zhou, *Uniform asymptotics for polynomials orthogonal with respect to varying exponential weights and applications to universality questions in random matrix theory*, Comm. Pure Appl. Math. **52** (1999), no. 11, 1335–1425, DOI 10.1002/(SICI)1097-0312(199911)52:11<1335::AID-CPA1>3.0.CO;2-1. MR1702716 ↑111, 148

[43] P. Deift and X. Zhou, *A steepest descent method for oscillatory Riemann-Hilbert problems. Asymptotics for the MKdV equation*, Ann. of Math. (2) **137** (1993), no. 2, 295–368, DOI 10.2307/2946540. MR1207209 ↑97

[44] Percy Deift, *Integrable operators*, Differential operators and spectral theory, Amer. Math. Soc. Transl. Ser. 2, vol. 189, Amer. Math. Soc., Providence, RI, 1999, pp. 69–84, DOI 10.1090/trans2/189/06. MR1730504 ↑146

[45] Percy Deift, Alexander Its, and Igor Krasovsky, *Asymptotics of Toeplitz, Hankel, and Toeplitz+Hankel determinants with Fisher-Hartwig singularities*, Ann. of Math. (2) **174** (2011), no. 2, 1243–1299, DOI 10.4007/annals.2011.174.2.12. MR2831118 ↑99, 139

[46] Percy Deift, Alexander Its, and Igor Krasovsky, *Toeplitz matrices and Toeplitz determinants under the impetus of the Ising model: some history and some recent results*, Comm. Pure Appl. Math. **66** (2013), no. 9, 1360–1438, DOI 10.1002/cpa.21467. MR3078693 ↑104, 136

[47] A. S. Fokas, A. R. Itš, and A. V. Kitaev, *The isomonodromy approach to matrix models in 2D quantum gravity*, Comm. Math. Phys. **147** (1992), no. 2, 395–430. MR1174420 ↑97, 99

[48] P. J. Forrester and J. R. Ipsen, *Selberg integral theory and Muttalib-Borodin ensembles*, Adv. in Appl. Math. **95** (2018), 152–176, DOI 10.1016/j.aam.2017.11.004. MR3759215 ↑92

[49] Peter J. Forrester, Satya N. Majumdar, and Grégory Schehr, *Non-intersecting Brownian walkers and Yang-Mills theory on the sphere*, Nuclear Phys. B **844** (2011), no. 3, 500–526, DOI 10.1016/j.nuclphysb.2010.11.013. MR2747559 ↑104, 107

[50] Markus Fulmek and Michael Kleber, *Bijective proofs for Schur function identities which imply Dodgson's condensation formula and Plücker relations*, Electron. J. Combin. **8** (2001), no. 1, Research Paper 16, 22, DOI 10.37236/1560. MR1855857 ↑126

[51] Ira Gessel and Gérard Viennot, *Binomial determinants, paths, and hook length formulae*, Adv. in Math. **58** (1985), no. 3, 300–321, DOI 10.1016/0001-8708(85)90121-5. MR815360 ↑92

[52] Roozbeh Gharakhloo, *Strong Szegő limit theorems for multi-bordered, framed, and multi-framed Toeplitz determinants*, SIGMA Symmetry Integrability Geom. Methods Appl. **20** (2024), Paper No. 062, 51, DOI 10.3842/SIGMA.2024.062. MR4771221 ↑92, 95, 97, 101, 103, 104, 136, 138, 139, 140, 143

[53] Roozbeh Gharakhloo and Alexander Its, *A Riemann-Hilbert approach to asymptotic analysis of Toeplitz+Hankel determinants*, SIGMA Symmetry Integrability Geom. Methods Appl. **16** (2020), Paper No. 100, 47, DOI 10.3842/SIGMA.2020.100. MR4158685 ↑92

[54] Roozbeh Gharakhloo and Nicholas S. Witte, *Modulated bi-orthogonal polynomials on the unit circle: the $2j-k$ and $j-2k$ systems*, Constr. Approx. **58** (2023), no. 1, 1–74, DOI 10.1007/s00365-022-09604-2. MR4636253 ↑92, 94, 101, 126, 130

[55] Vadim Gorin and Karl Liechty, *Boundary statistics for the six-vertex model with DWBC*, Preprint, `arXiv:2310.12735`, 2023. 105, 124, 125

[56] Mark C. Ho, *Properties of slant Toeplitz operators*, Indiana Univ. Math. J. **45** (1996), no. 3, 843–862, DOI 10.1512/iumj.1996.45.1973. MR1422109 ↑92, 94

[57] Mark C. Ho, *Adjoints of slant Toeplitz operators*, Integral Equations Operator Theory **29** (1997), no. 3, 301–312, DOI 10.1007/BF01320703. MR1477322 ↑92, 94

[58] Mark C. Ho, *Spectra of slant Toeplitz operators with continuous symbols*, Michigan Math. J. **44** (1997), no. 1, 157–166, DOI 10.1307/mmj/1029005627. MR1439675 ↑92, 94

[59] Mark C. Ho, *Adjoints of slant Toeplitz operators. II*, Integral Equations Operator Theory **41** (2001), no. 2, 179–188, DOI 10.1007/BF01295304. MR1847171 ↑92, 94

[60] Mark Chung-Hsuan Ho, *Spectral properties of slant Toeplitz operators*, ProQuest LLC, Ann Arbor, MI, 1996. Thesis (Ph.D.)–Purdue University. MR2695217 ↑92, 94

[61] B. Q. Jin and V. E. Korepin, *Entanglement entropy for disjoint subsystems in xx spin chain*, Preprint, `arXiv:1104.1004`, 2011. 97

[62] Kurt Johansson, *Discrete polynuclear growth and determinantal processes*, Comm. Math. Phys. **242** (2003), no. 1-2, 277–329, DOI 10.1007/s00220-003-0945-y. MR2018275 ↑109

[63] Kurt Johansson and Eric Nordenstam, *Eigenvalues of GUE minors*, Electron. J. Probab. **11** (2006), no. 50, 1342–1371, DOI 10.1214/EJP.v11-370. MR2268547 ↑125

[64] Samuel Karlin and James McGregor, *Coincidence probabilities*, Pacific J. Math. **9** (1959), 1141–1164. MR114248 ↑112, 132

[65] I. V. Krasovsky, *Correlations of the characteristic polynomials in the Gaussian unitary ensemble or a singular Hankel determinant*, Duke Math. J. **139** (2007), no. 3, 581–619, DOI 10.1215/S0012-7094-07-13936-X. MR2350854 ↑92, 98

[66] A. B. J. Kuijlaars, K. T.-R. McLaughlin, W. Van Assche, and M. Vanlessen, *The Riemann-Hilbert approach to strong asymptotics for orthogonal polynomials on $[-1,1]$*, Adv. Math. **188** (2004), no. 2, 337–398, DOI 10.1016/j.aim.2003.08.015. MR2087231 ↑98

[67] Karl Liechty, *Nonintersecting Brownian motions on the half-line and discrete Gaussian orthogonal polynomials*, J. Stat. Phys. **147** (2012), no. 3, 582–622, DOI 10.1007/s10955-012-0485-y. MR2923331 ↑104, 107, 108, 111, 115

[68] Karl Liechty, Gia Bao Nguyen, and Daniel Remenik, *Airy process with wanderers, KPZ fluctuations, and a deformation of the Tracy-Widom GOE distribution* (English, with English and French summaries), Ann. Inst. Henri Poincaré Probab. Stat. **58** (2022), no. 4, 2250–2283, DOI 10.1214/21-aihp1229. MR4492978 ↑109, 110, 111

[69] Karl Liechty and Dong Wang, *Nonintersecting Brownian motions on the unit circle*, Ann. Probab. **44** (2016), no. 2, 1134–1211, DOI 10.1214/14-AOP998. MR3474469 ↑115

[70] Bernt Lindström, *On the vector representations of induced matroids*, Bull. London Math. Soc. **5** (1973), 85–90, DOI 10.1112/blms/5.1.85. MR335313 ↑92

[71] Barry M. McCoy and Tai Tsun Wu, *The two-dimensional Ising model*, Harvard University Press, Cambridge, MA, 1973, DOI 10.4159/harvard.9780674180758. MR3618829 ↑92

[72] Madan Lal Mehta, *Random matrices*, 3rd ed., Pure and Applied Mathematics (Amsterdam), vol. 142, Elsevier/Academic Press, Amsterdam, 2004. MR2129906 ↑92

[73] Gábor Szegő, *Orthogonal polynomials*, 4th ed., American Mathematical Society Colloquium Publications, Vol. XXIII, American Mathematical Society, Providence, RI, 1975. MR372517 ↑100

[74] G. Szegö, *On certain Hermitian forms associated with the Fourier series of a positive function*, Comm. Sém. Math. Univ. Lund [Medd. Lunds Univ. Mat. Sem.] **1952** (1952), no. Tome Supplémentaire, 228–238. MR51961 ↑136

[75] Craig A. Tracy and Harold Widom, *Level-spacing distributions and the Airy kernel*, Comm. Math. Phys. **159** (1994), no. 1, 151–174. MR1257246 ↑113

[76] Craig A. Tracy and Harold Widom, *On orthogonal and symplectic matrix ensembles*, Comm. Math. Phys. **177** (1996), no. 3, 727–754. MR1385083 ↑108

[77] Craig A. Tracy and Harold Widom, *Nonintersecting Brownian excursions*, Ann. Appl. Probab. **17** (2007), no. 3, 953–979, DOI 10.1214/105051607000000041. MR2326237 ↑109

[78] Harold Widom, *Asymptotic behavior of block Toeplitz matrices and determinants*, Advances in Math. **13** (1974), 284–322, DOI 10.1016/0001-8708(74)90072-3. MR409511 ↑136

[79] P. Zinn-Justin, *Six-vertex model with domain wall boundary conditions and one-matrix model*, Phys. Rev. E (3) **62** (2000), no. 3, 3411–3418, DOI 10.1103/PhysRevE.62.3411. MR1788950 ↑105, 123

Department of Mathematics, University of California, Santa Cruz, California
Email address: roozbeh@ucsc.edu

Department of Mathematical Sciences, DePaul University, Chicago, Illinois
Email address: kliechty@depaul.edu

Contemporary Mathematics
Volume **822**, 2025
https://doi.org/10.1090/conm/822/16476

Two useful facts about generating functions

Alex Kasman and Robert Milson

ABSTRACT. Sequences are often conveniently encoded in the form of a generating function depending on a formal variable. This note presents two observations that allow one to draw conclusions about the generated sequence from the generating function. The first constructively produces "recursion relations" for the sequence from differential operators in the formal variable having the generating function as an eigenfunction. The second allows one to determine whether the sequence is orthogonal with respect to some inner product by considering the result of taking the inner product of the generating function with itself. Examples presented to demonstrate the use and value of these methods include a sequence of numbers, a family of Exceptional Hermite Polynomials, and an example illustrating the result in a non-commutative setting.

1. Introduction

It is often convenient to encode an infinite sequence $\{a_0, a_1, a_2, \ldots\}$ in the form of a generating function $\psi(z)$ which has the sequence as the coefficients in a power series representation. In particular, when $\psi(z)$ is an ordinary generating function which can be written in a closed form that is analytic at $z = 0$ it is then possible to recover elements of the sequence as

$$a_n = \frac{1}{n!}\frac{d^n}{dz^n}\psi(z)\bigg|_{z=0}.$$

Generating functions are common in many areas of pure and applied mathematics, with the variable z usually seen as nothing more than a formal variable with no function aside from encoding the sequence. As George Pólya wrote

> *A generating function is a device somewhat similar to a bag. Instead of carrying many little objects detachedly, which could be embarrassing, we put them all in a bag, and then we have only one object to carry, the bag* [**5**].

However, the formal variable in the generating function can play a more useful and interesting role.

The main result of this note (Theorem 2.3) is a correspondence between differential operators in z having the generating function $\psi(z)$ as an eigenfunction and difference operators acting with the same eigenvalue on the generated sequence.

2020 *Mathematics Subject Classification.* Primary 47B39.

©2025 American Mathematical Society

A second result (Theorem 3.1) allows one to determine whether the sequence is orthogonal by applying that inner product to the generating function itself[1].

Examples demonstrating both of these techniques are presented in Section 4. It is our hope that others are able to apply these results to the generating functions that arise in their own research.

1.1. Notation and definitions. Let $\mathcal{R}$ be a ring and $\mathcal{R}_0 \subseteq \mathcal{R}$ be any subring. We will usually be interested in the case that the elements of $\mathcal{R}$ are functions of a variable x and $\mathcal{R}_0$ consists of just the constant functions. No assumptions are made about commutativity in $\mathcal{R}$ aside from the fact that the formal variable z introduced below commutes with all elements of the ring.

The $\mathcal{R}$-valued function $\psi(z)$ of the complex variable z is assumed to be analytic at $z = 0$ with power series representation

$$\psi(z) = \sum_{n=0}^{\infty} a_n z^n. \tag{1.1}$$

(This necessarily implies the existence of a suitable topology on $\mathcal{R}$, but it will play no role here besides allowing for the assumed convergence of the infinite sum in (1.1), and so no more will be said about it.)

Associated to the choice of the generating function $\psi(z)$ we have infinite sequence $\{a_0, a_1, a_2, \ldots\}$ with $a_n \in \mathcal{R}$ for each $n \in \mathbb{N} \cup \{0\}$ define by the formula $\psi^{(n)}(0) = n!a_n$. For later convenience, we define $a_n = 0$ to be the zero element of $\mathcal{R}$ when $n < 0$.

The monomial differential operator $z^i\partial_z^j$ for $i \in \mathbb{Z}$ and $j \in \mathbb{N} \cup \{0\}$ acts on a sufficiently differentiable $\mathcal{R}$-valued function $\gamma(z)$ according to the formula

$$z^i\partial_z^j(\gamma(z)) = z^i \frac{\partial^j}{\partial z^j}(\gamma(z)) = z^i\gamma^{(j)}(z). \tag{1.2}$$

The ring $\mathcal{R}_0[z, z^{-1}, \partial_z]$ is made up of all finite linear combinations of such monomial differential operators with "constant" coefficients from $\mathcal{R}_0$ and their action can be extended linearly from the action of those basis elements in (1.2).

If there exist $\lambda \in \mathcal{R}$ and $L \in \mathcal{R}_0[z, z^{-1}, \partial_z]$ such that

$$L(\psi(z)) = \lambda\psi(z) \tag{1.3}$$

$$\text{or} \quad L(\psi(z)) = \psi(z)\lambda. \tag{1.4}$$

then we say that $\psi(z)$ is an eigenfunction for the differential operator L with eigenvalue λ. More specifically, in the non-commutative case we say that λ is a left eigenvalue when (1.3) holds and that it is a right eigenvalue when (1.4) holds.

Consider any function $f : \mathbb{N} \cup \{0\} \to \mathcal{R}_0$ with values in the "ring of constants" $\mathcal{R}_0$ and any integer $i \in \mathbb{Z}$. The corresponding elementary difference operator $f\mathfrak{S}^i$ acts on the sequence $\{a_n\}$ turning it into a new sequence of elements from $\mathcal{R}$ according to the rule

$$f\mathfrak{S}^i(a)_n = f(n)a_{n+i}. \tag{1.5}$$

[1]After this paper was submitted, it was pointed out to the authors (by another participant in this workshop) that the results of this theorem were implicitly used in [**1**]. Consequently, we cannot claim this result as original. However, it appears not to be very well-known and we hope its inclusion here will help spread the word about this interesting way to demonstrate orthogonality.

(which is zero if $n+i<0$ by an earlier assumption). Any finite sum of these elementary difference operators is a difference operator whose action is extended linearly from the formula above. If Ω is a difference operator satisfying $\Omega(a)_n=\lambda a_n$ (or alternatively $\Omega(a)_n=a_n\lambda$) for every $n\in\mathbb{N}\cup\{0\}$ and some fixed $\lambda\in\mathcal{R}$ then we say that $\{a_n\}$ is an eigensequence for Ω with left (or alternatively right) eigenvalue λ.

2. Difference equations from generating eigenfunctions

This section concerns a certain difference operator Ω_L corresponding to the choice of any differential operator $L\in\mathcal{R}_0[z,z^{-1},\partial_z]$:

DEFINITION 2.1. Suppose $L\in\mathcal{R}_0[z,z^{-1},\partial_z]$ is an ordinary differential operator in z with singularities only at $z=0$. Then it can be written in the form

$$L=\sum_{i=\mu}^{\nu}\sum_{j=0}^{\kappa}c_{ij}z^i\partial_z^j,\qquad \mu<\nu\in\mathbb{Z},\ \kappa\in\mathbb{N}\cup\{0\},c_{ij}\in\mathcal{R}_0.$$

Associate to the choice of any such differential operator a corresponding *difference* operator Ω_L defined as

$$\begin{aligned}\Omega_L&=\sum_{i=\mu}^{\nu}\sum_{j=0}^{\kappa}c_{ij}\prod_{k=0}^{j-1}(n+j-i-k)\mathfrak{S}^{j-i}\\&=\sum_{i=\mu}^{\nu}\sum_{j=0}^{\kappa}c_{ij}(n+j-i)(n+j-i-1)\cdots(n-i+1)\mathfrak{S}^{j-i}.\end{aligned}$$

2.1. General lemma. Before the main result which applies in the special case that $\psi(z)$ is an eigenfunction of L, we first state a more general result showing a correspondence between the action of those two operators *without* making any assumptions about the generating function. As this lemma shows, the action of L on any generating function ψ can be elegantly described in terms of the corresponding difference operator Ω_L:

LEMMA 2.2. *For any differential operator* $L\in\mathcal{R}_0[z,z^{-1},\partial_z]$ *and the corresponding difference operator* Ω_L *as in Definition* 2.1 *we have*

$$L(\psi)=\sum_{n-\mu}^{\infty}\Omega_L(a)_nz^n.$$

PROOF. For $j,m\in\mathbb{N}\cup\{0\}$ one has

$$\partial_z^j(z^m)=m(m-1)\cdots(m-j+1)z^{m-j}.\tag{2.1}$$

Note that this formula is valid even in the case that $m<j$ because in that case z^m is in the kernel of the operator and a factor of $(m-m)$ appears in the product on the right side, so they are both zero.

It then follows by linearity that

$$\begin{aligned}
L(\psi) &= \sum_{i=\mu}^{\nu}\sum_{j=0}^{\kappa} c_{ij} z^i \partial_z^j \psi(z) \\
&= \sum_{i=\mu}^{\nu}\sum_{j=0}^{\kappa} c_{ij} z^i \partial_z^j \left(\sum_{m=0}^{\infty} a_m z^m\right) \quad \langle \textit{expanding } \psi \textit{ using the index } m\rangle \\
&= \sum_{i=\mu}^{\nu}\sum_{j=0}^{\kappa}\sum_{m=0}^{\infty} c_{ij} a_m z^i \left(\partial_z^j z^m\right) \\
&= \sum_{i=\mu}^{\nu}\sum_{j=0}^{\kappa}\sum_{m=0}^{\infty} c_{ij} a_m z^i \left(m(m-1)\cdots(m-j+1)z^{m-j}\right) \quad \langle \textit{by } (2.1)\rangle \\
&= \sum_{i=\mu}^{\nu}\sum_{j=0}^{\kappa}\sum_{m=0}^{\infty} c_{ij} a_m m(m-1)\cdots(m-j+1) z^{m-j+i} \\
&= \sum_{i=\mu}^{\nu}\sum_{j=0}^{\kappa}\sum_{n=i-j}^{\infty} c_{ij} a_{n+j-i}(n+j-i)(n+j-i-1)\cdots(n-i+1)z^n.
\end{aligned}$$

The final equality above is obtained by rewriting the inner-most sum in terms of the new variable $n = m - j + i$. After doing so, this sum begins at $n = i - j$. However, the equality would still be true if it started at a value *lower* than $i - j$ since $a_n = 0$ for $n < 0$. In addition, the smallest power of z which can appear in this expression with a non-zero coefficient is z^μ because ψ is expanded only in non-negative powers of z, their derivatives only involve non-negative powers of z, and those are multiplied by z^i for $i \geq \mu$. So, the sum in n can be taken to start at $n = \mu$. Then, since there are only a finite number of terms in these sums having z^n for any fixed value of n, there can be no issues about reordering that would prevent us from reordering the sums. Consequently, we have

$$\begin{aligned}
L(\psi) &= \sum_{i=\mu}^{\nu}\sum_{j=0}^{\kappa}\sum_{n=\mu}^{\infty} c_{ij} a_{n+j-i}(n+j-i)(n+j-i-1)\cdots(n-i+1)z^n \\
&= \sum_{n=\mu}^{\infty}\left(\sum_{i=\mu}^{\nu}\sum_{j=0}^{\kappa} c_{ij}(n+j-i)(n+j-i-1)\cdots(n-i+1)a_{n+j-i}\right) z^n.
\end{aligned}$$

The claim follows from the observation that the coefficient of each term z^n in the formula above is precisely $\Omega_L(a)_n$. $\square$

For example, note that $L = z\partial_z$ corresponds to the difference operator $\Omega_L = n$ which simply multiplies a_n by its index. That makes sense because

$$z\partial_z(\psi(z)) = z\left(\sum_{n=0}^{\infty} n a_n z^{n-1}\right) = \sum_{n=0}^{\infty} n a_n z^n.$$

This is true regardless of the series $\{a_n\}$. However, our primary interest below will be on the case in which $\psi(z)$ is an *eigenfunction* for a differential operator L, and whether that is true clearly depends not only on the choice of operator but also on the sequence.

2.2. When the generating functions is an eigenfunction of L. Again consider a differential operator $L \in \mathcal{R}_0[z, z^{-1}, \partial_z]$ and its corresponding difference operator Ω_L as in Definition 2.1. However, we now further assume that the generating function

$$\psi(z) = \sum_{n=0}^{\infty} a_n z^n$$

is an eigenfunction for L satisfying either $L(\psi) = \lambda\psi$ or $L(\psi) = \psi\lambda$ for some $\lambda \in \mathcal{R}$. As an easy corollary of Lemma 2.2 we then get:

THEOREM 2.3. *If the generating function $\psi(z)$ is an eigenfunction for L with (left or right) eigenvalue λ then the generated sequence $\{a_0, a_1, a_2, \ldots\}$ is an eigensequence for the difference operator Ω_L with (left or right) eigenvalue λ.*

PROOF. Suppose $L(\psi) = \lambda\psi$. Then

$$L(\psi) = \lambda\psi = \sum_{n=0}^{\infty} \lambda a_n z^n$$

and from the lemma we have

$$L(\psi) = \sum_{n=\mu}^{\infty} \Omega_L(a)_n z^n.$$

Setting the coefficients equal we get that $\Omega_L(a)_n = 0$ for $n < 0$ and $\Omega_L(a)_n = \lambda a_n$ for $n \geq 0$, which is precisely the definition of being an eigensequence. The same argument also works with the eigenvalue λ appearing on the right. $\square$

3. Determining orthogonality using generating functions

Suppose that $\mathcal{R}$ is actually a vector space with inner product $\langle \cdot, \cdot \rangle_{\mathcal{R}}$. A natural question to then ask about the sequence $\{a_0, a_1, a_2, \ldots\}$ is whether it is *orthogonal* in the sense that there exists some sequence $\{\omega_0, \omega_1, \ldots\}$ so that

$$\langle a_m, a_n \rangle_{\mathcal{R}} = \omega_n \delta_{mn} \tag{3.1}$$

for all $m, n \in \mathbb{N} \cup \{0\}$. The following result provides a means to answer that question using the generating function:

THEOREM 3.1. *Let $\psi(z)$ be a regular generating function for the sequence $\{a_0, a_1, \ldots\}$ as in* (1.1). *Then the sequence is orthogonal with respect to the inner-product $\langle \cdot, \cdot \rangle_{\mathcal{R}}$ if and only if there exists a function $\theta(z)$ such that*

$$\langle \psi(w), \psi(z) \rangle_{\mathcal{R}} = \theta(wz).$$

Furthermore, if that is the case, then the function $\theta(z)$ is a regular generating function for the sequence $\{\omega_0, \omega_1, \ldots\}$.

PROOF. Expanding the generating functions and distributing[2] the inner product we have

$$\begin{aligned}\langle \psi(w), \psi(z) \rangle_{\mathcal{R}} &= \left\langle \sum_{m=0}^{\infty} a_m w^m, \sum_{n=1}^{\infty} a_n z^n \right\rangle_{\mathcal{R}} \\ &= \sum_{m=0}^{\infty} \sum_{n=0}^{\infty} \langle a_m, a_n \rangle_{\mathcal{R}} w^m z^n.\end{aligned}$$

[2]To avoid worrying about complex conjugates, we assume $w, z \in \mathbb{R}$.

If the sequence satisfies (3.1) then the terms for which $m \neq n$ have a coefficient of zero and the result is

$$\langle \psi(w), \psi(z) \rangle_{\mathcal{R}} = \sum_{n=0}^{\infty} \omega_n (wz)^n = \theta(wz), \tag{3.2}$$

which as claimed is a function of the product wz that is a generating function for the sequence $\{\omega_0, \omega_1, \ldots\}$.

On the other hand, if the sequence $\{a_0, a_1, \ldots\}$ is *not* orthogonal, then the right hand side of the expression (3.2) involves a product of the form $w^m z^n$ where $m \neq n$ and thus is not of the form $\theta(wz)$ for any function θ. □

Section 4.2 below includes an example in which this result is used to demonstrate the orthogonality of a sequence of polynomials.

4. Examples

4.1. A sequence of numbers. The function $\psi(z) = (z-1)e^z$ is an eigenfunction for the differential operator $L = \partial_z^2 - \frac{2}{z}\partial_z$ with eigenvalue $\lambda = 1$. In this simplest example where $\mathcal{R} = \mathcal{R}_0 = \mathbb{R}$ we will just get from this eigenfunction a difference equation satisfied by a sequence of real numbers.

Expanding the function as a power series in z we see that

$$\psi(z) = -1 + \frac{z^2}{2} + \frac{z^3}{3} + \frac{z^4}{8} + \frac{z^5}{30} + \frac{z^6}{144} + \cdots$$

and hence this is a generating function for a sequence which begins $a_0 = -1$, $a_1 = 0$, $a_2 = \frac{1}{2}$, $a_3 = \frac{1}{3}$, $a_4 = \frac{1}{8}$, $a_5 = \frac{1}{30}$.

"Translating" the differential operator L into a difference operator according to Definition 2.1 we get

$$\Omega_L = (n^2 + n - 2)\mathfrak{S}^2.$$

Theorem 2.3 then assures us that applying this difference operator to the elements of the sequence will multiply them by $\lambda = 1$ and hence leave them unchanged. Indeed, we can verify that this is the case for the terms at the start of the sequence since:

$$\Omega_L(a)_0 = (0^2 + 0 - 2)a_2 = -2 \times \frac{1}{2} = -1 = a_0,$$

$$\Omega_L(a)_1 = (1^2 + 1 - 2)a_3 = 0 \times \frac{1}{3} = 0 = a_1,$$

$$\Omega_L(a)_2 = (2^2 + 2 - 2)a_4 = 4 \times \frac{1}{8} = \frac{1}{2} = a_2,$$

and so on.

4.2. Exceptional hermite polynomials. As readers of this volume of Contemporary Mathematics surely know, the classical orthogonal polynomials are important examples of sequences that are often written in terms of generating functions. Moreover, the three-term recursion relations that follow as a consequence of their orthogonality make them eigensequences for difference operators, in the terminology of this paper.

Exceptional orthogonal polynomials are a generalization of the classical orthogonal polynomials in which a finite number of positive integer degrees are missing [**3**]. The Exceptional Hermite Polynomials (XHPs) are a special class which was first studied in 2014 [**2**]. In fact, there is a family of XHPs associated to each

Young diagram. More recently, it was found that for every Young diagram there is a KP Wave Function from Gr_0 that is a generating function for the corresponding family of XHPs [**4**]. Moreover, the wave functions from Gr_0 are all eigenfunctions for differential operators in $\mathbb{C}[z, z^{-1}, \partial_z]$ [**6**]. Here we will consider just one example to illustrate the two main theorems of this note.

The ordinary generating function

$$\psi(x,z) = \left(2x^2z^2 - 4xz + z^2 + 4\right) e^{xz - \frac{z^2}{4}}$$

produces the sequence:

$$a_0 = 4,\ a_1 = a_2 = 0,\ a_3 = \frac{2x^3}{3} + x,\ a_4 = \frac{x^4}{2} + \frac{x^2}{2} - \frac{1}{8},\ a_5 = \frac{x^5}{5} - \frac{x}{4}, \cdots$$

which includes a polynomial of every degree *except* degrees 1 and 2.

Now, we will use Theorem 3.1 to demonstrate the orthogonality of this sequence of polynomials with respect to the inner product

$$\langle p(x), q(x)\rangle = \int_{-\infty}^{\infty} p(x)q(x)\frac{e^{-x^2}}{(x^2+1/2)^2}\,dx.$$

To that end, set

$$S = w + z, \quad P = \frac{1}{2}wz - 1$$

$$\phi(x,z) = \frac{z^2}{2} + \frac{1-xz}{x^2+1/2}.$$

An elementary calculation shows that

$$\text{(4.1)} \qquad \partial_x\left[\frac{2xP - S}{2x^2+1}e^{x(S-x)}\right] = -\left(\phi(x,w)\phi(x,z) - P^2 - 1\right)e^{x(S-x)}.$$

It follows that,

$$\int_{-\infty}^{\infty}\phi(x,w)\phi(x,z)e^{x(S-x)}\,dx = (P^2+1)\int_{-\infty}^{\infty} e^{x(S-x)}\,dx = (P^2+1)e^{\frac{S^2}{4}}\sqrt{\pi}.$$

Since

$$\frac{\psi(x,z)}{x^2+1/2} = 4\phi(x,z)e^{xz-z^2/4}$$

we have that

$$\psi(x,w)\psi(x,z)\frac{e^{-x^2}}{(x^2+1/2)^2} = 16\phi(x,w)\phi(x,z)e^{x(S-x)}e^{-\frac{w^2}{4}-\frac{z^2}{4}}$$

$$\text{and} \quad \int_{-\infty}^{\infty}\psi(x,w)\psi(x,z)\frac{e^{-x^2}}{(x^2+1/2)^2}\,dx = 4e^{\frac{wz}{2}}(w^2z^2 - 4wz + 8)\sqrt{\pi}.$$

That expression depends only on the product wz. Hence, it follows from Theorem 3.1 that

$$\langle a_m, a_n\rangle = \left.\frac{d^n}{dz^n}\frac{4e^{\frac{z}{2}}(z^2 - 4z + 8)\sqrt{\pi}}{n!}\right|_{z=0}\delta_{mn} = \frac{16(n-1)(n-2)}{2^n n!}\sqrt{\pi}\,\delta_{mn}.$$

Because of the orthogonality, it is not surprising that the polynomials satisfy a set of "recursion relations", and because of the missing degrees it is not surprising that it involves a commutative algebra of higher-order difference operators. But, it is not immediately clear what is the corresponding algebra of difference operators.

To elucidate the commutative algebra of difference operators and the corresponding eigenvalues, let us observe that

$$\partial_x \psi(x,z) = (1+2x^2)z^3 e^{xz-\frac{z^2}{4}}.$$

Let $\mathcal{R}$ denote the codimension-2 vector space of polynomials $\lambda(x)$ with the property that $\lambda'(x)$ is divisible by $1+2x^2$. In fact, $\mathcal{R}$ is a ring because if $f,g\in\mathcal{R}$, then $(fg)' = f'g+fg'$ is also divisible by $1+2x^2$. As a vector space, $\mathcal{R}$ is infinite-dimensional. However, as a commutative ring, $\mathcal{R}$ is generated by

$$\lambda_0 = 1,\quad \lambda_3 = x^3+\frac{3}{2}x,\quad \lambda_4 = x^4+x^2,\quad \lambda_5 = x^5-\frac{5}{4}x.$$

Thus, $\psi(x,z)$ may be regarded as a kind of generating function for $\mathcal{R}$ with the coefficients being a basis that is orthogonal with respect to the above inner-product. Since $\mathcal{R}$ is a ring, for every $n\neq 1,2$ the coefficients of $\lambda_n(x)\psi(x,z)$ satisfies the same defining property, namely their derivative is divisible by $1+2x^2$. It is therefore very plausible that the coefficients of $\lambda_n(x)\psi(x,z)$ can be obtained by applying a difference operator to the orthogonal sequence $a_n(x)$. However, the key to obtaining these difference operators is Theorem 2.3 together with the fact that for every $n\neq 1,2$ we have that

$$L_n(z,\partial_z)\psi(x,z) = \lambda_n(x)\psi(x,z)$$

where $L_n(z,\partial_z)$ is a differential operator of order n. This immediately implies that the L_n, $n\neq 1,2$ span a commutative algebra of differential operators.

The procedure for constructing the operators L_n and the corresponding difference operators is given in [**4**]. Here are the operators corresponding to eigenvalues $\lambda_3,\lambda_4,\lambda_5$, as defined above:

$$\begin{aligned}
L_3 &= \partial_z^3 + \left(\frac{3}{2}z - 6z^{-1}\right)\partial_z^2 + \left(\frac{3}{4}z^2 - 3 + 12z^{-2}\right)\partial_z + \frac{1}{8}z^3,\\
L_4 &= \partial_z^4 + \left(2z-8z^{-1}\right)\partial_z^3 + \left(\frac{3}{2}z^2 - 8 + 28z^2\right)\partial_z^2,\\
&\quad + \left(\frac{1}{2}z^3 - 2z + 12z^{-1} - 40z^{-3}\right)\partial_z + \frac{1}{16}z^4 + \frac{1}{4}\\
L_5 &= \partial_z^5 + \left(\frac{5}{2}z - 10z^{-1}\right)\partial_z^4 + \left(\frac{5}{2}z^2 - 15 + 50z^{-2}\right)\partial_z^3\\
&\quad + \left(\frac{5}{4}z^3 - \frac{15}{2}z + 45z^{-1} - 140z^{-3}\right)\partial_z^2\\
&\quad + \left(\frac{5}{16}z^4 - \frac{5}{4}z^2 + 10 - 60z^{-2} + 180z^{-4}\right)\partial_z + \frac{1}{32}z^5.
\end{aligned}$$

Each of the above differential operators L_k of order k corresponds to a difference operator $\Omega_k = \Omega_{L_k}$ of order $2k$. These, difference operators commute and generate a ring of recurrence relations for the exceptional Hermite polynomials:

$$\Omega_k(a)_n = \lambda_k a_n. \tag{4.2}$$

Using the construction described in Theorem 2.3 the difference operators in question are:

$$\begin{aligned}
\Omega_3 &= (n-2)(n-1)(n+3)\mathfrak{S}^3 + \frac{3}{2}(n-2)(n+1)\mathfrak{S}^1 + \frac{3}{4}(n-1)\mathfrak{S}^{-1} + \frac{1}{8}\mathfrak{S}^{-3},\\
\Omega_4 &= (n-2)(n-1)(n+1)(n+4)\mathfrak{S}^4 + 2(n-2)(n-1)(n+2)\mathfrak{S}^2\\
&\quad + \frac{1}{4}\left(6n^2 - 14n + 1\right) + \frac{1}{2}(n-2)\mathfrak{S}^{-2} + \frac{1}{16}\mathfrak{S}^{-4},\\
\Omega_5 &= (n-1)(n+1)(n+2)(n+5)(n-2)\mathfrak{S}^5 + \frac{5}{2}(n-1)n(n+3)(n-2)\mathfrak{S}^3\\
&\quad + \frac{5}{2}(n+1)(n-2)^2\mathfrak{S}^1 + \frac{5}{16}(n-3)\mathfrak{S}^{-3} + \frac{5}{4}(n-3)(n-1)\mathfrak{S}^{-1} + \frac{1}{32}\mathfrak{S}^{-5}.
\end{aligned}$$

The Classical Hermite polynomials $h_n(x),\ n = 0, 1, 2, \dots$ can also be defined as the coefficients of the classical generating function

$$\psi_0(x, z) = e^{xz - \frac{1}{4}z^2} = \sum_{n=0}^{\infty} h_n(x)\frac{z^n}{n!},$$

and by means of the classical recurrence relation

$$xh_n(x) = \frac{1}{2}h_{n+1}(x) + nh_{n-1}.$$

Analogously, the exceptional Hermite polynomials $a_n(x)$ can be recursively defined by means of the recurrence relations (4.2). Indeed, the recurrences

$$\begin{aligned}
\lambda_3 a_0 &= \Omega_3(a)_0 = 6a_3\\
\lambda_4 a_0 &= \Omega_4(a)_0 = 8a_4 + \frac{1}{4}a_0\\
\lambda_5 a_0 &= \Omega_5(a)_0 = 20a_5\\
\lambda_3 a_3 &= \Omega_3(a)_3 = 12a_6 + 6a_4 + \frac{1}{8}a_0
\end{aligned}$$

serve to define a_3, a_4, a_5, a_6. However, a_7 can be defined recursively in two ways:

$$\begin{aligned}
\lambda_4 a_3 &= \Omega_4(a)_3 = 56a_7 + 20a_5 + \frac{13}{4}a_3\\
\lambda_3 a_4 &= \Omega_3(a)_4 = 42a_7 + 15a_5 + \frac{9}{4}a_3.
\end{aligned}$$

The fact that these two definitions give the same value of a_7 is a consequence of the fact that the Ω_3, Ω_4 commute, which in turn is a consequence of the fact that L_3 and L_4 commute.

Although this procedure can be applied to any of the families of XHPs, we will do so here in just one case as an example. (For information about the general case, consult [**4**] or the paper by Luke Paluso in this same volume of Contemporary Mathematics.)

4.3. Non-commutative example. The function

$$\psi = \begin{pmatrix} xz - 1 & -z^2 \\ 0 & xz \end{pmatrix} e^{xz}$$

is a generating function for a sequence of matrix polynomials which begins

$$a_0 = \begin{pmatrix} -1 & 0 \\ 0 & 0 \end{pmatrix}, a_1 = \begin{pmatrix} 0 & 0 \\ 0 & x \end{pmatrix}, a_2 = \begin{pmatrix} \frac{x^2}{2} & -1 \\ 0 & x^2 \end{pmatrix}, \quad a_3 = \begin{pmatrix} \frac{x^3}{3} & -x \\ 0 & \frac{x^3}{2} \end{pmatrix}.$$

It happens to be an eigenfunction for the matrix differential operator

$$L = \begin{pmatrix} 1 & 0 \\ 0 & 1 \end{pmatrix} \partial_z^3 + \begin{pmatrix} -\frac{3}{z} & 0 \\ 0 & -\frac{3}{z} \end{pmatrix} \partial_z^2 + \begin{pmatrix} \frac{3}{z^2} & 3 \\ 0 & \frac{6}{z^2} \end{pmatrix} \partial_z + \begin{pmatrix} 0 & -\frac{6}{z} \\ 0 & -\frac{6}{z^3} \end{pmatrix}$$

satisfying the eigenvalue equation $L\psi = x^3\psi$.

Theorem 2.3 then assures us that the generated sequence is an eigensequence for the difference operator

$$\Omega_L = \begin{pmatrix} n^3 + 3n^2 - n - 3 & 0 \\ 0 & n\left(n^2 + 3n + 2\right) \end{pmatrix} \mathfrak{S}^3 + \begin{pmatrix} 0 & 3(n-1) \\ 0 & 0 \end{pmatrix} \mathfrak{S}$$

with the same eigenvalue x^3. For example, applying the difference operator to a_2 we get

$$\Omega_L(a)_2 = \begin{pmatrix} 15 & 0 \\ 0 & 24 \end{pmatrix} a_5 + \begin{pmatrix} 0 & 3 \\ 0 & 0 \end{pmatrix} a_3 = \begin{pmatrix} \frac{x^5}{2} & -x^3 \\ 0 & x^5 \end{pmatrix} = x^3 a_2.$$

5. Concluding remarks

In the paper [**4**] the operators having the generating function for Exceptional Hermite Polynomials as an eigenfunction were "translated" into the recursion relations. The main goal of this note is to explain and generalize the process by which that took place. We also presented a procedure by which one can demonstrate orthogonality of a sequence by using its generating function. (See [**1**] and the footnote in the introduction.) It is hoped that others may find it useful to apply Theorems 2.3 and 3.1 in their own research.

For the sake of clarity, some generalizations and tangential topics have been ignored in the body of the paper above. In particular:

- This paper considered only *ordinary* generating functions in which the coefficients in the power series expansion are precisely the sequence of interest. More generally, it is common to consider generating functions involving some other known sequence $\{c_0, c_1, c_2, \ldots\}$ with $c_i \neq 0$ so that

$$\psi(z) = \sum_{n=0}^{\infty} c_n a_n z^n.$$

  For instance, many sequences have generating functions of this form with $c_n = 1/(n!)$ or $c_n = 1/n$. The main results could be stated so as to be valid for these more general generating functions as well. One could simply rescale the coefficients of Ω_L in Theorem 2.3 in order to handle the general case.
- Similarly, the results could be generalized to handle the case of a generating function whose Laurent series representation (including negative powers of the formal variable) has the sequence of interest as coefficients.
- In the case of the Exceptional Orthogonal Polynomials, it is important that the terms in the sequence which are equal to zero do not arise in the recursion relations. In particular, one wants the additional property that all of the terms which appear in the difference equations obtained by

applying one of the operators Ω_L to a non-zero polynomial in the sequence area also non-zero polynomials. See [**4**] for a construction which takes this additional restriction into account.

- The generating functions used in Sections 4.2 and 4.3 also happen to be eigenfunctions for differential operators in the variable x. Although this *bispectrality* plays an essential role in the papers [**4**] and [**6**], it does not seem to be a fundamental feature of the main result in this paper and hence was not mentioned above.

Acknowledgments

We thank the organizers of this special session for their efforts and thank the participants for their talks and helpful feedback.

References

[1] Philip Feinsilver and René Schott, *Krawtchouk transforms and convolutions*, Bull. Math. Sci. **10** (2020), no. 3, 1950009, 18, DOI 10.1142/S1664360719500097. MR4179132

[2] David Gómez-Ullate, Yves Grandati, and Robert Milson, *Rational extensions of the quantum harmonic oscillator and exceptional Hermite polynomials*, J. Phys. A **47** (2014), no. 1, 015203, 27, DOI 10.1088/1751-8113/47/1/015203. MR3146977

[3] David Gómez-Ullate, Niky Kamran, and Robert Milson, *An extended class of orthogonal polynomials defined by a Sturm-Liouville problem*, J. Math. Anal. Appl. **359** (2009), no. 1, 352–367, DOI 10.1016/j.jmaa.2009.05.052. MR2542180

[4] Alex Kasman and Robert Milson, *The adelic Grassmannian and exceptional Hermite polynomials*, Math. Phys. Anal. Geom. **23** (2020), no. 4, Paper No. 40, 51, DOI 10.1007/s11040-020-09365-z. MR4173419

[5] G. Pólya, *Mathematics and plausible reasoning. Vol. I*, Princeton University Press, Princeton, NJ, 1990. Induction and analogy in mathematics; Reprint of the 1954 original. MR1109055

[6] George Wilson, *Bispectral commutative ordinary differential operators*, J. Reine Angew. Math. **442** (1993), 177–204, DOI 10.1515/crll.1993.442.177. MR1234841

DEPARTMENT OF MATHEMATICS, COLLEGE OF CHARLESTON, CHARLESTON, SOUTH CAROLINA 29401

Email address: kasmana@cofc.edu

DEPARTMENT OF MATHEMATICS AND STATISTICS, DALHOUSIE UNIVERSITY, HALIFAX, B3H 3J5, NOVA SCOTIA, CANADA

Email address: rmilson@dal.ca

Contemporary Mathematics
Volume **822**, 2025
https://doi.org/10.1090/conm/822/16481

Exceptional Hermite polynomials and Calogero-Moser pairs

Luke Paluso and Alex Kasman

ABSTRACT. There are two equivalent descriptions of George Wilson's adelic Grassmannian Gr^{ad}, one in terms of differential "conditions" and another in terms of Calogero-Moser Pairs. The former approach was used in the 2020 paper by Kasman-Milson which found that each family of Exceptional Hermite Polynomials has a generating function which lives in Gr^{ad}. This suggests that Calogero-Moser Pairs should also be useful in the study of Exceptional Hermite Polynomials, but no researchers have pursued that line of inquiry prior to the first author's thesis. The purpose of this note is to summarize highlights from that thesis, including a novel formula for Exceptional Hermite Polynomials in terms of Calogero-Moser Pairs and a theorem utilizing this correspondence to produce explicit finitely-supported distributions which annihilate them.

1. Introduction

The mathematical objects mentioned in the title have separately been the subjects of many papers. Section 2 below will provide additional details, but for the sake of this introduction let us just briefly recall:

- A pair of $N \times N$ constant complex matrices (X, Z) whose commutator differs from the identity by a rank one matrix are called a Calogero-Moser Pair (CM Pair). Interest in them first arose in the 1970s to explain the integrable dynamics of the Calogero-Moser particle system using Hamiltonian reduction [**7**]. In 1998, Wilson showed that the rank one bispectral wave functions of the KP Hierarchy can all be written in terms of CM Pairs. [**11**].
- Exceptional orthogonal polynomials, a generalization of the classical orthogonal polynomials sharing many of their important properties despite lacking polynomials of certain degrees, were first defined in 2009 [**3**] and the particular example of the Exceptional Hermite Polynomials (XHPs) was first considered in 2014 [**2**].

It is not clear from those descriptions that there is any meaningful connection between XHPs and CM Pairs and they have not appeared together in any prior papers. However, it was shown in 2020 that each family of XHPs has a generating function which is a rank one bispectral wave function of the KP Hierarchy [**6**]. Consequently, it follows from Wilson's observation that the generating functions for XHPs can be written in terms of CM Pairs.

2020 *Mathematics Subject Classification.* Primary 42C05.

©2025 American Mathematical Society

Finding a formula for XHPs in terms of CM Pairs and applications for the correspondence between them was the thesis project assigned to the first author by the second author. This note constitutes a research announcement aiming to convey the main results found in [**8**]. The thesis and its references should be consulted for details and complete proofs.

2. Background material

2.1. Partitions and Young Diagrams. For a fixed natural number $N \in \mathbb{N}$, a *partition of* N is a non-increasing sequence of non-negative integers $\lambda = \{\lambda_n\}_{n\in\mathbb{N}}$ such that

$$|\lambda| = \sum_{n=1}^{\infty} \lambda_n = N.$$

Since the elements of the sequence must eventually be zero, a partition can be equivalently written as a finite list by omitting all of the 0 entries. The length of the partition $\ell = \ell(\lambda)$, is the number of non-zero entries in the sequence.

For example, the partition $\lambda = \{4,3,2,0,0,\ldots\}$ is a partition of $|\lambda| = 9$ which has length $\ell(\lambda) = 3$ and hence could also be written $\lambda = \{4,3,2\}$.

Following [**6**] we define some sets of integers corresponding to the choice of a partition λ,

$$\mathcal{M}^{(\lambda)} = \{m_i = \lambda_i - i : i \in \mathbb{N}\} \tag{2.1}$$

$$\mathcal{K}_n^{(\lambda)} = \{m_i + n : 1 \le i \le n\} \tag{2.2}$$

$$\mathcal{I}^{(\lambda)} = \mathbb{Z} \setminus \mathcal{M}^{(\lambda)} + |\lambda| = \left\{ z + |\lambda| \ : \ z \in \mathbb{Z} \setminus \mathcal{M}^{(\lambda)} \right\}. \tag{2.3}$$

$\mathcal{M}^{(\lambda)}$ is known as the *Maya Diagram* of a partition λ. The set defined in (2.2) depends on the choice of an integer n and will be of special interest in the case that $n = \ell(\lambda)$ is the length of the partition: ${\mathcal{K}_\ell}^{(\lambda)}$ is known as the *Index Set* of a partition and we will just write it as $\mathcal{K}^{(\lambda)} = \{k_1, \ldots, k_\ell\}$. It is also noteworthy that $\mathcal{I}^{(\lambda)} \subset \mathbb{N}$.

A *Young Diagram* is a finite array of boxes arranged in left-justified rows, with the rows in non-increasing order according to their lengths. There is a natural bijection between partitions and Young Diagrams which associates to $\lambda = \{\lambda_1, \lambda_2, \ldots\}$ the diagram whose i^{th} row has λ_i boxes.

An *alternate* method for representing partitions is the *Frobenius notation.* Consider the Young Diagram corresponding to a partition λ. The number of boxes to the right of the box in the ith row and i^{th} column in the Young Diagram is denoted α_i. Similarly, β_i is the number of boxes below the box in the ith row and i^{th} column. The partition corresponding to that Young Diagram can then be denoted $\lambda = \{(\alpha_1, \ldots, \alpha_{N'})|(\beta_1, \ldots, \beta_{N'})\}$ where N' is the highest integer j so that $\lambda_j \ge j$ and where the Young Diagram has a box in position (j,j).

The Young Diagram for the partition $\lambda = \{4,3,2\}$ is shown in Figure 1. The three boxes to the right of the box in the first column and first row (shaded in light gray) tell us that $\alpha_1 = 3$. Similarly, the two boxes below the box in the first column (shaded in dark gray) correspond to $\beta_1 = 2$. Similarly we can conclude that $\alpha_2 = \beta_2 = 1$. Hence, in Frobenius notation the partition corresponding to that Young Diagram is $\lambda = \{(3,1) \mid (2,1)\}$.

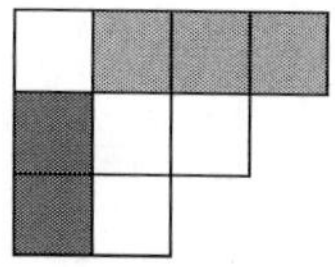

FIGURE 1. $\lambda = \{4,3,2\} = \{(3,1)|(2,1)\}$

2.2. Bivariate exceptional Hermite polynomials. Before we get to the XHPs, we begin by defining the classical Hermite polynomials. Readers may already be familiar with them and inclined to skip over this definition, but note that here we work with them in an unusual *bivariate* form.

For the purposes of this paper, the Hermite polynomials $H_n(x,y)$ are defined by the generating function

$$e^{xz+yz^2} = \sum_{n=0}^{\infty} H_n \frac{z^n}{n!}.$$

In the special case $y = -1/4$ they yield the standard univariate Hermite polynomials. Moreover, for any negative value of y they are orthogonal satisfying

$$\langle H_n, H_m \rangle_{H_y} = 2(-\pi y)^{\frac{1}{2}}(-2y)^n n! \delta_{n,m}.$$

with respect to the inner product

$$\langle f(x), g(x) \rangle_{H_y} = \int_{-\infty}^{\infty} f(x)g(x)e^{\frac{x^2}{4y}}\, dx.$$

To each partition λ, we associate a family of bivariate Exceptional Hermite Polynomials (briefly referred to as the λ-XHPs) [6]. This is the set $\left\{H_n^{(\lambda)} : n \in \mathcal{I}^{(\lambda)}\right\}$ of polynomials defined by the formula

$$H_n^{(\lambda)} = \frac{\mathrm{Wr}(H_{k_\ell}, H_{k_{\ell-1}} \dots, H_{k_1}, H_{n-N+\ell})}{\prod_{i<j}(k_i - k_j)\prod_i(n - N + \ell - k_i)} \qquad \text{for } n \in \mathcal{I}^{(\lambda)}, \tag{2.4}$$

where the Wronskian involves differentiation with respect to the variable x, $N = |\lambda|$, and where $\mathcal{I}^{(\lambda)}$ and index set $\mathcal{K}^{(\lambda)} = \{k_1, \dots, k_\ell\}$ are as in Section 2.1.

The polynomial $H_n^{(\lambda)}(x,y)$ is homogeneous of weighted degree n where $\deg(x) = 1$ and $\deg(y) = 2$. Consequently, the set includes polynomials of every non-negative degree *except* those in $\mathcal{K}_N^{(\lambda)}$.

Whenever λ is an *even* partition (which is a partition whose distinct entries all appear an even number times) and $y < 0$ is any negative number then the λ-XHPs are orthogonal with respect to a Hermite-like inner product:

$$\langle H_n^{(\lambda)}, H_m^{(\lambda)} \rangle_\lambda = \int_{-\infty}^{\infty} H_n^{(\lambda)} H_m^{(\lambda)} \frac{e^{\frac{x^2}{4y}}}{\left(\tau^{(\lambda)}(x,y)\right)^2}\, dx = \nu_{n-N}(y)\delta_{n,m} \tag{2.5}$$

where

$$\tau^{(\lambda)}(x,y) = \frac{\mathrm{Wr}(H_{k_\ell}, \dots, H_{k_1})}{\prod_{i<j<\ell}(k_i - k_j)} \text{ and } \nu_n(y) = 2(-\pi y)^{\frac{1}{2}}(-2y)^m \frac{(m+\ell)!}{\prod_{i=1}^{\ell}(m - m_i(\lambda))}.$$

REMARK 2.1. If the partition is not even, the λ-XHPs are still orthogonal under a modified version of the inner product above where the only change is that the path of integration in the complex plane is shifted in order to avoid the roots of $\tau^{(\lambda)}(x,y)$ [**4**].

2.3. The adelic Grassmannian Gr^{ad}. In his two papers that introduced and studied the so-called "adelic Grassmannian", George Wilson provided two seemingly different but entirely equivalent descriptions [**10, 11**]. They are each best described as a procedure for making functions of a certain type, in one case using Wronskians and linear functionals and in the other using matrix multiplication and rank one conditions. Remarkably, the set of functions produced by each of the procedures is the same.

2.3.1. *Gr^{ad} in terms of linear functionals.* For $n \in \mathbb{N} \cup \{0\}$ and $c \in \mathbb{C}$ define:

$$\delta_c^n(f(z)) = \partial_z^n\left(f(z)\right)\Big|_{z=c}.$$

to be the linear functional which differentiates n times in the variable z and evaluates the function at $z = c$, which is called the support of the functional.

Select n linear combinations $D = \{d_1, \ldots, d_n\}$ of these functionals with complex coefficients with the properties that:

- The set of all n of them is linearly independent (i.e. no linear combination of them is the functional that takes the constant value zero).
- Each one of them is supported at only one point in $\mathbb{C}$.

The second property assures us that there are n (not necessarily distinct numbers) $z_j \in \mathbb{C}$ such that

$$d_j = \sum_{q=0}^{m_j} \gamma_{jq}\delta_{z_j}^q \text{ for some } \gamma_{jq} \in \mathbb{C}.$$

Let the *vacuum* wave function

$$\mathbf{\Psi}_0(t_1, t_2, \ldots, z) = \exp\sum_{j=1}^{\infty} t_j z^j$$

be a function of the infinitely many "time variables" t_j (for $j \in \mathbb{N}$) and z.

From the selection of D, we produce a modified wave function according to the formula

$$\mathbf{\Psi}_D(t_1, t_2, \ldots, z) = \frac{\mathrm{Wr}(\phi_1, \ldots, \phi_n, \mathbf{\Psi}_0)}{\mathrm{Wr}(\phi_1, \ldots, \phi_n)\prod_{j=1}^{n}(z - z_j)}$$

where the Wronskian involves differentiation with respect to t_1 and $\phi_j = d_j(\mathbf{\Psi}_0)$ is the result of applying the jth functional to the vacuum wave function. Notice that the z_j's appearing in the denominator are the supports of the distributions in D.

One definition of Gr^{ad} is that it is the collection of all wave functions $\mathbf{\Psi}_D$ which can be made in this way using *any* selection of functionals D meeting the criteria above.

REMARK 2.2. It should be noted that the map $D \mapsto \mathbf{\Psi}_D$ is *not* one to one. In particular, if D and D' span the same subspace then $\mathbf{\Psi}_D = \mathbf{\Psi}_{D'}$. But, the lack of uniqueness is much worse than that. In fact, it is always possible to start with a choice of D with n elements, "integrate" the elements in a certain way, and then add δ_c^0 to yield an alternative set containing $N + 1$ elements corresponding to the same wave function. However, we will not be elaborating further on this issue in this note.

REMARK 2.3. The wave function Ψ_D is connected to the KP Hierarchy in the following sense: Write

$$\Psi_D(t_1, t_2, \ldots, z) = \left(1 + \sum_{j=1}^{\infty} \alpha_j(t_1, t_2, \ldots) z^{-j}\right) \Psi_0.$$

Let $W = 1 + \sum \alpha_j \partial_{t_1}^{-j}$ be the pseudo-differential operator in the variable t_1 with the same coefficients and let $\mathcal{L} = W \circ \partial_{t_1} \circ W^{-1}$ be the result of conjugating the simple differential operator $\partial_{t_1} = \frac{d}{dt_1}$ by W. Then $\mathcal{L}$ satisfies the hierarchy of equations

$$\frac{\partial}{\partial t_j}\mathcal{L} = [(\mathcal{L}^j)_+, \mathcal{L}]$$

for every positive integer j (where the positive subscript indicates projection onto differential operators). Consequently, the coefficients of $\mathcal{L}$ satisfy integrable nonlinear partial differential equations, including many from mathematical physics such as the KP Equation that models the behavior of water waves. (For more information about these objects and their role in soliton theory, see [**5**] or [**9**].)

REMARK 2.4. Following standard convention in the context of the KP hierarchy, this paper will identify the variables x (appearing in the orthogonal polynomials) and t_1 (the first time variable of the KP Hierarchy), considering them to be two different expressions representing the same variable.

2.3.2. *Gr^{ad} and CM Pairs.* When the commutator of two operators is equal to the identity operator, they are said to satisfy the "canonical commutation relation". This identity is of fundamental importance in quantum mechanics. However, it is easy to see that finite-dimensional matrices over $\mathbb{C}$ cannot satisfy the canonical commutation relation[1]. Since the commutator of two such matrices cannot differ from the identity by the zero matrix, one can argue that the closest that they can get to that situation is if their commutator differs from the identity by a rank *one* matrix. Matrices which satisfy this "almost canonical commutation relation" are called *Calogero-Moser Pairs* and they arise in the study of integrable particle systems [**7**]. We define them formally by saying:

DEFINITION 2.5. Let $X, Z \in M_{NN}(\mathbb{C})$ be constant $N \times N$ matrices. The ordered Pair (X, Z) is said to be a Calogero-Moser Pair (CM Pair) if and only if $\mathrm{Rank}([X, Z] - I) = \mathrm{Rank}(XZ - ZX - I) = 1$.

Let (X, Z) be any CM Pair. Then, since $[X, Z] - I$ is a rank one matrix, there exist a row vector $\boldsymbol{a}$ and a column vector $\boldsymbol{b}$ such that $[X, Z] - I = \boldsymbol{ba}$. Wilson [**11**] showed that the function

$$\Psi^{(X,Z)}(t_1, t_2, t_3, \ldots, z) = (1 + \boldsymbol{a}\tilde{X}^{-1}\tilde{Z}^{-1}\boldsymbol{b})\Psi_0$$

is a KP wave function in Gr^{ad} (cf. Remark 2.3) where

$$\tilde{X} = -X + \sum_{j=1}^{\infty} jt_j Z^{j-1} \qquad \text{and} \qquad \tilde{Z} = zI - Z.$$

[1]Since the properties of the trace imply that $\mathrm{tr}[X, Z] = 0$, the commutator of two $n \times n$ matrices can only equal the identity if the scalar field has positive characteristic so that $n \cdot 1 = 1 + 1 + \cdots + 1 = 0$.

Moreover, running over all CM Pairs this produces exactly the same set of functions as one gets by computing $\mathbf{\Psi}_D$ as above and running through all choices of functionals D. It is in that sense that Gr^{ad} can be equivalently described in terms of functionals or CM Pairs.

REMARK 2.6. Although there is not a unique choice of vectors $\boldsymbol{a}$ and $\boldsymbol{b}$ corresponding to a fixed (X, Z), the wave function depends only on the choice of the CM Pair and not on the vectors chosen. Moreover, the map $(X, Z) \mapsto \mathbf{\Psi}^{(X,Z)}$ is not one-to-one in the sense that for any invertible $N \times N$ matrix S, the matrices SXS^{-1} and SZS^{-1} are also a CM Pair and they correspond to the exact same wave function.

2.3.3. *Stationary wave functions and the bispectral involution.* The main question of interest in Wilson's papers [**10**] and [**11**] was the *bispectrality* of the wave functions in Gr^{ad}. In that context, the infinitely many time variables t_i for $i \geq 2$ are unnecessary. For a wave function $\mathbf{\Psi}(t_1, t_2, t_3, \ldots, z)$ in Gr^{ad}, the result of setting $t_1 = x$ and $t_i = 0$ for $i \geq 2$ is called the corresponding stationary wave function. Since the map

$$\mathbf{\Psi}(t_1, t_2, t_3, \ldots, z) \mapsto \mathbf{\Psi}(x, 0, 0, \ldots, z)$$

is one-to-one, no information is lost by considering only stationary wave functions. Furthermore, an interesting and important symmetry of Gr^{ad} which was discovered by Wilson is most naturally written in terms of CM Pairs and their stationary wave functions: Note that if (X, Z) is a CM Pair then so is $(Z^\top, X^\top)$. The corresponding stationary wave functions are related by nothing more than an exchange of the two variables:

$$\mathbf{\Psi}^{(Z^\top, X^\top)}(x, 0, 0, \ldots, z) = \mathbf{\Psi}^{(X,Z)}(z, 0, 0, \ldots, x). \tag{2.6}$$

2.4. XHPs hiding in Gr^{ad}. It was a surprise in 2020 when it was realized that the Exceptional Hermite Polynomials have a generating function that is a KP wave function from Gr^{ad} [**6**]. In the terminology and notation of this paper, this result can be most simply stated in the following way:

Choose any partition λ and let $D = \{\delta_0^{k_1}, \cdots, \delta_0^{k_\ell}\}$ be the set of linear functionals supported at $z = 0$ having orders determined by the set $\mathcal{K}^{(\lambda)} = \{k_1, \ldots, k_\ell\}$ from Section 2.1. Then define

$$\mathbf{\Psi}^{(\lambda)}(x, y, z) = \mathbf{\Psi}_D(x, y, 0, 0, \ldots, z) \tag{2.7}$$

to be the corresponding wave function with $t_1 = x$, $t_2 = y$ and $t_i = 0$ for $i \geq 3$. The function $\mathbf{\Psi}^{(\lambda)}$ is a generating function in the sense that

$$\mathbf{\Psi}^{(\lambda)}(x, y, z) = \sum_{m=-\ell}^{\infty} \frac{\prod_{i=1}^{\ell}(m - m_i(\lambda))}{(m+\ell)!} \frac{H_{m+N}^{(\lambda)}(x, y)}{\tau^{(\lambda)}(x, y)} z^m \tag{2.8}$$

or equivalently

$$z^N \tau^{(\lambda)}(x, y) \mathbf{\Psi}^{(\lambda)}(x, y, z) = \sum_{n=0}^{\infty} \frac{\prod_{k \in \mathcal{K}_N}(n-k)}{n!} H_n^{(\lambda)}(x, y) z^n. \tag{2.9}$$

The usefulness of this observation in the paper [**6**] stemmed mostly from the bispectrality of this wave function. Here, however, we have a different focus: what is the CM Pair corresponding to this wave function and how can it be used to further study the Exceptional Hermite Polynomials?

3. The CM Pair for the λ-XHPs

It follows from Equation (2.7) that the generating function $\Psi^{(\lambda)}(x, y, z)$ of the λ-XHPs is a KP wave function from Gr^{ad} (with the time variables set appropriately). However, we also know that every KP wave function in Gr^{ad} can be written in terms of CM Pairs. So, the question arises: What CM Pair corresponds to the generating function for the λ-XHPs?

Before we answer that question, we introduce the following notation:

DEFINITION 3.1. The elementary matrix $E_{\mu\nu}(i, j)$ is defined to be the $\mu \times \nu$ matrix whose only non-zero entry is a 1 in the ith row and j^{th} column. If the size of the matrix is clear from context, the subscript may be omitted so that it is written simply as $E(i, j)$.

Now, we are ready to give a matrix pair which answers the question, but it is necessary to write the partition using Frobenius notation (cf. Section 2.1).

DEFINITION 3.2. Fix a partition $\lambda = \{(\alpha_1, \ldots, \alpha_J)|(\beta_1, \ldots, \beta_J)\}$ of N. Let $N_k = \alpha_k + \beta_k + 1$ for $1 \leq k \leq J$ and define X_{ij}^{λ} and Z_{ij}^{λ} to be the $N_i \times N_j$ matrices

$$
\begin{aligned}
X_{ii}^{\lambda} &= -\sum_{n=1}^{\beta_i} nE(n+1, n) + \sum_{n=1}^{\alpha_i} nE(N_i + 1 - n, N_i - n) \\
X_{ij}^{\lambda} &= N_j \sum_{n=0}^{\alpha_j} E(\beta_i + 2 + n, \beta_j + 1 + n) && \text{for} \quad i < j \\
X_{ij}^{\lambda} &= -N_j \sum_{n=0}^{\beta_i} E(\beta_i + 1 - n, \beta_j - n) && \text{for} \quad j < i \\
Z_{ii}^{\lambda} &= \sum_{n=1}^{N_i - 1} E(n, n+1) \\
Z_{ij}^{\lambda} &= \mathbf{0} && \text{for} \quad i \neq j.
\end{aligned}
$$

The $N \times N$ matrices X^{λ} and Z^{λ} with block decomposition

$$
X^{\lambda} = \begin{bmatrix} X_{11}^{\lambda} & \cdots & X_{1J}^{\lambda} \\ \vdots & \ddots & \vdots \\ X_{J1}^{\lambda} & \cdots & X_{JJ}^{\lambda} \end{bmatrix} \quad \text{and} \quad Z^{\lambda} = \begin{bmatrix} Z_{11}^{\lambda} & \cdots & Z_{1J}^{\lambda} \\ \vdots & \ddots & \vdots \\ Z_{J1}^{\lambda} & \cdots & Z_{JJ}^{\lambda} \end{bmatrix}
$$

are called the CM Pair associated to λ.

Of course, it is not yet clear that these matrices are related in any useful way to the λ-XHPs, or even that they are actually a CM Pair. The next two results from [**8**] show that they are:

THEOREM 3.3. *The matrices X^{λ} and Z^{λ} satisfy the rank one condition*

$$[X^{\lambda}, Z^{\lambda}] - I = \boldsymbol{ba}$$

where $\boldsymbol{b}$ *and* $\boldsymbol{a}$ *are the* $N \times 1$ *and* $1 \times N$ *matrices with block decompositions*

$$\boldsymbol{b} = \begin{bmatrix} E_{N_1,1}(\beta_1+1,1) \\ \vdots \\ E_{N_i,1}(\beta_i+1,1) \\ \vdots \\ E_{N_J,1}(\beta_J+1,1) \end{bmatrix} \qquad \boldsymbol{a}^\top = \begin{bmatrix} -N_1E_{N_1,1}(\beta_1+1,1) \\ \vdots \\ -N_jE_{N_j,1}(\beta_j+1,1) \\ \vdots \\ -N_JE_{N_J,1}(\beta_J+1,1) \end{bmatrix}. \tag{3.1}$$

and hence (X^λ, Z^λ) *is a CM Pair.*

Furthermore, not only is (X^λ, Z^λ) *some* CM Pair, it is one in terms of which the generating function can be written:

THEOREM 3.4. *The generating function* $\boldsymbol{\Psi}^{(\lambda)}(x,y,z)$ *of the* λ*-XHPs can be written in terms of the CM Pair* (X^λ, Z^λ) *as*

$$\boldsymbol{\Psi}^{(\lambda)}(x,y,z) = \boldsymbol{\Psi}^{(X^\lambda,Z^\lambda)}(x,y,0,0,\ldots,z) = (1+\boldsymbol{a}\tilde{X}^{-1}\tilde{Z}^{-1}\boldsymbol{b})\boldsymbol{\Psi}_0 \tag{3.2}$$

with $\boldsymbol{a}$ *and* $\boldsymbol{b}$ *as in* (3.1) *and where*

$$\tilde{X} = -X^\lambda + xI + 2yZ^\lambda \qquad \text{and} \qquad \tilde{Z} = zI - Z^\lambda.$$

REMARK 3.5. The proof of Theorem 3.3 is nothing more than a long and tedious computation. The proof of Theorem 3.4 given in [**8**], on the other hand, involves some interesting ideas. In particular, it utilizes the fact that wave functions in Gr^{ad} can be identified by the finitely-supported linear functionals which annihilate them. It is relatively easy to see that in the case of $\boldsymbol{\Psi}_D$ those would be precisely the ones which multiply by $\prod(z-z_j)$ and then apply some linear combination of the functionals in D. On the other hand, Lemma 3.1 from the paper [**1**] provides a method involving residues for finding the linear functionals which annihilate a wave function $\boldsymbol{\Psi}^{(X,Z)}$ written in terms of a CM Pair. Confirming that these collections of functionals are equal in the special case $D = D_\lambda$ and $(X,Z) = (X^\lambda, Z^\lambda)$ was the proof method utilized in [**8**]. (Another proof method of an equivalent result can be found in the appendix by Macdonald to Wilson's 1998 paper [**11**]. We became aware of this alternate proof only afterwards.)

EXAMPLE 3.6. We will illustrate this procedure for associating a CM Pair to a partition with an example. Choose the partition $\lambda = \{2,2,1,1\}$. Using (2.1), (2.2), and (2.3) we can conclude that:

$$\mathcal{K}_N^{(\lambda)} = \{7,6,4,3,1,0\} \qquad \text{and} \qquad \mathcal{I} = \{2,5,8,9,\ldots\}.$$

Now we use Definition 3.2 in order to construct:

$$X = \begin{bmatrix} 0 & 0 & 0 & 0 & 0 & 0 \\ -1 & 0 & 0 & 0 & 0 & 0 \\ 0 & -2 & 0 & 0 & 0 & 0 \\ 0 & 0 & -3 & 0 & 0 & 0 \\ 0 & 0 & 0 & 1 & 0 & 1 \\ 0 & 0 & -5 & 0 & 0 & 0 \end{bmatrix} \qquad Z = \begin{bmatrix} 0 & 1 & 0 & 0 & 0 & 0 \\ 0 & 0 & 1 & 0 & 0 & 0 \\ 0 & 0 & 0 & 1 & 0 & 0 \\ 0 & 0 & 0 & 0 & 1 & 0 \\ 0 & 0 & 0 & 0 & 0 & 0 \\ 0 & 0 & 0 & 0 & 0 & 0 \end{bmatrix}.$$

Then $[X,Z] - I = \boldsymbol{ba}$ where:

$$\boldsymbol{b}^\top = \begin{bmatrix} 0 & 0 & 0 & 1 & 0 & 1 \end{bmatrix} \text{ and } \boldsymbol{a} = \begin{bmatrix} 0 & 0 & 0 & -5 & 0 & -1 \end{bmatrix}.$$

Now we can form

$$\tilde{X} = -X + xI + 2yZ = \begin{bmatrix} x & 2y & 0 & 0 & 0 & 0 \\ 1 & x & 2y & 0 & 0 & 0 \\ 0 & 2 & x & 2y & 0 & 0 \\ 0 & 0 & 3 & x & 2y & 0 \\ 0 & 0 & 0 & -1 & x & -1 \\ 0 & 0 & 5 & 0 & 0 & x \end{bmatrix}.$$

The corresponding KP wave function is then given by (3.2). This computation will be continued below in Example 4.3 where the exceptional orthogonal polynomials associated to the partition λ will be produced directly from these matrices.

4. Formula for λ-XHPs in terms of their CM Pair

The main result of the thesis is the following formulas which give any Exceptional Hermite Polynomial as a linear combination of classical Hermite Polynomials with coefficients written in terms of the corresponding CM Pair:

THEOREM 4.1. *Let λ be a partition of N, $(X,Z) = (X^\lambda, Z^\lambda)$ from Definition 3.2, $\tilde{X} = -X + xI + 2yZ$, and $\boldsymbol{a}, \boldsymbol{b}$ be the vectors from* (3.1). *The λ-XHPs $\{H_n^{(\lambda)}(x,y) \ : \ n \in \mathcal{I}^{(\lambda)}\}$ can be computed using the formulas:*

$$H_n^{(\lambda)} = \frac{\det(\tilde{X})}{s(n)} \left(\sum_{k=0}^{n} \frac{n!}{(n-k)!} H_{n-k}(x,y) \boldsymbol{a} \tilde{X}^{-1} Z^{N-k-1} \boldsymbol{b} \right)$$

$$H_n^{(\lambda)} = \frac{\det(\tilde{X})}{s(n)} \left(\sum_{k=0}^{N-1} \frac{n!}{(n-k)!} H_{n-k}(x,y) \boldsymbol{a} \tilde{X}^{-1} Z^{N-k-1} \boldsymbol{b} + \frac{n!}{(n-N)!} H_{n-N}(x,y) \right)$$

when $n < N$ and $n \geq N$ respectively and where $s(n) = \prod_{k \in \mathcal{K}_N^{(\lambda)}} (n-k)$.

REMARK 4.2. The proof of Theorem 4.1 begins with the observation that applying δ_0^n to (2.9) yields

$$\delta_0^n \left(z^N \tau^{(\lambda)}(x,y) \Psi^{(\lambda)} \right) = s(n) H_n^{(\lambda)}(x,y) \tag{4.1}$$

Since Equation (3.2) gives the generating function in terms of matrix products, it is relatively easy to apply the functional δ_0^n to it. In addition, two identities were used to further simplify the result: The rank one condition satisfied by the CM Pair implies that

$$ZX + \boldsymbol{ba} = XZ - I$$

and the Matrix Determinant Lemma [12] further allows one to translate back and forth between matrix products and determinants. Additionally, we made use of the correspondence that $\tau^{(\lambda)}(x,y) = \det(\tilde{X})$.

EXAMPLE 4.3. Recall Example 3.6 where specific matrices X, Z, $\tilde{X}$, $\boldsymbol{a}$, and $\boldsymbol{b}$ were associated to the choice of partition $\lambda = \{2,2,1,1\}$. Here we will illustrate the use of those matrices in producing the corresponding orthogonal polynomials $H_n^{(\lambda)}(x,y)$. According to Theorem 4.1 we can calculate $H_n^{(\lambda)}(x,y)$ using matrix products and the determinant of $\tilde{X}$ for all $n \in \mathcal{I}$. Here, for example, are the cases

$n = 2$, 5 and 8:

$$
\begin{aligned}
H_2^{(\lambda)}(x,y) &= \frac{\det\left(\tilde{X}\right)}{80}\left(\sum_{k=0}^{2}\frac{2!}{(2-k)!}H_{2-k}(x,y)\boldsymbol{a}\tilde{X}^{-1}Z^{6-k-1}\boldsymbol{b}\right)\\
&= \frac{\det\left(\tilde{X}\right)}{80}\left(2H_0(x,y)\boldsymbol{a}\tilde{X}^{-1}Z^3\boldsymbol{b}\right)\\
&= x^2-2y\\
H_5^{(\lambda)}(x,y) &= \frac{\det\left(\tilde{X}\right)}{80}\left(\sum_{k=0}^{5}\frac{5!}{(5-k)!}H_{5-k}(x,y)\boldsymbol{a}\tilde{X}^{-1}Z^{6-k-1}\boldsymbol{b}\right)\\
&= \frac{\det\left(\tilde{X}\right)}{80}(20H_3(x,y)\boldsymbol{a}\tilde{X}^{-1}Z^3\boldsymbol{b}+60H_2(x,y)\boldsymbol{a}\tilde{X}^{-1}Z^2\boldsymbol{b}\\
&\quad+120H_1(x,y)\boldsymbol{a}\tilde{X}^{-1}Z\boldsymbol{b}+120H_0(x,y)\boldsymbol{a}\tilde{X}^{-1}I\boldsymbol{b})\\
&= x^5-20x^3y+60xy^2\\
H_8^{(\lambda)}(x,y) &= \frac{\det(\tilde{X})}{2240}\left(\sum_{k=0}^{5}\frac{8!}{(8-k)!}H_{8-k}(x,y)\boldsymbol{a}\tilde{X}^{-1}Z^{5-k}\boldsymbol{b}+\frac{8!}{2!}H_2(x,y)\right)\\
&= x^8-20x^6y+120x^4y^2-240x^2y^3-240y^4
\end{aligned}
$$

5. Finitely-supported distributions annihilating λ-XHPs

For $n \in \mathbb{N}$ and $c \in \mathbb{C}$, let Δ_c^n be the linear functional

$$\Delta_c^n(f(x)) = f^{(n)}(c)$$

which differentiates n times in the variable x and then evaluates at $x = c$. From an abstract perspective, this is the same functional as δ_c^n introduced earlier, with the only difference being that the this new one acts on functions of the variable x while the one we saw earlier acted on functions of z. (Since we will be applying these finitely supported distributions to functions that depend on both variables, this notation allows us to indicate on which variable they are acting. For example, if $\psi(x,z) = x^2z^3$ then $\Delta_1^1(\psi(x,z)) = 2z^3$ while $\delta_1^1(\psi(x,z)) = 3x^2$.)

For any fixed partition λ it is known that there are functionals which can be written as linear combinations of the Δ_c^n that annihilate the entire family of λ-XHPs. In particular, [**6**] explains how one can find them using Wilson's bispectral involution on Gr^{ad}. However, the bispectral involution is not defined in an explicit form when Gr^{ad} is described using functionals in z and so [**6**] was only able to give an implicit description.

In contrast, the bispectral involution is not only explicitly described but also very simple in terms of CM Pairs. Combining (2.6) together with Theorem 3.4 allowed us to prove the following result giving formulas for those functionals annihilating the λ-XHPs for almost all values of y:

THEOREM 5.1. *Let λ be a partition of N and let $(X,Z) = (X^\lambda, Z^\lambda)$ be the corresponding CM Pair as in Definition 3.2. Now, suppose y is chosen so that the matrix $X - 2yZ$ has distinct eigenvalues $\{\gamma_1,\ldots,\gamma_N\}$. Then, let S be a matrix such* *that* $S^{-1}(X^\top - 2yZ^\top)S = \operatorname{diag}(\gamma_1, \dots, \gamma_N)$ *and define*

$$X' = S^{-1}Z^\top S,\ Z' = \operatorname{diag}(\gamma_1, \dots, \gamma_N),\ \mathbf{b}' = S^{-1}\mathbf{a}^\top,\ \textit{and } \mathbf{a}' = \mathbf{b}^\top S$$

where $\mathbf{a}$ *and* $\mathbf{b}$ *are as in* (3.1). *Finally, for* $1 \le i \le N$ *define the functions*

$$g_i(x) = \frac{\mathbf{a}'_i \mathbf{b}'_i + X'_{ii}(x - \gamma_i)}{\prod_{n \ne i}(x - \gamma_n)}.$$

Then for $1 \le i \le N$ *the functionals*

$$\theta_i = g_i(\gamma_i)\Delta^1_{\gamma_i} + g'_i(\gamma_i)\Delta^0_{\gamma_i}$$

annihilate each of the λ*-XHPs. Furthermore, a finitely-supported functional* d *satisfies* $d(H_n^{(\lambda)}) = 0$ *for all* $n \in \mathcal{I}^{(\lambda)}$ *if and only if* $d \in \operatorname{span}(\theta_1, \dots, \theta_N)$.

EXAMPLE 5.2. Consider the partition $\lambda = \{1, 1\} = \{(0)|(1)\}$. The corresponding CM pair is

$$X = \begin{bmatrix} 0 & 0 \\ -1 & 0 \end{bmatrix} \qquad Z = \begin{bmatrix} 0 & 1 \\ 0 & 0 \end{bmatrix}$$

which satisfy

$$[X, Z] - I = \begin{bmatrix} 0 & 0 \\ 0 & -2 \end{bmatrix} = \mathbf{ba} = \begin{bmatrix} 0 \\ 1 \end{bmatrix} \begin{bmatrix} 0 & -2 \end{bmatrix}.$$

From these one can make the KP wave function

$$\psi^{(X,Z)}(x, y, z) = e^{xz + yz^2}\left(\frac{2}{z^2(x^2 - 2y)} - \frac{2x}{z(x^2 - 2y)} + 1\right)$$

with corresponding τ-function

$$\tau(x, y) = x^2 - 2y.$$

Clearing the denominator of the wave function and expanding as a series in z yields:

$$\tau(x,y)z^2\psi(x,y,z) = 2 + 0z + 0z^2 + \frac{1}{3}\left(x^3 - 6xy\right)z^3 + \frac{1}{4}\left(x^4 - 4x^2y - 4y^2\right)z^4 + \frac{1}{10}\left(x^5 - 20xy^2\right)z^5 + O\left(z^6\right).$$

The non-zero coefficients in that expression are (up to constant scalar multiples) the family of Exceptional Hermite Polynomials corresponding to λ.

This section explains how we can use the CM pair to produce differential conditions in the variable x which are satisfied by all of the polynomials in that sequence whenever y is chosen so that $\tau(x, y)$ has distinct roots. Allow us to demonstrate this procedure.

To keep things simple, let us begin by choosing $y = -2$ so that

$$\tau(x, -2) = x^2 + 4 = (x - 2\boldsymbol{i})(x + 2\boldsymbol{i})$$

has the distinct roots $\pm 2\boldsymbol{i} \in \mathbb{C}$. It is no coincidence that these same numbers are the eigenvalues of the matrix $X - 2yZ$ (and its transpose) We will focus here only on this value of y and on the eigenvalue $2\boldsymbol{i}$, though the procedure works the same for any y for which the eigenvalues are distinct and for any one of those eigenvalues.

It is easy to see that

$$X^\top - 2yZ^\top = \begin{bmatrix} 0 & -1 \\ -4\boldsymbol{i} & 0 \end{bmatrix}$$

and we know that this matrix is diagonalizable (because of how y was chosen). In fact,

$$Z' = S^{-1}(X^\top - 2yZ^\top)S = \text{diag}(-2\boldsymbol{i}, 2\boldsymbol{i}) \text{ where } S = \begin{bmatrix} -\boldsymbol{i} & \boldsymbol{i} \\ 2 & 2 \end{bmatrix}.$$

We use that same matrix S to define

$$X' = S^{-1}Z^\top S, \ \boldsymbol{a}' = \boldsymbol{b}^\top S, \text{ and } \boldsymbol{b}' = S^{-1}\boldsymbol{a}^\top.$$

They turn out to be:

$$X' = \begin{bmatrix} -\frac{\boldsymbol{i}}{4} & \frac{\boldsymbol{i}}{4} \\ -\frac{\boldsymbol{i}}{4} & \frac{\boldsymbol{i}}{4} \end{bmatrix}, \ \boldsymbol{a}' = \begin{bmatrix} 2 & 2 \end{bmatrix}, \text{ and } \boldsymbol{b}' = \begin{bmatrix} -\frac{1}{2} \\ -\frac{1}{2} \end{bmatrix}.$$

These form a new CM pair because

$$[X', Z'] - I = \boldsymbol{b}'\boldsymbol{a}'.$$

(Essentially, what we have done is shift the pair (X, Z) using the second flow of the KP hierarchy and then applied the bispectral involution).

These matrices give us the values we need to construct those differential conditions satisfied by the λ-XHPs. Since the eigenvalue of interest here is $2\boldsymbol{i}$ which is in the second row and column of Z', we will use $i = 2$ in the formulas. (If instead we were looking at the eigenvalue $-2\boldsymbol{i}$ we would have been using $i = 1$.) In particular, we compute the function $g_i(x)$ from Equation (5.1)

$$g_2(x) = \frac{\boldsymbol{a}'_2\boldsymbol{b}'_2 + X'_{22}(x - 2\boldsymbol{i})}{x + 2\boldsymbol{i}} = \frac{-1 + \frac{1}{4}\boldsymbol{i}(x - 2\boldsymbol{i})}{x + 2\boldsymbol{i}}$$

(where the subscripts on $\boldsymbol{a}'$, $\boldsymbol{b}'$ and X' just indicate that we should select the corresponding entry from the constant matrices of those names listed above).

The theorem then tells us that the linear functional

$$g_2(2\boldsymbol{i})\Delta^1_{2i} + g'_2(2\boldsymbol{i})\Delta^0_{2i} = \frac{\boldsymbol{i}}{4}\Delta^1_{2i}$$

will annihilate all of the λ-XHPs. Indeed, one can check that each of the coefficient functions in the series expansion above has the property that its derivative with respect to x evaluated at $x = 2\boldsymbol{i}$ is zero.

6. Concluding remarks

6.1. Relationship to the thesis. This opportunity to report on the main results in [**8**] also has given us a chance to correct some of the errors that thesis contained. For example, the definition of $g_i(x)$ given in Theorem 5.1 was written incorrectly in the thesis, with γ_i being written in place of X'_{ii} by mistake. Furthermore, the statement of Theorem 4.1 in the thesis unfortunately confused n and N. In general, wherever statements in this research announcement differ from those in the thesis, it should be understood that a small error has been corrected.

On the other hand, the proof methods utilized in [**8**] remain essentially valid (with those small corrections implemented), and so one should still consult that original document for more details on the proofs.

The thesis also falsely claims that it is *always* possible to choose y so that the roots of $\tau^{(\lambda)}(x, y)$ (equivalently the eigenvalues of $X - 2yZ$) are distinct. If the corresponding solution to the KP Hierarchy happens to be a *KdV* solution then $\tau^{(\lambda)}(x, y)$ will be independent of y and the only roots will be $x = 0$.

6.2. Future outlook. This is the first investigation about the connections between Exceptional Hermite Polynomials and Calogero-Moser Pairs. The results above conclusively demonstrate that this correspondence can be fruitful. However, there is more to do.

One of the main goals we were unable to achieve was to create an alternate proof for orthogonality of the XHPs using CM Pairs. As stated in Section 2.2, it is already well known that the λ-XHPs are orthogonal with respect to the inner product (2.5). However, the proof of this fact requires some higher level analytic techniques. As a result we felt it would be fruitful to give a proof using Theorem 4.1, elementary linear algebra, and some basis properties of integrals. Unfortunately we were not able to devise such a proof.

There are many CM Pairs other than the special ones (X^λ, Z^λ) corresponding to partitions that appear in all of these results. There may be a way to generalize these results to a larger subset of them, or even perhaps to all of Gr^{ad}, though we did not find anything useful to say in that more general case.

It would be nice to know whether or not Theorem 5.1 works for all $y \neq 0$. As is currently stated, we only have that it works everywhere except on a set of Lebesgue measure 0. The problem amounts to showing whether or not the matrix $-X + 2yZ$ is diagonalizable. Furthermore, it ought to be possible to generalize Theorem 5.1 so that it works also in the case that the eigenvalues are *not* distinct. The same method should work, but it gets very complicated.

Acknowledgments

Both authors thank the organizers for the opportunity to present at this fantastic special session and the participants for their talks and feedback. The first author would also like to thank the other members of his thesis committee – Annalisa Calini and Rob Milson – as well as Dan Maroncelli for their assistance and support. Finally, we are grateful to the anonymous referees for their helpful suggestions.

References

[1] Maarten Bergvelt, Michael Gekhtman, and Alex Kasman, *Spin Calogero particles and bispectral solutions of the matrix KP hierarchy*, Math. Phys. Anal. Geom. **12** (2009), no. 2, 181–200, DOI 10.1007/s11040-009-9058-y. MR2497322

[2] David Gómez-Ullate, Yves Grandati, and Robert Milson, *Rational extensions of the quantum harmonic oscillator and exceptional Hermite polynomials*, J. Phys. A **47** (2014), no. 1, 015203, 27, DOI 10.1088/1751-8113/47/1/015203. MR3146977

[3] David Gómez-Ullate, Niky Kamran, and Robert Milson, *An extended class of orthogonal polynomials defined by a Sturm-Liouville problem*, J. Math. Anal. Appl. **359** (2009), no. 1, 352–367, DOI 10.1016/j.jmaa.2009.05.052. MR2542180

[4] William A. Haese-Hill, Martin A. Hallnäs, and Alexander P. Veselov, *Complex exceptional orthogonal polynomials and quasi-invariance*, Lett. Math. Phys. **106** (2016), no. 5, 583–606, DOI 10.1007/s11005-016-0828-8. MR3490949

[5] Alex Kasman, *Glimpses of soliton theory—the algebra and geometry of nonlinear PDEs*, Student Mathematical Library, vol. 100, American Mathematical Society, Providence, RI, [2023] ©2023. Second edition [of 2731261]. MR4566329

[6] Alex Kasman and Robert Milson, *The adelic Grassmannian and exceptional Hermite polynomials*, Math. Phys. Anal. Geom. **23** (2020), no. 4, Paper No. 40, 51, DOI 10.1007/s11040-020-09365-z. MR4173419

[7] D. Kazhdan, B. Kostant, and S. Sternberg, *Hamiltonian group actions and dynamical systems of Calogero type*, Comm. Pure Appl. Math. **31** (1978), no. 4, 481–507, DOI 10.1002/cpa.3160310405. MR478225

[8] Luke Paluso, *Exceptional Hermite polynomials and Calogero-Moser matrices, Master's thesis, College of Charleston*, ProQuest Dissertations Publishing, 2023.

[9] Graeme Segal and George Wilson, *Loop groups and equations of KdV type*, Inst. Hautes Études Sci. Publ. Math. **61** (1985), 5–65. MR783348

[10] George Wilson, *Bispectral commutative ordinary differential operators*, J. Reine Angew. Math. **442** (1993), 177–204, DOI 10.1515/crll.1993.442.177. MR1234841

[11] George Wilson, *Collisions of Calogero-Moser particles and an adelic Grassmannian*, Invent. Math. **133** (1998), no. 1, 1–41, DOI 10.1007/s002220050237. With an appendix by I. G. Macdonald. MR1626461

[12] *Matrix determinant lemma*, https://en.wikipedia.org/wiki/Matrix_determinant_lemma, Jul 2021, Last visited on: 04.18.2023.

Department of Mathematics, College of Charleston, Charleston, South Carolina 29401

Email address: palusol@g.cofc.edu

Department of Mathematics, College of Charleston, Charleston, South Carolina 29401

Email address: kasmana@cofc.edu

Contemporary Mathematics
Volume **822**, 2025
https://doi.org/10.1090/conm/822/16477

Reverse orthogonal polynomials

Steven H. Weintraub

ABSTRACT. In our earlier work, we defined, studied, and determined the reverse Legendre polynomials. We continue that work here by studying reverse orthogonal polynomials more generally. We extend our results on the reverse Legendre polynomials and determine the reverse Chebyshev polynomials of the first and second kinds as well. An ingredient in our proof is a new expression for the (classical) Legendre polynomials and the Chebyshev polynomials of the first and second kinds.

Let $\mathcal{P}$ be the vector space of all polynomials, and let $w(x)$ be a function such that $\langle f(x), g(x)\rangle = \int_{-1}^{1} f(x)g(x)w(x)dx$ defines a (positive definite) inner product on $\mathcal{P}$. We may apply the Gram-Schmidt procedure to the ordered basis $\mathcal{B} = \{1, x, x^2, \ldots\}$ of $\mathcal{P}$, with appropriate normalization, to obtain a family of orthogonal polynomials $\{F_0(x), F_1(x), F_2(x), \ldots\}$. This is classical, and three of the most prominent families of polynomials obtained in this way are (see [**1**, Chapter 22]):
(a) the Legendre polynomials $\{P_n(x)\}$, with $w(x) = 1$ and with normalization $P_n(1) = 1$ for every n,
(b) the Chebyshev polynomials of the first kind $\{T_n(x)\}$, with $w(x) = 1/\sqrt{1-x^2}$ and with normalization $T_n(1) = 1$ for every n,
(c) the Chebyshev polynomials of the second kind $\{U_n(x)\}$, with $w(x) = \sqrt{1-x^2}$ and with normalization $U_n(1) = n + 1$.

Instead, for a nonnegative integer n, let $\mathcal{P}_n$ be the vector space of polynomials of degree at most n, equipped with the same inner product. Let $\mathcal{B} = \{x^n, x^{n-1}, \ldots, 1\}$, an ordered basis of $\mathcal{P}_n$. Let $\{\overleftarrow{F}^n_n(x), \overleftarrow{F}^n_{n-1}(x), \ldots, \overleftarrow{F}^n_0(x)\}$ be the orthogonal basis of $\mathcal{P}_n$ obtained by applying the Gram-Schmidt procedure to $\mathcal{B}$, with the appropriate normalization. We call these polynomials reverse F-polynomials, as we obtain these by applying the Gram-Schmidt procedure in the reverse order. To be precise, we have:
(a) the reverse Legendre polynomials $\{\overleftarrow{P}^n_k(x)\}$, with normalization $\overleftarrow{P}^n_k(1) = 1$,
(b) the reverse Chebyshev polynomials of the first kind $\{\overleftarrow{T}^n_k(x)\}$, with normalization $\overleftarrow{T}^n_k(1) = 1$,

2020 *Mathematics Subject Classification.* 42C05, Secondary 33C45, 05A10, 15A99, 41A55.
Key words and phrases. Reverse orthogonal polynomials, orthogonal polynomials, Gram-Schmidt procedure.

©2025 American Mathematical Society

(c) the reverse Chebyshev polynomials of the second kind $\{\overleftarrow{U}_k^n(x)\}$, with normalization $\overleftarrow{U}_k^n(1) = 2[(n-k)/2] + 2$.

(The normalizations in (a) and (b) are the obvious ones, which turn out to be best. The normalization in (c), while not obvious, also turns out to be best, as we will see below.)

In our earlier work [**5**], we determined the reverse Legendre polynomials $\{\overleftarrow{P}_k^n(x)\}$.

In this work, we consider reverse orthogonal polynomials more generally. We extend our earlier work on the reverse Legendre polynomials and we determine the reverse Chebyshev polynomials of the first and second kinds as well.

We proceed as follows.

In Section 1 we give results on $\{\overleftarrow{F}_k^n(x)\}$ for a general even "weight" function $w(x)$.

In Section 2 we give new expressions for the classical polynomials $\{P_n(x)\}$, $\{T_n(x)\}$, and $\{U_n(x)\}$.

In Section 3 we give our main results, the determination of the reverse polynomials $\{\overleftarrow{P}_k^n(x)\}$, $\{\overleftarrow{T}_k^n(x)\}$, and $\{\overleftarrow{U}_k^n(x)\}$.

In [**5**] we gave tables of values of $\{\overleftarrow{P}_k^n(x)\}$. In Section 4 we give tables of values of $\{\overleftarrow{T}_k^n(x)\}$ and $\{\overleftarrow{U}_k^n(x)\}$.

1. General results

We assume henceforth that $w(x)$ is an even function. We let $\{c_k^n\}$ be an arbitrary set of nonzero real numbers and we normalize the functions $\{\overleftarrow{F}_k^n(x)\}$ by setting $\overleftarrow{F}_k^n(1) = c_k^n$.

THEOREM 1.1. *Let m, n and k be nonnegative integers with $k \le n$.*

(a) *$\overleftarrow{F}_k^n(x)$ is a polynomial of degree at most n whose low-order term is a nonzero multiple of x^k.*

(b) *$\overleftarrow{F}_k^n(x)$ is an even polynomial if k is even and an odd polynomial if k is odd.*

(c) $\overleftarrow{F}_k^{2m+1+k}(x) = (c_k^{2m+1+k}/c_k^{2m+k})\overleftarrow{F}_k^{2m+k}(x)$.

(d) *$\overleftarrow{F}_k^{2m+k}(x)$ is a polynomial of degree $2m+k$, with a zero of order k at $x=0$ and $2m$ simple zeroes at nonzero values of x, all of them real numbers symmetrically located around the origin and lying in the open interval $(-1,1)$.*

(e) *$\overleftarrow{F}_k^n(x)$ is a polynomial of degree n if $n-k$ is even and of degree $n-1$ if $n-k$ is odd.*

(f) *The polynomial $\overleftarrow{F}_k^n(x)$ is uniquely determined by condition (a) above and by the conditions that $\langle \overleftarrow{F}_j^n(x), \overleftarrow{F}_k^n(x)\rangle = 0$ for $j > k$ and that $\overleftarrow{F}_k^n(1) = c_k^n$.*

PROOF. The proof of [**5**, Theorem 1.3] goes through almost unchanged to show this general result. □

REMARK 1.2. Since $\mathrm{Span}(\{x^{k+1}, \dots, x^n\}) = \mathrm{Span}(\{\overleftarrow{F}_{k+1}^n(x), \dots, \overleftarrow{F}_n^n(x)\})$, a polynomial is orthogonal to all polynomials in the first of these sets if and only if it is orthogonal to all the polynomials in the second of these sets. Thus we can replace $\overleftarrow{F}_j^n(x)$ by x^j in Theorem 1.1(f). Also, since $\overleftarrow{F}_j^n(x)$, or x^j, has opposite parity to $\overleftarrow{F}_k^n(x)$ for $j \not\equiv k \pmod 2$, we can restrict our attention to the case $j \equiv k \pmod 2$ there.

There are some cases in which we can readily determine $\overleftarrow{F}_k^n(x)$. Note that in the statement of this theorem, $F_{2m+1}(x)$ is an "ordinary" F-polynomial.

THEOREM 1.3.

(a) $\overleftarrow{F}_n^n(x) = c_n^n x^n$ *for all* $n \geq 0$.

(b) $\overleftarrow{F}_0^{2m}(x) = (c_0^{2m}/F_{2m+1}(1))F_{2m+1}(x)/x$ *for all* $m \geq 0$.

PROOF. The proof of [**5**, Theorem 1.4 (a) and (d)] goes through almost unchanged to show this general result. □

It is well-known (see [**1**, Chapter 22]) that orthogonal polynomials satisfy a three-term recurrence relation. We have a similar result for reverse orthogonal polynomials, which plays a key role.

THEOREM 1.4. *For any integer* $n \geq 2$ *and any integer* k *with* $0 \leq k \leq n-2$, *there are unique real numbers* α_k^n *and* β_k^n *such that*

$$\overleftarrow{F}_k^n(x) = \alpha_k^n \overleftarrow{F}_k^{n-2}(x) + \beta_k^n x \overleftarrow{F}_{k+1}^{n-1}(x).$$

PROOF. From Theorem 1.1(c) we see that it suffices to prove this in the case $k \equiv n \pmod 2$, so we make that restriction.

We note that the low-order terms of $\overleftarrow{F}_k^n(x)$ and $\overleftarrow{F}_k^{n-2}(x)$ are of degree k while the low-order term of $x\overleftarrow{F}_{k+1}^{n-1}(x)$ is of degree $k+2$, and the high-order terms of $\overleftarrow{F}_k^n(x)$ and $x\overleftarrow{F}_{k+1}^{n-1}(x)$ are of degree n while the high-order term of $\overleftarrow{F}_k^{n-2}(x)$ is of degree $n-2$. Hence if such recursion holds, we must have

$$\alpha_k^n = \frac{\text{trailing coefficient of } \overleftarrow{F}_k^n(x)}{\text{trailing coefficient of } \overleftarrow{F}_k^{n-2}(x)},$$

$$\beta_k^n = \frac{\text{leading coefficient of } \overleftarrow{F}_k^n(x)}{\text{leading coefficient of } \overleftarrow{F}_{k+1}^{n-1}(x)} x.$$

We set α_k^n and β_k^n equal to these values. Then we see that

$$\overleftarrow{F}_k^n(x) = \alpha_k^n \overleftarrow{F}_k^{n-2}(x) + \beta_k^n x \overleftarrow{F}_{k+1}^{n-1}(x) + g(x)$$

where $g(x)$ is a polynomial in x whose terms have degree x^j for $j \in J = \{k+2 \leq j \leq n-2,\ j \equiv k \pmod 2\}$. Now any such power of x is in $\mathrm{Span}(\{\overleftarrow{F}_j^{n-2}(x) \mid j \in J\})$. Thus we may rewrite the above equation as

$$\overleftarrow{F}_k^n(x) = \alpha_k^n \overleftarrow{F}_k^{n-2}(x) + \beta_k^n x \overleftarrow{F}_{k+1}^{n-1}(x) + \sum_{j \in J} \gamma_j \overleftarrow{F}_j^{n-2}(x)$$

for some constants γ_j.

Let j_0 be any element of J. Then

$$\begin{aligned}
&\langle \overleftarrow{F}^{n-2}_{j_0}(x), \overleftarrow{F}^n_k(x)\rangle \\
&= \langle \overleftarrow{F}^{n-2}_{j_0}(x), \alpha^n_k \overleftarrow{F}^{n-2}_k(x)\rangle + \langle \overleftarrow{F}^{n-2}_{j_0}(x), \beta^n_k x \overleftarrow{F}^{n-1}_{k+1}(x)\rangle + \langle \overleftarrow{F}^{n-2}_{j_0}(x), \sum_{j\in J} \gamma_j \overleftarrow{F}^{n-2}_j(x)\rangle \\
&= \alpha^n_k \langle \overleftarrow{F}^{n-2}_{j_0}(x), \overleftarrow{F}^{n-2}_k(x)\rangle + \beta^n_k \langle \overleftarrow{F}^{n-2}_{j_0}(x), x\overleftarrow{F}^{n-1}_{k+1}(x)\rangle + \sum_{j\in J} \gamma_j \langle \overleftarrow{F}^{n-2}_{j_0}(x), \overleftarrow{F}^{n-2}_j(x)\rangle \\
&= \alpha^n_k \langle \overleftarrow{F}^{n-2}_{j_0}(x), \overleftarrow{F}^{n-2}_k(x)\rangle + \beta^n_k \langle x\overleftarrow{F}^{n-2}_{j_0}(x), \overleftarrow{F}^{n-1}_{k+1}(x)\rangle + \sum_{j\in J} \gamma_j \langle \overleftarrow{F}^{n-2}_{j_0}(x), \overleftarrow{F}^{n-2}_j(x)\rangle.
\end{aligned}$$

Let us examine this sum from left to right. As for the term on the left-hand side, we see that $\overleftarrow{F}^{n-2}_{j_0}(x) \in \operatorname{Span}(\{x^j \mid j \in J\}) = \operatorname{Span}(\{\overleftarrow{F}^{n-2}_j(x) \mid j \in J\})$ so the value of this term is 0, by orthogonality. The value of the first term on the right-hand side is 0, by orthogonality. As for the next term on the right-hand side, we see that $x\overleftarrow{F}^{n-2}_{j_0}(x) \in \operatorname{Span}(\{x^{j+1} \mid j \in J\}) = \operatorname{Span}(\{\overleftarrow{F}^{n-1}_{j+1}(x) \mid j \in J\})$ so the value of this term is 0, by orthogonality. As for the last term on the right-hand side, we see that its value is $\gamma_{j_0}\langle \overleftarrow{F}^{n-2}_{j_0}(x), \overleftarrow{F}^{n-2}_{j_0}(x)\rangle$, also by orthogonality. Thus we conclude that $\gamma_{j_0} = 0$, so $g(x) = 0$ and we have established our claimed recursion. □

We can give an alternate expression for α^n_k and β^n_k.

COROLLARY 1.5. *Let α^n_k and β^n_k be as in Theorem 1.4. Then they are determined by the linear system*

$$\begin{aligned}
\alpha^n_k \langle x^n, \overleftarrow{F}^{n-2}_k(x)\rangle + \beta^n_k \langle x^{n+1}, \overleftarrow{F}^{n-1}_{k+1}(x)\rangle &= 0 \\
\alpha^n_k \overleftarrow{F}^{n-2}_k(1) \qquad + \beta^n_k \overleftarrow{F}^{n-1}_{k+1}(1) \qquad &= \overleftarrow{F}^n_k(1)
\end{aligned}$$

PROOF. From Theorem 1.1(c) we see that it suffices to prove this in the case $k \equiv n \pmod 2$, so we make that restriction.

We prove this result by using the fact that $\{\overleftarrow{F}^n_k(x)\}$ satisfy the conditions of Theorem 1.1(f).

By Remark 1.2, it suffices to consider the cases $j \in J$, with J as above, and the case $j = n$.

First suppose that $j = n$. Note that $\overleftarrow{F}^n_n(x)$ is a multiple of x^n so we may instead consider x^n. Then we obtain

$$\begin{aligned}
0 = \langle x^n, \overleftarrow{F}^n_k(x)\rangle &= \langle x^n, \alpha^n_k \overleftarrow{F}^{n-2}_k(x) + \beta^n_k x\overleftarrow{F}^{n-1}_{k+1}(x)\rangle \\
&= \alpha^n_k \langle x^n, \overleftarrow{F}^{n-2}_k(x)\rangle + \beta^n_k \langle x^n, x\overleftarrow{F}^{n-1}_{k+1}(x)\rangle \\
&= \alpha^n_k \langle x^n, \overleftarrow{F}^{n-2}_k(x)\rangle + \beta^n_k \langle x^{n+1}, \overleftarrow{F}^{n-1}_{k+1}(x)\rangle.
\end{aligned}$$

Next suppose that $j \in J$, so that $j \leq n-2$. Note that $\overleftarrow{F}^n_j(x) - cx^n + g(x)$ for some constant c and some polynomial $g(x)$ whose terms have degrees between

j and $n-2$. We then have the equation

$$\begin{aligned}\langle \overleftarrow{F}_j^n(x), \overleftarrow{F}_k^n(x)\rangle &= \langle cx^n + g(x), \overleftarrow{F}_k^n(x)\rangle = c\langle x^n, \overleftarrow{F}_k^n(x)\rangle + \langle g(x), \overleftarrow{F}_k^n(x)\rangle \\ &= \langle g(x), \overleftarrow{F}_k^n(x)\rangle \\ &= \langle g(x), \alpha_k^n \overleftarrow{F}_k^{n-2}(x) + \beta_k^n x \overleftarrow{F}_{k+1}^{n-1}(x)\rangle \\ &= \alpha_k^n \langle g(x), \overleftarrow{F}_k^{n-2}(x)\rangle + \beta_k^n \langle g(x), x\overleftarrow{F}_{k+1}^{n-1}(x)\rangle \\ &= \alpha_k^n \langle g(x), \overleftarrow{F}_k^{n-2}(x)\rangle + \beta_k^n \langle xg(x), \overleftarrow{F}_{k+1}^{n-1}(x)\rangle = 0 + 0 = 0\end{aligned}$$

by orthogonality (and hence none of these equations imposes any additional conditions on α_k^n or β_k^n).

Also, setting $x=1$ we obtain the equation

$$\overleftarrow{F}_k^n(1) = \alpha_k^n \overleftarrow{F}_k^{n-2}(1) + \beta_k^n \overleftarrow{F}_{k+1}^{n-1}(1).$$

Thus we have a system of two linear equations in the two unknowns α_k^n and β_k^n. While a priori such a system may not have a unique solution, in this case we know that it does, as we know that α_k^n and β_k^n exist and are unique by Theorem 1.4. □

THEOREM 1.6. *Each of the following sets determines the others:*

(a) $\{\overleftarrow{F}_k^n(x)\}$ *for all* n, k.

(b1) $\{\alpha_k^n\}$ *and* $\{\beta_k^n\}$ *for all* n, k, *and* $\{\overleftarrow{F}_n^n(x)\}$ *for all* n.

(b2) $\{\alpha_k^n\}$ *and* $\{\beta_k^n\}$ *for all* n, k, *and* $\{\overleftarrow{F}_0^n(x)\}$ *for all* n.

PROOF. The proof of Theorem 1.4 shows that $\{\overleftarrow{F}_k^n(x)\}$ for all n,k determines $\{\alpha_k^n\}$ and $\{\beta_k^n\}$ for all n,k.

By induction on $n-k$, Theorem 1.4, solving for the left-hand side in terms of the right-hand side, shows that $\{\alpha_k^n\}$ and $\{\beta_k^n\}$ for all n,k, and $\{\overleftarrow{F}_n^n(x)\}$ for all n, determines $\{\overleftarrow{F}_k^n(x)\}$ for all n,k.

By induction on k, Theorem 1.4, solving for the second term on the right-hand in terms of the left-hand side and the first term on the right-hand side, shows that $\{\alpha_k^n\}$ and $\{\beta_k^n\}$ for all n,k, and $\{\overleftarrow{F}_0^n(x)\}$ for all n, determines $\{\overleftarrow{F}_k^n(x)\}$ for all n,k. □

2. Legendre and Chebyshev polynomials

We recall the following explicit formulas for the Legendre and Chebyshev polynomials ([**1**, (22.3.6), (22.3.7), (22.3.8)]).

THEOREM 2.1. *Let* n *be a nonnegative integer and let* $m=[n/2]$.
(a) The Legendre polynomial $P_n(x)$ *is given by*

$$P_n(x) = \frac{1}{2^n}\sum_{p=0}^{m}(-1)^p\binom{2n-2p}{n}\binom{n}{p}x^{n-2p}$$

(b) The Chebyshev polynomial of the first kind $T_n(x)$ *is given by*

$$T_n(x) = \frac{n}{2}\sum_{p=0}^{m}(-1)^p 2^{n-2p}\frac{1}{n-p}\binom{n-p}{p}x^{n-2p}$$

(b) The Chebyshev polynomial of the second kind $U_n(x)$ is given by

$$U_n(x) = \sum_{p=0}^{m} (-1)^p 2^{n-2p} \binom{n-p}{p} x^{n-2p}$$

We have the following apparently new alternate explicit formulas for these polynomials.

For an integer t, we denote the product of m consecutive integers, the largest of which is t, by $\Gamma_m(t)$, and for an odd integer t, we denote the product of m consecutive odd integers, the largest of which is t, by $\Pi_m(t)$, i.e.,

$$\Gamma_0(t) = 1 \quad \text{and} \quad \Gamma_m(t) = \prod_{r=1}^{m} (t - (r-1)) \quad \text{for} \quad m > 0,$$

$$\Pi_0(t) = 1 \quad \text{and} \quad \Pi_m(t) = \prod_{r=1}^{m} (t - 2(r-1)) \quad \text{for} \quad m > 0.$$

THEOREM 2.2. *Let n be a nonnegative integer and let $m = [n/2]$.*
(a) The Legendre polynomial $P_n(x)$ is given by

$$P_n(x) = \frac{1}{2^m} \frac{1}{m!} \sum_{p=0}^{m} (-1)^p \binom{m}{p} \Pi_m(2n-1-2p) x^{2n-2p}$$

(a) The Chebyshev polynomial of the first kind $T_n(x)$ is given by

$$T_n(x) = \frac{1}{\Pi_m(2m-1)} \sum_{p=0}^{m} (-1)^p 2^{m-p} \binom{m}{p} \Gamma_{m-p}(n-1-p) \Pi_p(2n-2m-1) x^{n-2p}$$

(b) The Chebyshev polynomial of the second kind $U_n(x)$ is given by

$$U_n(x) = \frac{n+1}{\Pi_{m+1}(2m+1)} \sum_{p=0}^{m} (-1)^p 2^{m-p} \binom{m}{p} \Gamma_{m-p}(n-p) \Pi_p(2n-2m-1) x^{n-2p}$$

PROOF. (a) It is straightforward to check that

$$\frac{1}{2^m} \frac{1}{m!} \binom{m}{p} \Pi_m(2n-1-2p) = \frac{1}{2^n} \binom{2n-2p}{n} \binom{n}{p}$$

(b) It is straightforward to check that

$$\frac{1}{\Pi_m(2m-1)} 2^{m-p} \binom{m}{p} \Gamma_{m-p}(n-1-p) \Pi_p(2n-2m-1) = \frac{n}{2} 2^{n-2p} \frac{1}{n-p} \binom{n-p}{p}$$

(c) It is straightforward to check that

$$\frac{n+1}{\Pi_{m+1}(2m+1)} 2^{m-p} \binom{m}{p} \Gamma_{m-p}(n-p) \Pi_p(2n-2m-1) = 2^{n-2p} \binom{n-p}{p}$$

□

3. Reverse Legendre and Chebyshev polynomials

In [**5**] we determined the reverse Legendre polynomials. The proof we gave there does not generalize to the reverse Chebyshev polynomials. Here we give a unified proof that handles all these cases. This also gives a different (of course equivalent, though not obviously so) expression for the reverse Legendre polynomials than we obtained in [**5**].

LEMMA 3.1.

(a) *For any integer* $n \geq 2$ *and any integer* k *with* $0 \leq k \leq n-2$, $k \equiv n \pmod 2$

$$\overleftarrow{P}^n_k(x) = \alpha^n_k \overleftarrow{P}^{n-2}_k(x) + \beta^n_k x \overleftarrow{P}^{n-1}_{k+1}(x).$$

for

$$\alpha^n_k = -\frac{n+k+1}{n-k} \quad \text{and} \quad \beta^n_k = \frac{2n+1}{n-k}.$$

(b) *For any integer* $n \geq 2$ *and any integer* k *with* $0 \leq k \leq n-2$, $k \equiv n \pmod 2$,

$$\overleftarrow{T}^n_k(x) = \alpha^n_k \overleftarrow{T}^{n-2}_k(x) + \beta^n_k x \overleftarrow{T}^{n-1}_{k+1}(x).$$

for

$$\alpha^n_k = -\frac{n+k+1}{n-k-1} \quad \text{and} \quad \beta^n_k = \frac{2n}{n-k-1}.$$

(c) *For any integer* $n \geq 2$ *and any integer* k *with* $0 \leq k \leq n-2$, $k \equiv n \pmod 2$

$$\overleftarrow{U}^n_k(x) = \alpha^n_k \overleftarrow{U}^{n-2}_k(x) + \beta^n_k x \overleftarrow{U}^{n-1}_{k+1}(x).$$

for

$$\alpha^n_k = -\frac{(n-k+2)(n+k+1)}{(n-k)(n-k+1)} \quad \text{and} \quad \beta^n_k = \frac{2(n-k+2)(n+1)}{(n-k)(n-k+1)}.$$

THEOREM 3.2.

(a) *For any integer* $n \geq 0$ *and any integer* k *with* $0 \leq k \leq n$, $k \equiv n \pmod 2$, *let* $m = (n-k)/2$. *Then the reverse Legendre polynomial* $\overleftarrow{P}^n_k(x)$ *is given by*

$$\overleftarrow{P}^n_k(x) = \frac{1}{2^m}\frac{1}{m!}\left(\sum_{p=0}^{m}(-1)^p \binom{m}{p} \Pi_m(2n-2p+1)x^{n-2p}\right)$$

(b) *For any integer* $n \geq 0$ *and any integer* k *with* $0 \leq k \leq n$, $k \equiv n \pmod 2$, *let* $m = (n-k)/2$. *Then the reverse Chebyshev polynomial of the first kind* $\overleftarrow{T}^n_k(x)$ *is given by*

$$\overleftarrow{T}^n_k(x) = \frac{1}{\Pi_m(2m-1)}\left(\sum_{p=0}^{m}(-1)^p 2^{m-p}\binom{m}{p}\Gamma_{m-p}(n-p)\Pi_p(2n-2m+1)x^{n-2p}\right)$$

(c) *For any integer* $n \geq 2$ *and any integer* k *with* $0 \leq k \leq n$, $k \equiv n \pmod 2$, *let* $m = (n-k)/2$. *Then the reverse Chebyshev polynomial of the second kind* $\overleftarrow{U}^n_k(x)$ *is given by*

$$\overleftarrow{U}^n_k(x)$$

$$= \frac{2(m+1)}{\Pi_{m+1}(2m+1)}\left(\sum_{p=0}^{m}(-1)^p 2^{m-p}\binom{m}{p}\Gamma_{m-p}(n+1-p)\Pi_p(2n-2m+1)x^{n-2p}\right)$$

PROOF. Theorem 3.2 is a consequence of Lemma 3.1 via a straightforward inductive computation begining with the values $\overleftarrow{P}^n_n(x) = x^n$, $\overleftarrow{T}^n_n(x) = x^n$ and $\overleftarrow{U}^n_n(x) = 2x^n$ given by Theorem 1.3(a). But we see from Theorem 1.6 that in fact Theorem 3.2 and Lemma 3.1 are equivalent, given those values and also the values $\overleftarrow{P}^{2m+1}_0(x) = \overleftarrow{P}^{2m}_0(x) = P_{2m+1}(x)/x$, $\overleftarrow{T}^{2m+1}_0(x) = \overleftarrow{T}^{2m}_0(x) = T_{2m+1}(x)/x$, $\overleftarrow{U}^{2m+1}_0(x) = \overleftarrow{U}^{2m}_0(x) = U_{2m+1}(x)/x$, given by Theorem 1.1(c) and Theorem 1.3(b). Thus we shall prove Lemma 3.1 and Theorem 3.2 simultaneously.

Let $\{\overleftarrow{G}^n_k(x)\}$ be the family of polynomials given by any of the recursions in Lemma 3.1. It is easy to check that $\overleftarrow{G}^n_k(x)$ satisfies the condition of Theorem 1.1(a). We show that $\overleftarrow{G}^n_k(x)$ satisfies the conditions of Theorem 1.1(f) as well, so we may conclude that $\overleftarrow{G}^n_k(x) = \overleftarrow{F}^n_k(x) = \overleftarrow{P}^n_k(X)$, $\overleftarrow{T}^n_k(x)$, or $\overleftarrow{U}^n_k(x)$ as the case may be.

There are two conditions in Theorem 1.1(f). The first condition is $\langle \overleftarrow{G}^n_j(x), \overleftarrow{G}$-${}^n_k(x)\rangle = 0$ for $j > k$. We prove this by induction on k.

For $k = 0$, we observe that the formula for $\overleftarrow{G}^n_0(x)$ in Theorem 3.2 agrees with the formula obtained for $\overleftarrow{P}^n_0(x)$, $\overleftarrow{T}^n_0(x)$, or $\overleftarrow{U}^n_0(x)$ from Theorem 2.2 and Theorem 1.3(b).

For the inductive step, assume that $\overleftarrow{G}^m_k(x)$ is orthogonal to $\{x^{k+1}, \ldots, x^m\}$ for any m (see Remark 1.2). We must show that $\overleftarrow{G}^m_{k+1}(x)$ is orthogonal to $\{x^{k+2}, \ldots, x^m\}$. Let j be any integer between $k+2$ and $m = n-1$. We then have

$$\begin{aligned}\beta^n_k \langle x^j, \overleftarrow{G}^{n-1}_{k+1}(x)\rangle &= \beta^n_k \langle x^{j-1}, x\overleftarrow{G}^{n-1}_{k+1}(x)\rangle \\ &= \langle x^{j-1}, \overleftarrow{G}^n_k(x)\rangle - \alpha^n_k \langle x^{j-1}, \overleftarrow{G}^{n-2}_k(x)\rangle = 0 + 0 = 0\end{aligned}$$

as required.

The second condition in Theorem 1.1(f) is $\overleftarrow{G}^n_k(1) = c^n_k$. We prove this by induction on $n-k$. For $n-k=0$ we immediately see from Theorem 3.2 that $\overleftarrow{G}^n_n(1)$ has the correct value (1, 1, or 2 in each of the three cases). Then from Lemma 3.1 we obtain the three equations for $n-k$ even

$$-\frac{n+k+1}{n-k}1 + \frac{2n+1}{n-k}1 = 1$$

$$-\frac{n+k+1}{n-k-1}1 + \frac{2n}{n-k-1}1 = 1$$

$$-\frac{(n-k+2)(n+k+1)}{(n-k)(n-k+1)}(n-k) + \frac{2(n-k+2)(n+1)}{(n-k)(n-k+1)}(n-k) = n-k+2$$

as required, and this also gives the value of $\overleftarrow{G}^n_k(1)$ for $n-k$ odd, by Theorem 1.1(c). □

REMARK 3.3. This proof may at first glance appear to be a bit of a sleight of hand, since we used almost nothing about α^n_k and β^n_k in the first inductive step.

But note that in fact we had to use all of these values to begin with $\overleftarrow{G}^n_n(x)$ and obtain from that the values of $\overleftarrow{G}^n_0(x)$, which were necessary to start the induction.

We conjectured the following result in [**5**], and it was first proved by Amdeberhan and Moll in [**2**]. Our work here gives a very direct proof of it.

COROLLARY 3.4. *For any integer $n \geq 0$ and for any integer k with $0 \leq k \leq n$ and $k \equiv n \pmod 2$, $2^{n-k}\overleftarrow{P}^n_k(x)$ is a polynomial with integer coefficients.*

PROOF. It is an elementary fact that for any integer $m \geq 0$ and any odd integer t, $2^m\Pi_m(t)$ is divisible by $m!$. □

There are a number of identities relating the Chebyshev polynomials of the first and second kinds. We have ([**1**, (22.5.6), (22.5.7), (22.5.8), (22.5.10)]): $T_n(x) = U_n(x) - xU_{n-1}(x)$, $T_n(x) = xU_{n-1}(x) - U_{n-2}(x)$, $T_n(x) = (1/2)(U_n(x) - U_{n-2}(x))$, $(1-x^2)U_{n-1}(x) = xT_n(x) - T_{n+1}(x)$. We have analogous identities relating the reverse Chebyshev polynomials of the first and second kinds, with the first three having direct analogues while the analogue for the fourth is more involved.

COROLLARY 3.5. *For any integer $n \geq 2$ and for any integer k with $0 \leq k \leq n-2$ and $k \equiv n \pmod 2$:*

$$(a) \qquad \overleftarrow{T}^n_k(x) = \frac{n-k+1}{n-k+2}\overleftarrow{U}^n_k(x) - x\overleftarrow{U}^{n-1}_{k+1}(x)$$

$$(b) \qquad \overleftarrow{T}^n_k(x) = \frac{1}{n-k}\left((n+k+2)x\overleftarrow{U}^{n-1}_{k+1}(x) - (n+k+1)\overleftarrow{U}^{n-2}_k(x)\right)$$

$$(c) \qquad \overleftarrow{T}^n_k(x) = \frac{1}{2(n+1)}\left(\frac{(n-k+1)(n+k+2)}{n-k+2}\overleftarrow{U}^n_k(x) - (n+k+1)\overleftarrow{U}^{n-2}_k(x)\right)$$

For any integer $n \geq 3$ and for any integer k with $1 \leq k \leq n-2$ and $k \equiv n \pmod 2$:

$$(d) \qquad (n-k+1)(n+k+1)\overleftarrow{U}^{n-1}_{k-1}(x) - (n-k+2)(n+k+2)x^2\overleftarrow{U}^{n-1}_{k+1}(x) = (n-k+2)(n+k+2)x\overleftarrow{T}^n_k(x) - (n-k+1)(n-k+2)\overleftarrow{T}^{n+1}_{k-1}(x)$$

PROOF. (a) follows from Theorem 3.2(b) and (c) by direct calculation, with the final step being the identity, valid for any m and p with $0 \leq p \leq m-1$,

$$\binom{m}{p}(m-p) = m\binom{m-1}{p},$$

and then (b) and (c) follow from (a) and Lemma 3.1(c) by direct calculation. Now if we multiply (a) by x and take (b) for $\overleftarrow{T}^{n+1}_{k-1}$, we obtain two equations, each of which has a term in $x\overleftarrow{U}^n_k$. Taking a linear combination of these two equations that eliminates that term, and clearing denominators, we obtain (d). □

4. Tables of values

Table of $\overleftarrow{T}{}^n_k(x)$

	0	1	2	3	4	5	6
6							x^6
5						x^5	x^5
4					x^4	x^4	$12x^6 - 11x^4$
3				x^3	x^3	$10x^5 - 9x^3$	$10x^5 - 9x^3$
2			x^2	x^2	$8x^4 - 7x^2$	$8x^4 - 7x^2$	$40x^6 - 60x^4 + 21x^2$
1		x	x	$6x^3 - 5x$	$6x^3 - 5x$	$\frac{1}{3}(80x^5 - 112x^3 + 35x)$	$\frac{1}{3}(80x^5 - 112x^3 + 35x)$
0	1	1	$4x^2 - 3$	$4x^2 - 3$	$16x^4 - 20x^2 + 5$	$16x^4 - 20x^2 + 5$	$64x^6 - 112x^4 + 56x^2 - 7$

	7	8
8		x^8
7	x^7	x^7
6	x^6	$16x^8 - 15x^6$
5	$14x^7 - 13x^5$	$14x^7 - 13x^5$
4	$12x^6 - 11x^4$	$\frac{1}{3}(224x^8 - 364x^6 + 143x^4)$
3	$56x^7 - 88x^5 + 33x^3$	$56x^7 - 88x^5 + 33x^3$
2	$40x^6 - 60x^4 + 21^2$	$\frac{1}{5}(896x^8 - 1848x^6 + 1188x^4 - 231x^2)$
1	$112x^7 - 216x^5 + 126x^3 - 21x$	$112x^7 - 216x^5 + 126x^3 - 21x$
0	$64x^6 - 112x^4 + 56x^2 - 7$	$256x^8 - 576x^6 + 432x^4 - 120x^2 + 9$

Table of $\overleftarrow{U}{}^n_k(x)$

	0	1	2	3	4	5	6
6							$2x^6$
5						$2x^5$	$2x^5$
4					$2x^4$	$2x^4$	$\frac{4}{3}(14x^6 - 11x^4)$
3				$2x^3$	$2x^3$	$16x^5 - 12x^3$	$16x^5 - 12x^3$
2			$2x^2$	$2x^2$	$\frac{4}{3}(10x^4 - 7x^2)$	$\frac{4}{3}(10x^4 - 7x^2)$	$\frac{1}{5}(336x^6 - 216x^4 + 63x^2)$
1		$2x$	$2x$	$\frac{4}{3}(8x^3 - 5x)$	$\frac{4}{3}(8x^3 - 5x)$	$48x^5 - 56x^3 + 14x$	$48x^5 - 56x^3 + 14x$
0	2	2	$8x^2 - 4$	$8x^2 - 4$	$32x^4 - 32x^2 + 6$	$32x^4 - 32x^2 + 6$	$128x^6 - 192x^4 + 80x^2 - 8$

	7	8
8		$2x^8$
7	$2x^7$	$2x^7$
6	$2x^6$	$24x^8 - 20x^6$
5	$\frac{4}{3}(16x^7 - 13x^5)$	$\frac{4}{3}(16x^7 - 13x^5)$
4	$\frac{4}{3}(14x^6 - 11x^4)$	$\frac{1}{5}(576x^8 - 832x^6 + 286x^4)$
3	$\frac{1}{5}(448x^7 - 616x^5 + 198x^3)$	$\frac{1}{5}(448x^7 - 616x^5 + 198x^3)$
2	$\frac{1}{5}(336x^6 - 216x^4 + 63x^2)$	$\frac{1}{5}(1536x^8 - 2816x^6 + 1584x^4 - 264x^2)$
1	$\frac{1}{5}(1024x^7 - 1728x^5 + 864x^3 - 120x)$	$\frac{1}{5}(1024x^7 - 1728x^5 + 864x^3 - 120x)$
0	$128x^6 - 192x^4 + 80x^2 - 8$	$512x^8 - 1024x^6 + 672x^4 - 160x^2 + 10$

REMARK 4.1. The coefficients of $\overleftarrow{P}{}^n_k(x)$, $\overleftarrow{T}{}^n_k(x)$, and $\overleftarrow{U}{}^n_k(x)$ are rational numbers. We see from the above that, on the one hand, the coefficients of $\overleftarrow{P}{}^n_k(x)$ have denominators a power of 2, while, one the other hand, the coefficients of $\overleftarrow{T}{}^n_k(x)$ and $\overleftarrow{U}{}^n_k(x)$ have denominators odd integers.

5. Application to quadrature

We gave an application to quadrature of the reverse Legendre polynomials in Section 5 of [**5**]. Theorem 5.1 of [**5**] generalizes to the reverse Chebyshev polynomials (indeed, generalizes further) with no essential change to either its statement or proof, so we shall not restate it here. But we note that Chelyshkov considered polynomials more general than ours in [**3**] and [**4**], though his goal was not to find explicit expressions for them but rather to apply them to questions in quadrature.

Acknowledgments

We thank the referee for calling our attention to Chelyshkov's work.

References

[1] Milton Abramowitz and Irene A. Stegun, *Handbook of mathematical functions with formulas, graphs, and mathematical tables*, National Bureau of Standards Applied Mathematics Series, No. 55, U. S. Government Printing Office, Washington, DC, 1964. For sale by the Superintendent of Documents. MR167642

[2] T. Amdeberhan, and V. Moll, *Integers associated to the reverse Legendre polynomials*, to appear.

[3] Vladimir S. Chelyshkov, *Alternative orthogonal polynomials and quadratures*, Electron. Trans. Numer. Anal. **25** (2006), 17–26. MR2280360

[4] V. S. Chelyshkov, *Alternative Jacobi polynomials and orthogonal exponentials*, arXiv:1105.1838, 2011.

[5] Steven H. Weintraub, *Reverse Legendre polynomials*, Arch. Math. (Basel) **118** (2022), no. 6, 593–604, DOI 10.1007/s00013-022-01740-2. MR4423453

Department of Mathematics, Lehigh University, Bethlehem, Pennsylvania 18015
Email address: shw2@lehigh.edu

Contemporary Mathematics
Volume **822**, 2025
https://doi.org/10.1090/conm/822/16478

The integrality of reverse Legendre polynomials

Tewodros Amdeberhan and Victor H. Moll

ABSTRACT. In the study of reverse Legendre polynomials, certain coefficients are conjectured to be integers. We apply generating functions to confirm this claim.

1. Introduction

The classical Legendre polynomials $P_n(x)$ appear from the Gram-Schmidt orthogonalization process applied to the ordered basis $\mathcal{B} = \{1,\, x,\, x^2,\, \ldots\}$ with the usual inner product ($\langle f, g\rangle$ is the integral over $[-1,1]$) subject to the normalization $P_n(1) = 1$. In recent work S. Weintraub **[2]** constructs a family of polynomials applying the same orthogonalization procedure and normalization, but now starting with the ordered basis $\overline{\mathcal{B}} = \{x^n,\, x^{n-1},\, \ldots,\, x, 1\}$. These are the reverse Legendre polynomials. Several properties of this family are established in **[2]**. The author conjectures that the expression

$$\alpha(m,k,j) = 2\binom{m}{j} \times \beta(m,k,j) \tag{1.1}$$

where

$$\beta(m,k,j) = \frac{\binom{2m+2k+2j+1}{2m}\binom{2m}{m}}{\binom{m+k+j}{m}} \tag{1.2}$$

is always an integer. This was presented as a problem in the special session on Orthogonal Polynomials and their Applications.

The goal of this short note is to prove a slightly stronger result:

THEOREM 1.1. *The expression $\beta(m,k,j)$ is an integer.*

2. The proof

Make a notational simplicity by writing $\ell = k + j$ to obtain

$$\gamma(m,\ell) := \beta(m,k,j) = \frac{\binom{2m+2\ell+1}{2m}\binom{2m}{m}}{\binom{m+\ell}{m}}. \tag{2.1}$$

The power series coming from the identity

$$(1-4x)^{-(\ell+1)} \times (1-4x)^{-1/2} = (1-4x)^{-(\ell+3/2)} \tag{2.2}$$

2020 *Mathematics Subject Classification.* Primary 33C45, 11A99.

©2025 American Mathematical Society

yields

$$\left[\sum_{a=0}^{\infty} 4^a \binom{a+\ell}{a} x^a\right] \times \left[\sum_{b=0}^{\infty} \binom{2b}{b} x^b\right] = \sum_{m=0}^{\infty} 4^m \binom{m+\ell+\frac{1}{2}}{m} x^m. \tag{2.3}$$

Cauchy's formula for the product of the two series on the left and matching corresponding powers of x give the identity

$$\sum_{i=0}^{m} 4^i \binom{i+\ell}{i}\binom{2m-2i}{m-i} = 4^m \binom{m+\ell+\frac{1}{2}}{m}. \tag{2.4}$$

The integrality of $\gamma(m,\ell)$ follows once we show

$$\gamma(m,\ell) = 4^m \binom{m+\ell+\frac{1}{2}}{m}. \tag{2.5}$$

Since the left-hand side of (2.4) is clearly an integer, Theorem 1.1 is established.

The proof of (2.5) follows directly by elementary manipulation or by writing factorials in terms of the gamma function $\Gamma(x)$ and using the duplication formula [**1**, 8.335.1]

$$\Gamma(x+\tfrac{1}{2}) = \frac{\Gamma(2x)\sqrt{\pi}}{2^{2x-1}\Gamma(x)} \tag{2.6}$$

to write

$$\binom{r+\frac{1}{2}}{m} = \frac{(2r+1)!\,(r-m)!}{2^{2m} r! m! (2r-2m+1)!}. \tag{2.7}$$

Then (2.5) is immediate.

References

[1] I. S. Gradshteyn and I. M. Ryzhik, *Table of integrals, series, and products*, 8th ed., Elsevier/Academic Press, Amsterdam, 2015. Translated from the Russian; Translation edited and with a preface by Daniel Zwillinger and Victor Moll; Revised from the seventh edition [MR2360010]. MR3307944 ↑194

[2] Steven H. Weintraub, *Reverse Legendre polynomials*, Arch. Math. (Basel) **118** (2022), no. 6, 593–604, DOI 10.1007/s00013-022-01740-2. MR4423453 ↑193

Department of Mathematics, Tulane University, New Orleans, Louisiana 70118
Email address: tamdeberhan@tulane.edu

Department of Mathematics, Tulane University, New Orleans, Louisiana 70118
Email address: vhm@tulane.edu

Contemporary Mathematics
Volume **822**, 2025
https://doi.org/10.1090/conm/822/16479

On an identity by Ercolani, Lega, and Tippings

Maxim L. Yattselev

ABSTRACT. In this note we prove that

$$j!\,2^N\binom{N+j-1}{j}{}_2F_1\left({-j,-2j \atop -N-j+1};-1\right)=\sum_{l=0}^N\binom{N}{l}\prod_{i=0}^{j-1}2(2i+1+l),$$

where N and j are positive integers, which resolves a question posed by Ercolani, Lega, and Tippings.

In [**2**, Theorem 2.1], Ercolani, Lega, and Tippings have shown that the number of 2ν-valent maps with j vertices and two legs that can be embedded in a surface of genus $g\geq 1$ is given by

$$j!\left[2\nu(\nu-1)\binom{2\nu-1}{\nu-1}\right]^j\sum_{l=0}^{3g-1}a_l(g,\nu)\binom{2g-2+l+j}{j}{}_2F_1\left({-j,-\nu j \atop 2-2g-l-j};\frac{1}{1-\nu}\right)$$

for some coefficients $a_l(g,\nu)$. Based on numerical computations, they subsequently conjectured, see [**2**, Conjecture 4.2] or [**1**, Conjecture 5.1], that when $\nu=2$ the summands in the above expression can be stated without hypergeometric functions with the help of identity (1) further below. In this note we provide a proof of this fact.

THEOREM 1. *Let N,j be positive integers. Then*

$$j!2^N\binom{N+j-1}{j}{}_2F_1\left({-j,-2j \atop -N-j+1};-1\right)=\sum_{l=0}^N\binom{N}{l}\prod_{i=0}^{j-1}2(2i+1+l). \tag{1}$$

We use the following notation for the falling and rising factorials:

$$(a)^{\underline{0}}=(a)^{\overline{0}}:=1,\quad (a)^{\underline{n}}:=a(a-1)\cdots(a-n+1),\quad\text{and}\quad (a)^{\overline{n}}:=a(a+1)\cdots(a+n-1)$$

for $n\geq 1$. We prove (1) in three steps formulated as separate lemmas.

LEMMA 2. *The right-hand side of (1) is equal to $2^NR_j(N)$, where $R_j(x)=\sum_{i=0}^jR(i,j)(x)^{\underline{i}}$ and $R(i,j)$ are numbers uniquely determined by the relations*

$$R(0,j)=2^j(2j-1)!!,\quad R(j,j)=1,\quad R(i,j+1)=2(2j+i+1)R(i,j)+R(i-1,j),$$

with the recurrence relation holding for $i\in\{1,2,\ldots,j\}$.

2020 *Mathematics Subject Classification.* Primary 05C30, 33C05.

The research was supported in part by a grant from the Simons Foundation, CGM-706591.

©2025 American Mathematical Society

PROOF. Let $C_{i,j}$ be the following coefficients:

$$\prod_{i=0}^{j-1}(2i+1+x) = \sum_{k=0}^{j} C_{k,j}x^k.$$

Since

$$\prod_{i=0}^{j}(2i+1+x) = (2j+1+x)\sum_{k=0}^{j} C_{k,j}x^k = \sum_{k=0}^{j}(2j+1)C_{k,j}x^k + \sum_{k=1}^{j+1} C_{k-1,j}x^k,$$

it holds that

$$C_{0,j} = (2j-1)!!, \quad C_{j,j} = 1, \quad C_{k,j+1} = (2j+1)C_{k,j} + C_{k-1,j}, \;\; k \in \{1,2,\dots,j\}.$$

Exactly as in the case of the coefficients $R(i,j)$, the coefficients $C_{k,j}$ are uniquely defined by the above relations since the knowledge of all the coefficients on the level j allows one to compute all the coefficients on the level $j+1$ with the base case $C_{0,1} = C_{1,1} = 1$. Recall [**3**, Equation (26.8.10)] that

$$x^k = \sum_{i=1}^{k} \left\{ {k \atop i} \right\} (x)^{\underline{i}},$$

where $\left\{ {k \atop i} \right\}$ are Stirling numbers of the second kind. Therefore,

$$\prod_{i=0}^{j-1}(2i+1+x) = C_{0,j} + \sum_{i=1}^{j} \left(\sum_{k=i}^{j} C_{k,j} \left\{ {k \atop i} \right\} \right) (x)^{\underline{i}}.$$

Observe that

$$\begin{aligned} \sum_{l=0}^{N} \binom{N}{l} (l)^{\underline{i}} &= \sum_{l=0}^{N} \frac{N!}{l!(N-l)!}(l)(l-1)\cdots(l-i+1) \\ &= \sum_{l=i}^{N} \frac{N!}{(l-i)!(N-l)!} = \sum_{l=0}^{N-i} \frac{N!}{l!(N-i-l)!} = 2^{N-i}(N)^{\underline{i}}. \end{aligned}$$

Therefore, the right-hand side of (1) is equal to

$$2^j \sum_{l=0}^{N} \binom{N}{l} \left(C_{0,j} + \sum_{i=1}^{j} \left(\sum_{k=i}^{j} C_{k,j} \left\{ {k \atop i} \right\} \right) (l)^{\underline{i}} \right) = 2^N R_j(N),$$

where $R_j(x) := \sum_{i=0}^{j} R(i,j)(x)^{\underline{i}}$ with

$$R(0,j) := 2^j(2j-1)!! \quad \text{and} \quad R(i,j) := 2^{j-i} \sum_{k=i}^{j} C_{k,j} \left\{ {k \atop i} \right\}, \quad i \in \{1,2,\dots,j\}.$$

Since $\left\{ {j \atop j} \right\} = 1$, see [**3**, Equation (26.8.4)], it indeed holds that $R(j,j) = 1$. Thus, we only need to establish the recurrence relation. The recurrence relation for $C_{k,j}$

yields that

$$\begin{aligned} R(i,j+1) &= 2^{j+1-i}\sum_{k=i}^{j+1} C_{k,j+1}\left\{{k \atop i}\right\} \\ &= 2^{j+1-i}\sum_{k=i}^{j} C_{k,j+1}\left\{{k \atop i}\right\} + 2^{j+1-i}\left\{{j+1 \atop i}\right\} \\ &= 2(2j+1)R(i,j) + 2^{j+1-i}\sum_{k=i}^{j+1} C_{k-1,j}\left\{{k \atop i}\right\} \end{aligned}$$

for any $i \in \{1,2,\dots,j\}$, where we also used the fact that $C_{j,j} = 1$. Furthermore, we get from [**3**, Equation (26.8.22)] that

$$\begin{aligned} 2^{j+1-i}\sum_{k=i}^{j+1} C_{k-1,j}\left\{{k \atop i}\right\} &= 2^{j+1-i}\sum_{k=i-1}^{j} C_{k,j}\left\{{k+1 \atop i}\right\} \\ &= 2^{j+1-i}\sum_{k=i-1}^{j} C_{k,j}\left(i\left\{{k \atop i}\right\} + \left\{{k \atop i-1}\right\}\right) \\ &= 2iR(i,j) + R(i-1,j), \end{aligned}$$

where, by convention, $\left\{{i-1 \atop i}\right\} = 0$. This finishes the proof of the lemma. □

LEMMA 3. *The left-hand side of* (1) *is equal to* $2^N L_j(N)$, *where* $L_j(x) = \sum_{i=0}^{j} L(i,j)(x)^{\underline{i}}$ *with*

$$L(0,j) = \frac{(2j)!}{j!} \quad \textit{and} \quad L(i,j) = \frac{j!}{i!}\sum_{k=i}^{j}\binom{2j}{j+k}\binom{k-1}{i-1}, \quad i \in \{1,2,\dots,j\}.$$

PROOF. It follows from [**3**, Equation (15.2.1)] that the left-hand side of (1) is equal to $2^N L_j(N)$, where

$$L_j(x) := (x)^{\overline{j}}\sum_{k=0}^{j}\frac{(-j)^{\overline{k}}(-2j)^{\overline{k}}}{(-x-j+1)^{\overline{k}}}\frac{(-1)^k}{k!}$$

(we take this formula as the definition of $L_j(x)$ and show that this polynomial also admits an expression as in the statement of the lemma). Since

$$(-a)^{\overline{k}} = (-1)^k (a)^{\underline{k}} \quad \text{and} \quad \frac{(a)^{\overline{j}}}{(a+j-1)^{\underline{k}}} = (a)^{\overline{j-k}},$$

we have that

$$\begin{aligned} L_j(x) &= \sum_{k=0}^{j}(j)^{\underline{k}}(2j)^{\underline{k}}\frac{(x)^{\overline{j-k}}}{k!} = \sum_{k=0}^{j}\frac{j!}{(j-k)!}\frac{(2j)!}{(2j-k)!}\frac{(x)^{\overline{j-k}}}{k!} \\ &= \sum_{k=0}^{j}\binom{j}{k}\frac{(2j)!}{(2j-k)!}(x)^{\overline{j-k}} = \sum_{k=0}^{j}\binom{j}{k}\frac{(2j)!}{(j+k)!}(x)^{\overline{k}}, \end{aligned}$$

where we replaced $j-k$ by k to get the last equality. It is known [**4**] that

$$(x)^{\overline{k}} = \sum_{i=1}^{k} \binom{k-1}{i-1} \frac{k!}{i!} (x)^{\underline{i}}, \quad k \geq 1.$$

Hence,

$$\begin{aligned} L_j(x) &= \frac{(2j)!}{j!} + \sum_{k=1}^{j} \binom{j}{k} \frac{(2j)!}{(j+k)!} \sum_{i=1}^{k} \binom{k-1}{i-1} \frac{k!}{i!} (x)^{\underline{i}} \\ &= \frac{(2j)!}{j!} + \sum_{i=1}^{j} \left(\sum_{k=i}^{j} \binom{j}{k} \frac{(2j)!}{(j+k)!} \binom{k-1}{i-1} \frac{k!}{i!} \right) (x)^{\underline{i}}, \end{aligned}$$

which finishes the proof of the lemma. □

LEMMA 4. *It holds that $L(i,j) = R(i,j)$ for all $i \in \{0,1,\ldots,j\}$ and $j \geq 1$. In particular,* (1) *is true.*

PROOF. Clearly, $L(0,j) = R(0,j)$ and $L(j,j) = R(j,j) = 1$. Thus, we only need to show that

$$L(i,j+1) = 2(2j+i+1)L(i,j) + L(i-1,j), \quad i \in \{1,2,\ldots,j\},$$

since this recurrence relation and the marginals $L(0,j)$, $L(j,j)$ uniquely determine the whole table $L(i,j)$. In what follows, we agree that binomial coefficients with out-of-range indices are set to zero. In what follows we repeatedly use the elementary identity

$$\binom{n+1}{l} = \binom{n}{l} + \binom{n}{l-1}.$$

Using this identity twice and our convention concerning binomial coefficients with out-of-range indices, we get that

$$L(i,j+1) = 2(j+1)L(i,j) + \frac{(j+1)!}{i!} \sum_{k=i}^{j+1} \left[\binom{2j}{j+k+1} + \binom{2j}{j+k-1} \right] \binom{k-1}{i-1}$$

$$= 2(j+1)L(i,j) + \frac{(j+1)!}{i!} \left[\sum_{k=i+1}^{j} \binom{2j}{j+k} \binom{k-2}{i-1} + \sum_{k=i-1}^{j} \binom{2j}{j+k} \binom{k}{i-1} \right],$$

where the second row is obtained simply by a change of summation indices. Therefore, when $i=1$, we get that

$$\begin{aligned} L(1,j+1) &= 4(j+1)L(1,j) + (j+1)! \left[\binom{2j}{j} - \binom{2j}{j+1} \right] \\ &= 4(j+1)L(1,j) + L(0,j)\big[j+1-j\big], \end{aligned}$$

which establishes the desired recurrence relation for $i=1$. On the other hand, when $i > 1$, the above sum in square brackets can further be rewritten with the help of the elementary identity as

$$\sum_{k=i+1}^{j}\binom{2j}{j+k}\left[\binom{k-1}{i-1}-\binom{k-2}{i-2}\right]+\sum_{k=i-1}^{j}\binom{2j}{j+k}\left[\binom{k-1}{i-1}+\binom{k-1}{i-2}\right]$$
$$=2\sum_{k=i}^{j}\binom{2j}{j+k}\binom{k-1}{i-1}+\sum_{k=i-1}^{j}\binom{2j}{j+k}\left[\binom{k-1}{i-2}-\binom{k-2}{i-2}\right],$$

where the last sum simply reduced to $\binom{2j}{j+1}$ when $i=2$. Thus, using the elementary identity once more, we get that

$$L(i,j+1)=4(j+1)L(i,j)+\frac{(j+1)!}{i!}\sum_{k=i-1}^{j}\binom{2j}{j+k}\binom{k-2}{i-3},$$

which holds even for $i=2$ if we understand that $\binom{-1}{-1}=1$ and $\binom{k-2}{-1}=0$, $k\geq 2$. That is, to prove the lemma we need to show that

$$2(i-1)L(i,j)+L(i-1,j)=\frac{(j+1)!}{i!}\sum_{k=i-1}^{j}\binom{2j}{j+k}\binom{k-2}{i-3},$$

or equivalently that

$$\sum_{k=i-1}^{j}\binom{2j}{j+k}\left[2(i-1)\binom{k-1}{i-1}+i\binom{k-1}{i-2}-(j+1)\binom{k-2}{i-3}\right]=0.$$

When $i=2$, the left-hand side above becomes

$$\sum_{k=2}^{j}2k\binom{2j}{j+k}-(j-1)\binom{2j}{j+1}=\sum_{k=2}^{j}(j+k)\binom{2j}{j+k}-\sum_{k=1}^{j-1}(j-k)\binom{2j}{j+k}.$$

Similarly, when $i>2$, the left-hand side in question is equal to

$$\sum_{k=i-1}^{j}\binom{2j}{j+k}\frac{(k-2)!}{(k-i+1)!(i-2)!}\left[2(k-1)(k-i+1)+i(k-1)-(j+1)(i-2)\right]$$
$$=\sum_{k=i-1}^{j}\binom{2j}{j+k}\frac{(k-2)!}{(k-i+1)!(i-2)!}\left[(k-i+1)(j+k)-(j-k)(k-1)\right],$$

which is the same as

$$\sum_{k=i}^{j}(j+k)\binom{2j}{j+k}\binom{k-2}{i-2}-\sum_{k=i-1}^{j-1}(j-k)\binom{2j}{j+k}\binom{k-1}{i-2}$$

(as shown just before, this formula is valid for $i=2$ as well). This difference is indeed equal to zero as the second sum above can be rewritten as

$$\sum_{k=i-1}^{j-1}(j+k+1)\binom{2j}{j+k+1}\binom{k-1}{i-2}=\sum_{l=i}^{j}(j+l)\binom{2j}{j+l}\binom{l-2}{i-2}.\qquad\square$$

References

[1] Nicholas Ercolani, Joceline Lega, and Brandon Tippings, *Map enumeration from a dynamical perspective*, Recent progress in special functions, Contemp. Math., vol. 807, Amer. Math. Soc., [Providence], RI, [2024] ©2024, pp. 85–110, DOI 10.1090/conm/807/16167. MR4818103

[2] Nicholas Ercolani, Joceline Lega, and Brandon Tippings, *Non-recursive counts of graphs on surfaces*, Enumer. Comb. Appl. **3** (2023), no. 3, Paper No. S2R20, 24, DOI 10.54550/eca2023v3s3r20. MR4607276

[3] F. W. J Olver et al. editors, *NIST digital library of mathematical functions*, `http://dlmf.nist.gov`.

[4] I. Lah, *Eine neue Art von Zahlen, ihre Eigenschaften und Anwendung in der mathematischen Statistik* (German), Mitteilungsbl. Math. Statist. **7** (1955), 203–212. MR74435

Department of Mathematical Sciences, Indiana University Indianapolis, 402 North Blackford Street, Indianapolis, Indiana 46202

Email address: `maxyatts@iu.edu`

Selected Published Titles in This Series

822 **Ahmad Barhoumi, Roozbeh Gharakhloo, and Andrei Martinez-Finkelshtein, Editors,** Recent Developments in Orthogonal Polynomials, 2025

818 **Howard S. Cohl, Roberto S. Costas-Santos, and Robert S. Maier, Editors,** Classical Hypergeometric Functions and Generalizations, 2025

816 **Luca F. Di Cerbo and Laurenţiu G. Maxim, Editors,** Geometry and Topology of Aspherical Manifolds, 2025

815 **Daniel Orr and Joshua Jeishing Wen, Editors,** Macdonald Theory and Beyond, 2025

814 **Susan Montgomery and Siu-Hung Ng, Editors,** Quantum Groups, Hopf Algebras, and Applications, 2025

813 **Thomas Creutzig, Julia Plavnik, and David Ridout, Editors,** Quantum Symmetries, 2025

812 **Ross Geoghegan, Craig R. Guilbault, and Kim Ruane, Editors,** Topology at Infinity of Discrete Groups, 2025

811 **Krishnendu Gongopadhyay and Shashikant A. Katre, Editors,** Geometry, Groups and Mathematical Philosophy, 2025

810 **Robert J. Berman and Yanir A. Rubinstein, Editors,** Convex and Complex: Perspectives on Positivity in Geometry, 2025

809 **Fabrice Baudoin and Luca Rizzi, Editors,** New Trends in Sub-Riemannian Geometry, 2025

808 **Loring W. Tu, Editor,** Group Actions and Equivariant Cohomology, 2024

807 **Galina Filipuk, Editor,** Recent Progress in Special Functions, 2024

806 **Maximilian Engel, Hildeberto Jardón-Kojakhmetov, and Cinzia Soresina, Editors,** Topics in Multiple Time Scale Dynamics, 2024

805 **Anthony Iarrobino, Pedro Macias Marques, Maria Evelina Rossi, and Jean Vallès, Editors,** Deformation of Artinian Algebras and Jordan Type, 2024

804 **Mahir Bilen Can and Jörg Feldvoss, Editors,** A Glimpse into Geometric Representation Theory, 2024

803 **Peter Gothen, Margarida Melo, and Montserrat Teixidor i Bigas, Editors,** Moduli Spaces and Vector Bundles—New Trends, 2024

802 **Ralph M. Kaufmann, Martin Markl, and Alexander A. Voronov, Editors,** Higher Structures in Topology, Geometry, and Physics, 2024

801 **K. A. Brown, T. J. Hodges, M. Vancliff, and J. J. Zhang, Editors,** Recent Advances in Noncommutative Algebra and Geometry, 2024

800 **Avinoam Mann, Louis H. Rowen, David J. Saltman, Aner Shalev, Lance W. Small, and Uzi Vishne, Editors,** Amitsur Centennial Symposium, 2024

799 **Alberto A. Condori, Elodie Pozzi, William T. Ross, and Alan A. Sola, Editors,** Recent Progress in Function Theory and Operator Theory, 2024

798 **Marat V. Markin, Igor V. Nikolaev, and Carsten Trunk, Editors,** Advances in Functional Analysis and Operator Theory, 2024

797 **Sangita Jha, Mrinal Kanti Roychowdhury, and Saurabh Verma, Editors,** Recent Developments in Fractal Geometry and Dynamical Systems, 2024

796 **John Cremona, John Jones, Jennifer Paulhus, Andrew V. Sutherland, and John Voight, Editors,** LuCaNT: LMFDB, Computation, and Number Theory, 2024

795 **Michael A. Jones, David McCune, and Jennifer M. Wilson, Editors,** Mathematical Analyses of Decisions, Voting and Games, 2024

794 **Jean-Pierre Magnot, Editor,** Recent Advances in Diffeologies and Their Applications, 2024

For a complete list of titles in this series, visit the
AMS Bookstore at **www.ams.org/bookstore/conmseries/**.